Nature's Wealth

The Economics of Ecosystem Services and Poverty

Increasing pressure from economic development and population growth has resulted in the degradation of ecosystems around the world and the loss of the essential services that they provide. Understanding the linkages between ecosystem service provisioning and human well-being is crucial for the establishment of effective environmental and economic development policies.

Presenting new insights into the relationship between ecosystem services and livelihoods in developing countries, this book takes up the challenge of assessing these links to demonstrate their importance in policy development. The book pays special attention to innovative management opportunities that improve local livelihoods and alleviate poverty while enhancing ecosystem protection. Based on 18 studies in more than 20 developing countries, the authors explore the role of biodiversity-, marine-, forest-, water- and land-related ecosystem services, making this an invaluable contribution to research on the role of ecosystems in supporting the livelihoods of the poor around the world.

PIETER J. H. VAN BEUKERING is an Associate Professor of environmental economics and ecosystem services at the Institute for Environmental Studies (IVM) of VU University Amsterdam. Most of his research work takes place in Asia, the Caribbean, the Pacific and Africa, with a strong focus on natural resource management, economic valuation and poverty alleviation.

ELISSAIOS PAPYRAKIS is a Senior Researcher in environmental economics and climate change at VU University Amsterdam, and a Senior Lecturer in economics at the University of East Anglia. His work involves both theoretical and empirical analysis in development and environmental economics, with a particular emphasis on the economics of climate change, sustainable development and environmental management.

JETSKE BOUMA is a Senior Environmental Economist at the IVM, VU University Amsterdam. She specializes in community-based natural resource management and the conditions for self-enforcement, with specific attention to conservation–development trade-offs, incentives for cooperative behaviour and effective policy design.

ROY BROUWER is Professor and Head of the Department of Environmental Economics at the IVM, VU University Amsterdam, and also Professor of Water Economics for the Dutch National Research Programme 'Living with Water'. He is a recipient of a National Research Flagship Fellowship Award from CSIRO, Australia, for his work on water economics.

ECOLOGY, BIODIVERSITY AND CONSERVATION

The world's biological diversity faces unprecedented threats. The urgent challenge facing the concerned biologist is to understand ecological processes well enough to maintain their functioning in the face of the pressures resulting from human population growth. Those concerned with the conservation of biodiversity and with restoration also need to be acquainted with the political, social, historical, economic and legal frameworks within which ecological and conservation practice must be developed. The new Ecology, Biodiversity and Conservation series will present balanced, comprehensive, up-to-date and critical reviews of selected topics within the sciences of ecology and conservation biology, both botanical and zoological, and both 'pure' and 'applied'. It is aimed at advanced final-year undergraduates, graduate students, researchers and university teachers, as well as ecologists and conservationists in industry, government and the voluntary sectors. The series encompasses a wide range of approaches and scales (spatial, temporal and taxonomic), including quantitative, theoretical, population, community, ecosystem, landscape, historical, experimental, behavioural and evolutionary studies. The emphasis is on science related to the real world of plants and animals rather than on purely theoretical abstractions and mathematical models. Books in this series will, wherever possible, consider issues from a broad perspective. Some books will challenge existing paradigms and present new ecological concepts, empirical or theoretical models, and testable hypotheses. Other books will explore new approaches and present syntheses on topics of ecological importance.

Ecology and Control of Introduced Plants
Judith H. Myers and Dawn Bazely

Invertebrate Conservation and Agricultural Ecosystems
T. R. New

Risks and Decisions for Conservation and Environmental Management
Mark Burgman

Nature's Wealth

The Economics of Ecosystem Services and Poverty

Edited by
PIETER J. H. VAN BEUKERING
ELISSAIOS PAPYRAKIS
JETSKE BOUMA
ROY BROUWER

Institute for Environmental Studies (IVM), VU University, Amsterdam

CAMBRIDGE
UNIVERSITY PRESS

CAMBRIDGE UNIVERSITY PRESS
Cambridge, New York, Melbourne, Madrid, Cape Town,
Singapore, São Paulo, Delhi, Mexico City

Cambridge University Press
The Edinburgh Building, Cambridge CB2 8RU, UK

Published in the United States of America by Cambridge University Press, New York

www.cambridge.org
Information on this title: www.cambridge.org/9781107027152

© Cambridge University Press 2013

First published 2013

Printed and bound in the United Kingdom by the MPG Books Group

A catalogue record for this publication is available from the British Library

Library of Congress Cataloging in Publication data
Nature's wealth : the economics of ecosystem services and poverty / edited by Pieter J. H. van
Beukering ... [et al.].
    p.   cm. – (Ecology, biodiversity, and conservation)
ISBN 978-1-107-02715-2 (hbk.) – ISBN 978-1-107-69804-8 (pbk.)   1. Human ecology –
Developing countries.   2. Biotic communities – Developing countries.   3. Environmental
degradation – Developing countries.   4. Environmental policy – Developing
countries.   I. Beukering, Pieter van.
GF900.N374   2013
333.709172′4–dc23
                                                                                              2012027480

ISBN 978-1-107-02715-2 Hardback
ISBN 978-1-107-69804-8 Paperback

Contents

Colour plate section between pages 398 and 399.

Contributors

GIRMAY GEBRESAMUEL ABRAHA
College of Dryland Agriculture and Natural
Resources, Mekelle University, Mekelle, Tigray, Ethiopia

ZENEBE ABRAHA
Department of Natural Resources, Economics and
Management, Mekelle University, Mekelle, Tigray, Ethiopia

BHIM ADHIKARI
International Development Research Center, Ottawa, Canada

SONIA AKTER
Helmholtz Center for Environmental Research –
UFZ, Leipzig, Germany

JANET A.R. AMPONIN
Resources, Environment and Economics Center for Studies, Quezon
City, Philippines

ROSEMARY ATIENO
Institute for Development Studies, University of Nairobi,
Nairobi, Kenya

VITHANARACHCHIGE D.N. AYONI
Department of Agriculture, Socio Economics and Planning
Centre, Peradeniya, Sri Lanka

MA. EUGENIA C. BENNAGEN
Resources, Environment and Economics Center for Studies, Quezon
City, Philippines

JAMES N. BLIGNANT
Department of Economics, University of Pretoria

JETSKE BOUMA
Institute for Environmental Studies (IVM), VU University
Amsterdam, Amsterdam, The Netherlands

LUKE BRANDER
Brander Ltd, Kornhill, Hong Kong

ROY BROUWER
Institute for Environmental Studies (IVM), VU University
Amsterdam, Amsterdam, The Netherlands

MUYEYE CHAMBWERA
International Institute for Environment and
Development, London, United Kingdom

MARTINUS P. DE WIT
School of Public Leadership and De Wit Sustainable Options (Pty)
Ltd, Stellenbosch University, Brackenfell, South Africa

ANTONIUS JOHANNES DIETZ
African Studies Centre, Leiden, The Netherlands

SABINA L. DI PRIMA
Center for International Cooperation, VU University
Amsterdam, Amsterdam, The Netherlands

THOMAS EMWANU
Uganda Bureau of Statistics, Kampala, Uganda

AUYRZANA ENKH–AMGALAN
Centre for Policy Research, Ulaanbaatar, Mongolia

TUMUR ERDENECHULUUN
Wageningen University and Research, Wageningen, The Netherlands

REYER GERLAGH
Department of Economics, Tilburg University, Tilburg,
The Netherlands

LOKUGAM H.P. GUNARATNE
Department of Agricultural Economics and Business
Management, University of Peradeniya, Kandy, Sri Lanka

SHREEKANT GUPTA
Delhi School of Economics, University of Delhi, Delhi, India; LKY
School of Public Policy, National University of Singapore, Singapore

FITSUM HAGOS
International Water Management Institute, Eastern Africa and Nile
Sub-regional Office, Addis Ababa, Ethiopia

WOLFGANG HAIDER
School of Resource and Environmental Management, Simon Fraser
University, Burnaby, British Columbia, Canada

A. K. ENAMUL HAQUE
School of Business and Economics, United International
University, Dhaka, Bangladesh

RASHID HASSAN
Center for Environmental Economics and Policy in
Africa, University of Pretoria, Hatfield, Pretoria, South Africa

SEBASTIAAN M. HESS
Hess Environmental Economic Analyst, Hilversum,
The Netherlands

JOHANNES HOOGEVEEN
The World Bank, Washington DC, USA

MARK HORRIDGE
Centre of Policy Studies, Monash University,
Melbourne, Australia

ANABETH INDAB–SAN GREGORIO
Resources, Environment and Economics Center for
Studies, Manila, Philippines

RON JANSSEN
Institute for Environmental Studies (IVM), VU University
Amsterdam, Amsterdam, The Netherlands

ALISON R. JOUBERT
Southern Waters Ecological Research and Consulting, Mill
Street, Cape Town, South Africa

K. J. JOY
Society for Promoting Participative Ecosystem Management
(SOPPECOM), Pune, India

JANE KABUBO–MARIARA
School of Economics, University of Nairobi, Nairobi, Kenya

GODIUS KAHYARARA
Environment for Development in Tanzania (EfDT), Department of
Economics, University of Dar Es Salaam

DUNCAN KNOWLER
School of Resource and Environmental Management, Simon Fraser
University, Burnaby, British Columbia, Canada

BAKARY KONE
Wetlands International, Bamako, Mali

GIDEON KRUSEMAN
LEI-Wageningen, The Hague, The Netherlands

CRAIG LEISHER
The Nature Conservancy, Millburn NJ, USA

VINCENT LINDERHOF
LEI-Wageningen, The Hague, The Netherlands

WIETZE LISE
Åf-Mercados Emi, Odtu-Teknokent Met Alani,
Eskischir Road, Ankara, Turkey

MARGARET MABUGU
Human Science Research Council, Pretoria, South Africa

RAMOS EMMANUEL MABUGU
Financial and Fiscal Commission, Midrand, South Africa

SAKIB MAHMUD
Business and Economics Department, University of Wisconsin-
Superior, Superior, Wisconsin, USA

VICTOR G. MAKUNDI
Centre for Environmental Economics and Development Research
(CEDR), Kijitonyama, Dar es Salaam, Tanzania

ERIC E. MASSEY
Institute for Environmental Studies (IVM), VU University
Amsterdam, Amsterdam, The Netherlands

AFEWORKI MULUGETA
Department of Public Health, College of Health Sciences, Mekelle
University, Mekelle, Tigray, Ethiopia

S. MANSOOB MURSHED
Institute of Social Studies (ISS), Erasmus University Rotterdam, The
Hague, The Netherlands and Coventry University, Coventry, UK

HIYARE P. L. K. NANAYAKKARA
United Nations World Food Programme, Colombo, Sri Lanka

URVASHI NARAIN
The World Bank, Washington DC, USA

PAUL O. OKWI
International Development Research Centre (IDRCC),
Nairobi, Kenya

ELISSAIOS PAPYRAKIS
Institute for Environmental Studies (IVM), VU University
Amsterdam, Amsterdam, The Netherlands and School of International
Development, University of East Anglia, Norwich, UK

LORENZO PELLEGRINI
International Institute of Social Studies, Erasmus University, The
Hague, The Netherlands

NAM PHAM KHANH
Faculty of Development Economics, University of Economics, Ho Chi
Minh City, Vietnam

MAHESH POUDYAL
Department of Forest Resource Management, Swedish University of
Agricultural Sciences (SLU), Umeå, Sweden

PRIYANGA K. PREMARATHNE
ProSEES Consultancies (Pvt) Ltd, Dambulla, Udagaladeniya,
Sri Lanka

BYAMBA PUREV
School of Economics and Business, Mongolian State University of
Agriculture, Ulaanbaatar, Mongolia

LEA M. SCHERL
School of Earth and Environmental Sciences, James Cook University of
North Queensland, Townsville, Australia

THEODOR J. STEWART
Department of Statistical Sciences, University of Cape Town,
Rondebosch, South Africa

MARONEL STEYN
Natural Resources and the Environment, CSIR, Stellenbosch, Western Cape, South Africa

RICHARD S. J. TOL
Department of Economics, University of Sussex, Falmer, Brighton, United Kingdom and VU University Amsterdam

PIETER J. H. VAN BEUKERING
VU University Amsterdam, Institute for Environmental Studies (IVM), Amsterdam, The Netherlands

KIM VAN DER LEEUW
Do-inc VOF, Amsterdam, The Netherlands

JAN H. VAN HEERDEN
Department of Economics, University of Pretoria, Mooikloof, Pretoria, South Africa

KLAAS VAN 'T VELD
Department of Economics & Finance, University of Wyoming, Laramie, Wyoming, USA

EYASU YAZEW
Mekelle University, Mekelle, Tigray, Ethiopia

MEKONNEN YOHANNES
Department of Microbiology, Immunology and Parasitology, Mekelle University, Mekelle, Tigray, Ethiopia

LEO ZWARTS
Altenburg and Wymenga Ecological Consultants, Feanwâlden, The Netherlands

Acknowledgements

We would like to express our sincere thanks to all individuals who provided direct or indirect support to accomplish this research. There are really too many people involved in the 18 case studies to list in their entirety. Still, we would like to thank a few people in particular for making this book possible. To begin with, we are very grateful to all those families, farmers and fishermen around the world who took the time to complete interviews and share their knowledge with our researchers. Without their hospitality, patience and wisdom, this book would not have been written. We also gratefully acknowledge the Dutch Ministry of Development Cooperation (DGIS), Poverty Reduction and Environmental Management Programme (PREM) and the Institute of Environmental Studies (IVM), VU University Amsterdam for providing generous financial support to undertake this research. A special thanks goes to the Steering Committee of the PREM Programme for carefully guiding the research in a clear, stable and stimulating manner. Therefore, we express our gratitude to Professor Harmen Verbruggen, Dean of the Economics Faculty of the VU University, Professor Mansoob Murshed of the Institute of Social Studies (ISS) in The Hague, Professor Rashid Hassan, Director of the Centre for Environmental Economics and Policy in Africa (CEEPA) in Pretoria and Piet Klop, former Coordinator of the Dutch Ministry of Foreign Affairs. Furthermore, we are grateful to Professor Erwin Bulte, Professor Ton Dietz, Professor Charles Perrings, Professor Ian Bateman, Danielle Hirsch and Kirsten Schuyt who kindly invested time as external reviewers of the PREM programme and provided invaluable insights and feedback. Our gratitude also extends to the filmmakers that documented six out of the 18 case studies in the form of beautiful documentaries that tell the story in a very appealing manner. Their work can be appreciated at the website of the PREM programme (www.prem-online.org). We would also like to thank Kim van der Leeuw for being a strict and sociable programme assistant for the full length of the PREM programme and Marjolijn Staarink, Annabel

Aish, Kate Gallop and Rayvon Lemmert for meticulously making the necessary finishing touches and improvements to the manuscript. We were particularly privileged to work with the people at Cambridge University Press, who provided continuous support and detailed comments for both the content of the manuscript as well as the overall presentation, production and marketing. We received extremely useful input from the series editor Professor Michael Usher, who repeatedly encouraged us to actually publish the book and who oversaw the whole process from start to finish. Dominic Lewis, Christopher Miller and Megan Waddington assisted us in numerous ways such as processing the reviewers' feedback, designing the book cover and providing help and guidance whenever called upon. Last but not least, our deepest thanks go to our immediate family and friends for their patience and continuous support. We dedicate the book to them.

1 · *The economics of ecosystem services and poverty*

PIETER J. H. VAN BEUKERING,
ELISSAIOS PAPYRAKIS, JETSKE BOUMA
AND ROY BROUWER

1.1 Introduction

Ecosystems play a crucial role in the survival and well-being of human beings. Increasing pressure on ecosystems resulting from economic development and population growth has resulted in degrading ecosystems and losses of the services ecosystems provide throughout the world. According to the UN (2010) 'as a consequence of human actions, species are being lost at a rate estimated to be 100 times the natural rate of extinction. In the past century, 35% of the mangroves, 40% of the forests and 50% of the wetlands have been lost ... action is urgently needed to avoid reaching critical thresholds that will lead to an irreversible loss of biodiversity and ecosystem services, with dangerous consequences for human well-being'.

The Millennium Ecosystem Assessment (MEA) (2005) was the first to explicitly underline the linkages between ecosystems and human well-being, coining the term 'ecosystem services' to stress the important benefits that people derive from ecosystems (MEA 2005). The term 'Ecosystem Services' serves as a catalyst to stress the importance of ecosystems for human well-being. As indicated in Figure 1.1, the number of publications using the term has increased exponentially since 2005. Figure 1.1 also shows that most of the publications are in the domain of the natural sciences, with the governance-based sciences somewhat lagging behind. The publication of the influential 'The Economics of Ecosystems and Biodiversity' (TEEB) report (2009) on the value of ecosystem services for human well-being changed this picture, but still social sciences research that assesses the linkages between ecosystem services and human well-being is limited, especially where questions

Nature's Wealth: The Economics of Ecosystem Services and Poverty, ed. P. J. H. van Beukering, E. Papyrakis, J. Bouma and R. Brouwer. Published by Cambridge University Press, © Cambridge University Press 2013.

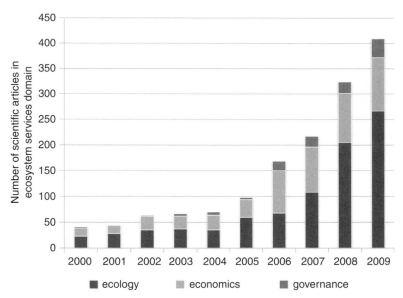

Figure 1.1 Number of journal papers in the 'ecosystem services' domain categorized by scientific discipline

relating to the governance of ecosystem services are concerned in the political science. Also, especially given the global nature of biodiversity and ecosystem service provision, the distribution of human well-being impacts needs to be assessed. The MEA (2005) and TEEB (2009) assume that ecosystem protection always benefits human well-being, but given the trade-offs arising between conservation and development, ecosystem protection can have adverse impacts on human well-being as well (Sunderlin *et al.* 2005). Understanding the linkages between ecosystem service provisioning and human well-being is crucial for avoiding adverse effects in environmental and economic development policy. Research assessing these linkages is limited, however, and constitutes an essential knowledge gap.

Nature's Wealth takes up this challenge and presents new insights into the relationship between ecosystem services and livelihoods in developing countries around the world. Based on 18 studies in more than 20 developing countries, evidence is presented on the role of ecosystems in supporting human well-being, especially in the livelihoods of the poor. *Nature's Wealth* pays special attention to innovative management opportunities that improve local livelihoods and alleviate poverty while at the same time enhancing ecosystem protection. To demonstrate the variety of ecosystem services and their linkages to local livelihoods, *Nature's Wealth* is organized in five parts, describing the role of biodiversity-, marine-, forest-, water- and land-related ecosystem services.

In the past decade, several books addressing poverty and the environment have been published. *The Environmentalism of the Poor* published in 2002 (Martinez-Alier, 2002) analyses several manifestations of the growing 'environmental justice movement' and the 'environmentalism of the poor'. It presents a systematic analysis of the clash between economy and environment, and discusses the attempts 'to take nature into account'. The book differs from *Nature's Wealth* in that it follows a socio-political framework instead of an environmental and ecological economics approach, and views the 'environment' in a more general, holistic manner, not specifically addressing the various types of ecosystem services it provides. A more recent book on poverty and ecosystem was written as part of the MEA. In the volume *Ecosystems and Human Well-Being* Chopra (2005) addresses the challenge of reversing the degradation of ecosystems while meeting increasing demands for their services through drastic policy and institutional changes. Although the aims of the book are similar to those of *Nature's Wealth*, the main difference is that it is not based on empirical case studies, but rather on the conceptual knowledge and views of a large number of experts. The book also has a more general macro-economic focus, while *Nature's Wealth* is more context-specific, based mainly on micro-economic research. The *Economics of Poverty, Environment and Natural-Resource Use* by Dellink and Ruijs (2008) also contributes to an improved understanding of the economic dimensions of environmental and natural resource management and poverty alleviation. The book differs from *Nature's Wealth* in that it focuses specifically on three themes: searching for explanations for the resource–poverty nexus; payments for and values of environmental and forestry resources; and sustainable land use. The book does not specifically have an ecosystem services perspective and considers the environment more generally. Finally, 'The economics of ecosystems and biodiversity' (TEEB) study (2009) generated various volumes specifically aimed at drawing attention to the global economic benefits of biodiversity, while highlighting the growing costs of biodiversity loss and ecosystem degradation. Although TEEB is currently the most extensive international initiative estimating the economic value of ecosystem services, it does not explicitly address poverty issues in contrast to *Nature's Wealth*.

In this chapter, we set the stage for the book by sketching a general framework for the assessment of the multiple dimensions of the relationship between ecosystem services and poverty. The background to the chapter is provided in Section 1.2, which describes the ecosystems–human well-being link. Section 1.3 more specifically addresses the link between ecosystems and poverty and highlights the relevant components underlying this complex relationship. Section 1.4 identifies various forms of interventions that can help to strengthen the link between ecosystem services and poverty alleviation. With this background in mind, Section 1.5 reflects on the main lessons learned from the 18 case studies covered in the book, following the central asset categories of security, capacity and ownership.

1.2 Ecosystem services and human well-being

The Millennium Ecosystem Assessment – MEA (2005) – built upon the concept of ecosystem services introduced in the 1970s (e.g. Ehrlich and Ehrlich 1981) to underline the inextricable linkages that exist between biodiversity, ecosystems and human well-being. Ecosystem services are 'the benefits people obtain from ecosystems' and by putting ecosystem services central to the debate on nature conservation, the MEA underlines the societal benefits of nature conservation and the need to align conservation and development goals. The MEA (2005) recognized four categories of services: supporting (e.g. nutrient cycling, soil formation and primary production); provisioning (e.g. food, fresh water, wood and fibre and fuel); regulating (e.g. climate regulation, flood and disease regulation and water purification); and cultural (aesthetic, spiritual, educational and recreational). Protecting the ecosystem helps to ensure the provisioning of these services in the long run and hence the associated flow of benefits.

Human beings benefit from ecosystem service provisioning through the access they have to it: for a global good such as climate regulation or biodiversity the whole world basically benefits, their public good character implies that people cannot be excluded from these benefits (Fisher *et al.* 2009). In practice, access may be limited by people's ability and capacity to access certain services, for example certain stretches of land providing the services, but the option and intrinsic value of climate regulation and biodiversity is often a benefit shared by all. For the different ecosystem services access to the benefits of service provision might be more confined to certain individuals or groups of people as property rights are assigned. In the case of food production, for example, land-use rights are usually with individual farmers, and access to food production is determined by these rights, and often rights to underlying groundwater resources to irrigate the land. Typically, regulating services like carbon uptake, water purification and climate regulation are public services, whereas many of the provisioning services have to a certain extent been privatized and submitted to the forces of market mechanisms. Especially in industrialized economies, water and food are usually provided through commercial companies, and ecological processes are being replaced by technological processes, but still depend on the ecological resource base. Finally, people benefit at different levels, provisioning services typically benefiting local and regional stakeholders and cultural and regulating services often benefiting stakeholders also at national and even global scale.

Protecting ecosystems thus serves human well-being, but this usually comes at a cost. Protecting ecosystems usually implies defining and enforcing resource-use restrictions, which may reduce the short-term extractive benefits that can be derived. Here too the distribution of benefits across spatial and temporal scales plays an important role. The benefits of protection may be regional, national or even global (biodiversity, carbon uptake, water purification), whereas the (opportunity) costs of conservation are felt locally. Here it is important to note, for example, that most of the world's biodiversity is located in developing countries, specifically in remote areas where economic development is low (Fisher and Christoph 2007). Depending on the approach taken to protect biodiversity, the opportunity costs of their protection will vary, but are still likely to be substantial. Ecosystems are often protected by completely banning the use and extraction of natural resources, i.e. fencing off an area and prohibiting human use. Cernea and Schmidt-Soltau (2006) showed that this approach may increase local poverty as people are relocated and denied access to the resource base on which their livelihoods depend. This increases vulnerability and makes them poorer, with sometimes adverse impacts on ecosystem conservation as well, due to the absence of resource maintenance and management.

Subsequently, integrated conservation–development approaches were developed, to improve local livelihoods and conservation at the same time (Salasky and Wollenberg 2000). Integrated conservation–development approaches try to substitute resource harvesting economically by creating alternative, higher use values of the ecosystem. These typically include approaches that complement the designation of protected areas with investments in alternative livelihoods, such as marketing of non-timber forest products (NTFP), community forestry and ecotourism. However, in those cases where there is no direct linkage between ecosystem conservation and the livelihood benefits people derive, integrated conservation–development approaches have not been very successful and the literature indicates that trade-offs are common and hard to avoid (Barrett et al. 2005, Sunderlin et al. 2005). For example, recent ecotourism studies indicate that income has increased in regions surrounding protected areas (Andam et al. 2010, Sims 2010), but Wittmeyer et al. (2008) suggest that ecotourism nevertheless results in increased ecosystem pressure due to increased population density and resource use. Similarly, the literature indicates that without additional investments in market access and services the economic feasibility of most NFTP projects is low (Belcher et al. 2005). Such investments, however, tend to increase resource exploitation and create conservation–development trade-offs.

Alternatively, more inclusive approaches to ecosystem protection have been developed together with approaches that directly link conservation and livelihood goals. Ferraro and Kiss (2002) indicate that direct linkage or incentive approaches are more effective in protecting biodiversity and enhancing livelihoods since they explicitly make local communities responsible for the conservation of the environmental resource base. Examples of direct linkage approaches are payments for ecosystem services (PES), as well as efforts to formalize the user rights of local communities and involve communities in protected area management, creating non-monetary incentives for sustainable use (see for example Maffi and Woodley 2010, Niesten and Milne 2009). For example, in PES schemes farmers are paid to conserve the forest and stop cutting trees (Ferraro and Kiss 2002). Thus, farmer income is improved and secured, and nature is conserved at the same time. Alternatively, decentralizing ecosystem management to local communities creates non-monetary incentives for conservation by partly transferring user rights: indigenous protected area or co-management approaches are examples of this approach where communities define, monitor and enforce resource-use restrictions themselves (Carlsson and Berkes 2005, Plummer and Fitzgibbon 2004). This has another advantage as local communities usually have more knowledge about the ecosystem and can better monitor and enforce sustainable resource use (Danielsen et al. 2008). This not only increases the effectiveness of conservation, but also lowers protected area monitoring and enforcement costs (Kubo and Supriyanto 2010, Somanathan et al. 2009). Community co-management of protected areas requires, however, that communities self-enforce resource use restrictions. Although the literature on common pool resource management has convincingly shown that communities are capable of doing this (Ostrom 1990, 2009), for effective self-enforcement certain conditions have to be met (Agrawal 2001).

Thus, even when it is possible to protect ecosystems at low opportunity and transaction costs, it is important to understand how ecosystem protection improves human well-being at different scales. This book explores the options for combining ecosystem protection with livelihood improvement for the poor. In the next section we further elaborate how the characteristics of poverty, ecosystem services and their protection relate.

1.3 Poverty, local livelihoods and ecosystem protection

In a world of more than 7 billion people, about a billion live on the estimated equivalent of less than a dollar per day (World Bank 2009). The overwhelming majority of these people live in rural areas in South Asia and sub-Saharan Africa, and due to population growth their number is expected to increase. Because poor people tend to have few assets, they depend for an important part of their livelihood on the natural resource base (Chen and Ravallion 2007). For example, poor people often depend on collective resources for livestock grazing (Kerr 2002), and consequently migrate to open access forest and wetland areas to improve their livelihood (Sunderlin et al. 2005). In fact, it is because of the dependence of poor people on common property resources that influential reports like MEA (2005) and TEEB (2009) argue that ecosystem protection will benefit the poor and that by improving the quality of ecosystem services, the benefits to poor people from the ecosystem will increase as well.

We view the relationship between poverty and ecosystem services in a slightly more critical light. First, there is increasing evidence that poor people might benefit less than non-poor people from improved resource management, and also bear most of the costs (see for example Adhikari et al. 2004, Kerr 2002). This is partly due to the fact that poor people are often not well represented in decision-making processes so that resource-use restrictions are not defined in their interests but in the interests of the better-off. Scholars such as Amartya Sen have pointed out that not being represented is an important determinant of poverty: people without a voice in decision-making lack the capacity and capability to pursue their needs (Anand and Sen 1997, Sen 1983, 1995). Although policymakers are becoming increasingly aware of the importance of ensuring the engagement of local communities in ecosystem management, getting the poor really on board to participate in decision-making is very difficult, and involves empowerment, awareness raising, education and other interventions as well (Bawa et al. 2007, Murphee 2009).

Second, the poverty–ecosystem nexus is not a one-way relationship but consists of a complex of interrelated factors. Poverty is often seen as one of the determinants of environmental degradation and loss of ecosystem services. At the same time, there is evidence suggesting that environmental degradation reinforces the extent of poverty. This suggests that the relationship between poverty and ecosystem services is often complex, since processes are often interlinked and several mediating factors influence the magnitude and sign of effects (Duraiappah 1998). This is illustrated in Figure 1.2.

1.3.1 How the poor affect ecosystems and their services

There are several potential mechanisms through which the poor influence available ecosystem services (summarized in the upper part of Figure 1.2). One should bear in mind that these mechanisms should always be analysed in combination with the effects of the environment on poverty (summarized in the lower part of Figure 1.2) rather than in isolation. Also the empirical evidence of the importance and validity of these mechanisms remains somewhat uncertain. We will start here by briefly describing some of the main driving forces behind ecosystem degradation and the pressures they exert on ecosystem services provision.

1.3.1.1 Depletion of natural resources

The poor are often accused of degrading their surrounding ecosystem, primarily through rapid depletion of natural resources (e.g. tree-cutting,

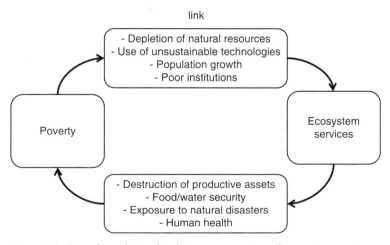

Figure 1.2 Complex relationship between poverty and ecosystem services

overfishing, etc.). Their livelihoods are largely dependent on their local natural resource base and environmental degradation is often the result of sole reliance on resource-intensive economic activities. While the poor may resort to unsustainable harvest patterns of their surrounding natural resource base, it is only fair to say that they are also often the victims of unsustainable consumption of the rich (Boyce 2002). Moreover, the poor often lack the skills and credit to diversify their economic activities and as a result they may rely exclusively on natural resources to sustain their livelihoods. Postponing consumption and use of natural resources for the sake of preventing environmental degradation is often not an option for the poor.

1.3.1.2 Population growth

A number of reasons explain why poorer countries and communities are traditionally characterized by higher fertility rates. In the absence of a well-established social security system parents tend to rely on their children for income transfers when old. Children often become productive assets of the household and participate in everyday economic activities. Due to the lack of formal employment and income for mothers, the opportunity cost of raising children is rather small. Population growth, coupled with inadequate means to increase production at the same rate, can lead to overexploitation of already fragile ecosystems.

1.3.1.3 Poor institutions

Poorer economies are generally characterized by weak institutional arrangements that often constrain further expansion of production. In poorer countries property rights are often poorly defined and enforced. This discourages households from investing in natural resources since there is high uncertainty about their future availability as household assets. Moreover, 'tragedy of the commons' scenarios often arise in the absence of well-defined property. Individuals fail to realize that private actions create negative external effects to other common-property users, which may result in pasture land degradation, water scarcity or extensive deforestation.

1.3.1.4 Use of unsustainable technologies

The lack of access to sustainable production technologies can lead to substantial disruption to ecosystem services (Asche 2008). Inappropriate fishing practices (e.g. in the form of unsuited fishing nets) often impact negatively on the stock of species and fish sizes not meant to be harvested. The lack of alternative environmentally friendly technology often restricts the poor to the unsustainable use of forest products for access to energy for cooking and heating. There are many reasons why the poor may resort to environmentally destructive technologies, such as lack of knowledge of alternative environmentally benign technologies and implementation, resistance to adopt unfamiliar practices and lack of credit to purchase alternative equipment.

1.3.2 How degraded ecosystems and their services affect the poor

The state of the ecosystem matters a lot to poor people, who are often hit the hardest by environmental degradation (Comim *et al.* 2009). Environmental damage can exacerbate poverty through destruction of productive assets, loss of food/water security, health impacts and exposure to natural disasters (see lower part of Figure 1.2).

1.3.2.1 Destruction of productive assets

Environmental damage often deprives the poor of productive assets. Pollution and soil erosion can reduce agricultural productivity of farmers and water pollution and water scarcity can reduce the availability of fish for dependent communities. Unsustainable deforestation can reduce the amount of forest products available to the poor and reduce energy access since they often rely excessively on locally available biomass for heating and cooking. Gradual environmental degradation may also imply that poor households need to travel longer distances to access ecosystem services that were previously locally located and directly available.

1.3.2.2 Food and water security

One of the most important provisioning services of ecosystems is in the form of food and water. The poor are often self-sufficient and dependent on locally produced food in order to meet their nutritional needs. They lack access to distant markets where food is traded and many of them resort to subsistence agriculture to meet their needs in food consumption. Significant declines in yields of staple food crops are often associated with soil contamination and erosion. The quantity and quality of water is also very much regulated by the conditions of local ecosystems. Water contamination restricts accessibility to safe drinking water for the poor. Deforestation often alters the local hydrological regime and significantly reduces water availability. Water scarcity is increasingly linked to migration, food security and conflict. With more than 1 billion people still lacking access to clean water sources, water poverty still remains a major challenge in the developing world (World Bank 2009).

1.3.2.3 Exposure to natural disasters

Ecosystems provide several regulating services to the poor. Flood and drought regulation are closely linked to sustainable soil and forest management. Landslides disrupt economic activity and may result in loss of life in areas that have been extensively deforested. In general, exposure to natural disasters is often larger for communities that live in a surrounding natural environment that is in a degraded state (Atker *et al.* 2011).

1.3.2.4 Human health

The health status of the poor is often dependent on the quality of ecosystem services in their vicinity. Almost 1.5 million children, primarily in the developing world, die every year as a result of inadequate access to safe drinking water and poor sanitation (WHO 2007). Poor people often live where environmental conditions are worst, since their lower purchasing power often restricts their mobility. As a result of poorer environmental health conditions they suffer disproportionately from diseases and face a higher risk of premature death. Poor individuals who face environmental health problems often experience a reduced ability to work and their family members sacrifice time to take care of them or work in their place. Furthermore, biodiversity loss may deprive poor communities of their immediate sources of medicine, and usually the only ones they can afford.

1.4 How to break the vicious cycle?

The poverty–environment nexus described in the previous section suggests that the poor can become trapped in a cycle of continuous environmental degradation and welfare decrease. The main challenge of policymakers operating in the domain of ecosystem management and poverty alleviation is to find ways to reverse the negative spiral and induce a mutually positive relationship between the poor and the environment.

The World Bank (2001) proposes a general framework for addressing the root causes of poverty by initiating action in three important areas:

- *Promoting opportunity*: expanding economic opportunity for poor people through a combination of market and non-market actions by expanding their assets and increasing the returns on these assets.
- *Facilitating empowerment*: strengthening the participation of poor people in political processes and local decision-making, and removing social barriers that result from distinctions of gender, ethnicity, race and social status.

- *Enhancing security*: reducing poor people's vulnerability to ill health, economic shocks, policy-induced dislocations, natural disasters and violence, as well as helping them cope with adverse shocks when they occur.

In adopting these three action categories as the basis for an effective poverty reduction strategy, the World Bank (2001) makes a number of nuances. First, the three areas for action are strongly interconnected. For example, action to expand opportunity is itself an important source of empowerment and improved security, and vice versa. Therefore, these actions cannot be considered in full isolation. Second, an effective poverty reduction strategy will require action on all three fronts, and needs to be implemented by all relevant agents in society – government, civil society, the private sector, and poor people themselves. Third, these areas of action are mainly targeted at the national or local level. Obviously, international actions are also required that harness global forces in favour of poor countries and poor people.

Although ecosystems are not a central factor, the above framework still provides a functional categorization of the potential interventions in the environment–poverty nexus. Therefore, we adopt the three areas of action and apply these to the ecosystem services domain. Before we describe the main lessons learned in the 18 case studies presented in the book, we explain the three areas of action in more detail.

1.4.1 Promoting opportunity and enhancing capacity

Poverty relates strongly to the capacity to influence decision-making as well as to the capacity and opportunity to invest in sustainable practices and alternative livelihoods. Poor health, for example, dramatically influences a person's capacity to make a living. Especially for the poor, health is closely related to ecosystem service provisioning, poor people usually not having access to purified drinking water and depending for medicine on the ecosystem as well (see above). In particular, the urban poor often live where environmental conditions are worst, suffering from bad health caused by environmental degradation since they cannot afford to move anywhere else. Capacity is related to underdevelopment and the socio-economic and institutional context in which many poor people live. Missing markets (for credit, inputs, insurance, etc.) reduce people's capacity to make efficient use of resources and sustain their livelihood. Lack of access to technology also reduces people's capacity to respond to changes and invest in sustainable resource use. Finally, capacity is related to education, knowledge and awareness, which help to diversify income strategies and facilitate more sustainable resource use. Alternatively, lack of capacity tends to increase degradation, making socio-economic development sometimes the main strategy for more sustainable resource use.

Poverty may arise due to a lack of access to multiple types of productive assets thereby limiting opportunities to escape poverty. The main forms of assets crucial for economic development include the following five categories (Hobley and Shields 2000):

- *Physical assets.* This mainly comprises already manufactured assets that can be used for productive purposes. Lack of access to capital assets lowers the productivity of resource use, but tends to reduce the environmental impacts of resource extraction as well.
- *Human assets.* The level of health and education of individuals are direct determinants of labour productivity. Healthy and educated workers are able to work longer and more productively. As discussed under capacity, lack of health and education are specific characteristics of the poor.
- *Financial assets.* Access to money (credit) is necessary for individuals in order to set up businesses and carry out investments, but also to secure and protect themselves against shocks. Without access to financial services, people tend to invest less and depend more on informal mechanisms to cope with vulnerabilities and stress.
- *Social assets.* Social capital is the trust, norms and networks that enable collective action (Bouma *et al.* 2008). Poor people do not necessarily own less social capital, but given their lack of access to the formal economy, they tend to depend more on social networks for ensuring their livelihoods. Social capital is especially important when considering community co-management arrangements, where poor households have to voluntarily cooperate in sustainable resource use.
- *Natural assets.* This comprises the stock of our natural ecosystems. The different types of services that our ecosystems provide to us (e.g. in terms of fuelwood, soil fertility, water availability, tourism, biodiversity) have a direct impact on our livelihoods and welfare levels. The role of natural capital and ecosystem service provision is the central question addressed in this book: Given that poor people have few assets of their own they tend to depend on collective resources to a relatively large extent, most notably natural capital.

Box 1.1 *Payments for Ecosystem Services (PES)*

These are economic incentives offered to individuals/firms – the 'potential providers' of ecosystem services – in exchange for a certain action that protects a valuable ecosystem service (Wunder 2008). Such schemes of monetary transfers have been used to a larger extent for climate change mitigation, protection of watershed services and biodiversity conservation. Although many such schemes exist, especially in developing countries in Central and South America and South East Asia, the factors that drive and explain their environmental performance are still poorly understood. Moreover, implementation of PES schemes can be problematic in a developing country context. The poor are often the most important beneficiaries of ecosystem services, but their low financial capacity prevents them from carrying out any payments. One should also bear in mind that in many cases the poor are both the potential providers as well as beneficiaries of ecosystem services. Bulte *et al.* (2008) focused on PES both as a mechanism for environmental protection and poverty reduction, and showed that tying PES and poverty reduction may result in lower efficiency in meeting either objective, thus it may be better to focus programmes that concentrate on one or the other objective separately.

The literature indicates that designing such mechanisms requires, amongst other things, a clear demand and supply of the ecosystem services, a condition that often cannot be met (Engel *et al.* 2008). Especially when considering biodiversity protection, where demand is global and unorganized and large incentives to free-ride exist (Pearce 2007) a payment mechanism may not be feasible for linking global nature conservation and local livelihoods. Proper monitoring of the impacts of PES schemes on the relevant ecosystem services provided at different scales is in many cases lacking. International monitoring guidelines are needed to facilitate comparisons between different scheme designs. Only in this way will relevant causal relationships between institutional design and environmental performance be understood and hence those design factors driving the success or failure of PES schemes in achieving environmental objectives (Brouwer *et al.* 2011).

Key in expanding opportunities for the poor is to build up assets and capabilities and try to increase the returns of these assets. A range of actions can support poor people in expanding their assets. The state has a central role, especially in providing infrastructure and social services such as health and education. Where access to land is highly unequal, there is a social and economic case for negotiated land reforms. Often the state's role in service provision can be complemented by market mechanisms, civil society and the private sector, increasing the benefits to the poor. The returns on these assets depend on access to markets and all the global, national and local influences on returns in these markets. A typical example of improving market access for poor people in the ecosystem services domain is the instrument PES, as described in Box 1.1.

1.4.2 Strengthen ownership and facilitate empowerment

Ecosystem services are valuable to the poor for the wide range of regulating and provisioning services that they provide. Unregulated access to natural resources often deprives poor communities of such services. This can happen because entities outside the community (e.g. industrial firms) have little interest in environmental externalities they impose on the poor. It can also happen because the community itself fails to regulate a sustainable provision for ecosystems services internally, as in the classic example of the tragedy of the commons. Strengthening ownership and facilitating empowerment are two ways to deal with the above-mentioned inefficiency.

Ownership of ecosystem services, ideally from their main beneficiaries, implies that identified owners have an increased incentive to manage their natural resources sustainably, as well as seek compensation whenever external forces impact negatively on them (Ostrom and Hess 2010). Ownership determines whether the externalities of ecosystem service provisioning can be internalized and, thus, whether people are willing to cooperate in sustainable resource use. Secure property rights and pricing of ecosystem services ensures that an economic value is attached to environmental damage, hence creating incentives for people to safeguard their valued natural resources: enforceable property rights (e.g. secure tenure) allow ownership of ecosystem services to be honoured and empower the poor (Araujo et al. 2009).

Empowerment means enhancing the capacity of the poor to influence institutions that affect their lives, by strengthening their participation in political processes and local decision-making (World Bank 2001). This implies the removal of political, legal and social barriers that work against poor people and building their assets to enable them to engage effectively in markets. Expanding economic opportunities for poor people indeed contributes to their empowerment. But efforts are needed to make state and social institutions work in the interests of poor people – to make them pro-poor. Formal democratic processes are part of empowerment.

A popular instrument, which is increasingly used to facilitate empowerment and strengthen ownership, is Community-based natural resource management (CBNRM) of ecosystems or natural areas (See Box 1.2).

Box 1.2   *Community-based natural resource management (CBNRM)*

Decentralizing ecosystem management to local communities creates non-monetary incentives for conservation by partly transferring user rights: indigenous protected area or co-management approaches are examples of this approach where communities define, monitor and enforce resource-use restrictions themselves (Plummer and Fitzgibbon 2004). For CBNRM to be successful, all relevant stakeholders need to come together and participate in the decision-making. Crucially, community-based management depends on whether the community succeeds in defining collective decision-making institutions and enforcing (in)formal agreements and rules (Agrawal and Gibson 1999, Ostrom 1990). Involving communities in natural resource management has another advantage as local communities usually have more knowledge about the ecosystem and can better monitor and enforce sustainable resource use (Danielsen *et al.* 2008). This not only increases the effectiveness of natural resource management, but also lowers protected area monitoring and enforcement costs (Somanathan *et al.* 2009). CBNRM requires, however, that communities self-enforce resource-use restrictions, a requirement which cannot always be met (Agrawal 2001). Also, CBNRM can be less suitable when macro-scale environmental issues extend beyond community borders and when the community is not integrated in formal decision-making at higher governance scales (Berkes 2006). It can also be less successful when the management design reflects to a larger extent the interests of a few influential stakeholders rather than the collective interest of the community.

1.4.3 Enhancing security

Security is an important contributing factor that can simultaneously alleviate poverty and enhance the provision of ecosystem services. Enhancing security for poor people means reducing their vulnerability to such risks as ill health, economic shocks and natural disasters and helping them cope with adverse shocks when they do occur (Armitage *et al.* 2008, World Bank 2009). Enhancing security can reduce the vulnerability of poor people through a range of approaches that provide the means for the poor to manage risk themselves, and strengthen institutions for risk management.

Because poor people cannot insure themselves against, for example, the risks of crop failure or flooding, they tend to be hit hardest by environmental degradation and climate change (Atker *et al.* 2011). Often ecosystems can have an important safety net function as well: in times of need poor people often resort to fishing and NTFP collection to fulfil their most imminent needs. Security also has to do with trade-offs and how redistribution of costs and benefits affects local livelihoods.

In that respect, ownership and security need to coexist in order to achieve poverty alleviation and environmental quality simultaneously. Income security more broadly allows poor households to achieve sustainable use of ecosystem services. Income volatility and uncertainty encourages a rapid depletion of natural resources and associated loss of ecosystem services (Andersson *et al.* 2011). How income security can be insured against climate change risks is illustrated in Box 1.3.

1.5 Lessons learned in the case studies

This chapter has presented an overall framework for actions in three areas – opportunity, empowerment and security – jointly to alleviate poverty and stimulate the provision of ecosystem services. Many questions still remain. For example, how can policymakers set priorities in practice, and do actions in all three areas have to be carried out at the same time? Obviously, there is no blueprint on how to alleviate poverty and improve ecosystem management since the mixture of actions needed depends on each country's or community's economic, socio-political, structural and cultural context. But even though choices depend on local conditions, it generally is necessary to consider scope for action in all three areas – opportunity, empowerment and security – because of their crucial complementarities (World Bank 2001).

Box 1.3   *Climate change risk micro insurance*

Insurance has been referred to as an effective tool for reducing, sharing and spreading climate change-induced disaster risks in both developed and developing countries (Botzen and van den Bergh 2008, Brouwer and Akter 2010). The body of literature related to catastrophe insurance and crop insurance, in particular, is vast and rapidly growing. Advocates argue that crop insurance can play a vital role as a risk management instrument to enable poor farmers in developing economies to cope with weather-related production risk, hence contributing to poverty alleviation (e.g. Hazell 1992). On the other hand, others describe catastrophic risks, including those related to agricultural crops, as uninsurable and unsustainable in the long run as the transfer of losses from affected groups to the community at large is not feasible at an affordable premium. Associated costs of providing insurance outweigh the gains from risk spreading.

Although practical experiences are limited, insurance as a mitigation strategy seems to have been unsuccessful based on standard commercial criteria throughout the world (Akter *et al.* 2009). Especially in developing countries where the poorest parts of the population often find themselves in a downward spiral of recurrent damages due to natural calamities, premiums for disaster insurance schemes fail to earn enough revenue to cover payouts as well as administrative implementation costs. As a result, new public–private partnerships are explored to overcome the institutional–economic barriers to the introduction of alternative risk mitigation schemes for the poor, encouraging them at the same time through risk premium differentiation to adapt to changing climate change conditions and associated risk exposure (Akter *et al.* 2011).

Adopting the three main areas of action, the most important lessons of the 18 case studies presented in the book are listed and categorized. This summary of lessons learned is not intended to give a complete overview of the main conclusions of the case studies, but aims to highlight similarities and differences that we can learn from.

1.5.1 Promoting opportunity and enhancing capacity

Lesson 1: Variation in capacity of the poor implies that there is no blueprint route to success

The analysis of human–elephant conflicts (Chapter 3), for example, indicates that whereas in some cases it is better to offer compensation, in other cases technical solutions are a better option and in other cases insurance is best. The analysis of livelihood–poverty–conservation linkages (Chapter 4) suggests that the dependence of the poor on the ecosystem also differs: in remote regions with subsistence economies the poor are often the food insecure households that depend on low productivity agriculture and the direct collection of natural resources for their survival, whereas in less remote areas they are often the income poor, landless households with non-diversified livelihood strategies that use the ecosystem to supplement their nutritional intake and/or for additional cash. Interventions that aim to conserve nature while avoiding negative livelihood impacts need to acknowledge the different linkages and the implications for conservation success.

Lesson 2: Higher educational and awareness attainments correlate with sustainable management

Education and awareness are important determinants of the capability of the poor to adopt alternative income-generating strategies. The case study on tenure security and ecosystem service provisioning in Kenya (Chapter 17) suggests that there is a strong positive correlation between the years of schooling of the household head and the amount of investment in water and soil conservation. This allows households to take better advantage to agricultural extension services, which tend to impact positively on average income levels. The case study on protected area management in Nepal (Chapter 2) also shows that households with relatively high levels of education are more willing to get involved in tourist-related activities. The marine protected area (MPA) case study (Chapter 5) demonstrated that with higher levels of education and awareness, the local support of protected area management increased, thereby making it easier to enforce its rules and regulations. The involvement of local knowledge institutes which educate the involved stakeholders about the underlying ecological processes while simultaneously monitoring the level of service provision of the protected area proved to be a crucial element in the schemes. At the same time, the desire for education may also harm the ecosystems involved. The study on PES in the Philippines (Chapter 9) reveals how many upstream forest dwellers, while being aware of possible detrimental effects, illegally cut the forest to generate the funds for their children to go to school.

Lesson 3: Time availability influences access to environmental assets
The case study on common pool resources and income poverty in rural India (Chapter 16) reveals that time constraints influence choices over resource use. In the Jhabua district, poorer households find it easier to engage in time-intensive flower/seed collection, since they suffer less from 'time poverty' relative to richer households. Richer households with less of a labour surplus (largely due to a smaller family size) prefer to engage in resource collection with a higher value:time ratio. Similarly, in rural Kenya (Chapter 17), households with a large number of children between 6 and 16 years of age find it easier to allocate time towards soil and water conservation. Likewise, different studies in this book indicate that local stakeholders prefer investments in alternative livelihoods as a way to compensate lost income caused by ecosystem management, since the opportunities for developing income-generating activities in many of the case study sites are low. Among others, this is the case in the Nepal and Sri Lanka studies on human–wildlife conflicts (Chapter 2 and 3) and the deforestation studies in Tanzania and Zambia (Chapters 8 and 10).

Lesson 4: Access to markets is a major contributing factor to poverty alleviation and sustainable ecosystem management
This is one of the key findings of the study on pastureland management and poverty among herders in Mongolia (Chapter 18). Making use of caloric terms of trade tables and corresponding conversion rates, the authors of the study conclude that access to urban markets allows rural herders to meet their nutritional needs with a smaller production of meat and milk. At the same time, this will allow dependent rural communities to cope better with risks associated with any temporary adverse weather conditions in their vicinity. Our case study on common pool resources and income poverty in rural India (Chapter 16) suggests that trading of common natural resources also takes place at a more localized level across rural communities, but this restricts the range of commodities that can be exchanged.

Lesson 5: A sustainable management of ecosystems requires good availability of census data on poverty and geographic information system (GIS) data on land-use changes

The case study on poverty and land-use changes (wetlands/forest cover) in rural Uganda (Chapter 19) combines extensive time-series census data on poverty with biophysical information on land-use changes of wetlands and forests. The overlapping of data series allows the authors to create maps that depict areas of excessive resource degradation and high poverty levels. Governments, particularly in developing nations, have limited resources to carry out interventions. The combined use of disaggregated datasets and mapping techniques allow policymakers to identify 'poverty/ environment hotspots' and design interventions accordingly. The analysis of human–elephant conflicts in Sri Lanka (Chapter 3) also illustrates the importance of geographical information and spatial mapping. Combining maps of rainfall, death from human–elephant conflicts and poverty indicated that conflicts were highest where people are poor. Similar conclusions can be drawn for all studies that involved watershed functions and river-basin-related ecosystem services such as the PES study in the Philippines (Chapter 9), the case study on the Niger in Mali (Chapter 13) and the flood-coping study in Bangladesh (Chapter 14).

Lesson 6: Because the role of ecosystem services is often ignored by the donor community, traditional development aid can do more harm than good

Development aid plays an important role in many case studies covered in the book. In some cases, development aid promotes sustainable management by explicitly taking into account the role of ecosystem services in local societies. Examples of such support include the role of the World Bank in Tanzania (Chapter 8) adopting the charcoal chain concept, which facilitates a more holistic approach in addressing the problem of deforestation in the Dar es Salaam region. However, the book also reports many detrimental impacts of development aid. The dam infrastructure in the Mali study (Chapter 13) and the flood defence measures in Bangladesh (Chapter 14), are typical examples of large infrastructural measures that insufficiently take into account possible environmental externalities caused by the interventions. On a smaller scale, donors may cause damage by pulling out in the middle of a transformation process. A good example is provided by the Pakistan case study (Chapter 11) where donor agencies were heavily involved in the decentralization process of forest management. Eventually, due to implementation slowness and especially because of the resistance at the provincial level to undertake the necessary steps to support the reform in due time, the support of donor agencies was withdrawn, leaving the heritage of unfinished reforms in several districts. Now trust in this type of intervention among local communities is lost and it will be substantially more difficult for new reforms to build confidence again.

1.5.2 Strengthen ownership and facilitate empowerment

Lesson 1: It is a misconception that only the poor rely on common-pool resources
The case study on common-pool resources and income poverty in rural India (Chapter 16) suggests that both the very poor as well as the relatively rich are largely dependent on (provisioning) common-property ecosystem services. Based on observations across 60 villages in the Jhabua district in Central India the study confirms that poverty alleviation alone is not sufficient to ease pressure on common resources. This contradicts earlier findings pointing to a strong substitutability between private assets and common-pool natural resources. Similar conclusions can be drawn from the supply chain study near the capital Dar es Salaam (Chapter 8) which found that charcoal in Tanzania serves the rich and the poor alike, implying that both groups in society are also responsible for the forest degradation that is caused by unsustainable charcoal production and consumption.

Lesson 2: Ownership encourages investment in natural resource conservation but is not necessarily a sufficient condition for enhanced household welfare
The case study on tenure security and ecosystem service provisioning in Kenya (Chapter 17) provides support to a strong positive link between tenure security and investment in soil and water conservation. Evidence from 18 villages in rural Kenya suggests that household income tends to increase as a result of land conservation investment. This is particularly the case when land is registered in the name of the household head rather than another member of the extended family. However, besides landownership, additional factors, such as investment in conservation measures, soil fertility and low population density, are equally important in ensuring that there is a positive ownership–income relationship. The case study on pastureland management and poverty in rural Mongolia (Chapter 18) emphasizes the role of market access as an additional complementary factor.

Lesson 3: Regulating access to common-pool resources prevents food shortages and improves the nutritional intake of households

This is a key finding of the case study on pastureland management and poverty among herders in Mongolia (Chapter 18). The study concludes that government policies that encourage productivity-oriented commercial herding can benefit herders both nutritionally as well as financially, when this is accompanied by increased access to markets. In this way, meat and milk products can be sustainably produced and exchanged for grain in urban markets. The case study on tenure security and ecosystem provisioning in Kenya (Chapter 16) emphasizes that secure land tenure can enhance agricultural productivity and food security, particularly in areas that experience high population growth. The MPA study in the Asia-Pacific region (Chapter 5) finds similar results. Villages managing MPAs effectively seem to be more resilient in times of crisis and are also better organized in terms of health care and education.

Lesson 4: The mediating role of ownership in the poverty–environment nexus can be qualitatively different between the short and the long term

This is the key conclusion of the case study on poverty and land-use changes (wetlands/forest cover) in rural Uganda (Chapter 19). The study combines welfare estimates with biophysical information to explore the links between poverty changes and resource degradation between 1992 and 1999. The authors of the study conclude that in the short run excessive degradation of local wetlands and forest can provide a temporary relief for impoverished communities. In the long term, the causality is likely to reverse and the corresponding loss of ecosystem services will exacerbate poverty levels. The study in Bangladesh (Chapter 14) on different coping strategies for flooding also finds that technical solutions, such as building dikes, provides short term protection against flooding, but in the longer term at the cost of the productivity of flood-dependent agriculture. Similarly, the study on the effects of dams in Mali (Chapter 13) shows that expanding dam capacity only leads to a transfer of benefits from flood-dependent downstream communities to the irrigation-based farmers upstream, rather than an increase of the overall welfare.

Lesson 5: Poorly designed formalization of ownership often deprives the poor from traditional access to ecosystem services

Often formalization of ownership is pursued with good intentions, yet due to ineffective implementation traditional rights systems are often disassembled without replacing them with a well-functioning alternative rights system managing ecosystems. Rather than slowing down, the degradation of ecosystems accelerates after such fruitless interventions. A classical example of such counterproductive interventions is provided by the study on the institutional reform in the forest sector in the Swat region of Pakistan (Chapter 11). Due to an incoherent set of external interventions and strategic reactions by different agents in the local communities, the emergent system of management is the one producing a dismal and unsustainable outcome. Such flawed interventions are also recorded by the study on the implementation of an MPA in Hon Mun in Vietnam (Chapter 6). This study revealed that almost half of the fishermen sacrificing fishing grounds for the creation of the MPA are now worse off than before the implementation of the MPA. Similar effects are also described in the study on fishing rights allocation in South Africa's Western Cape (Chapter 7), where strict quotas have been introduced for precisely the fisheries of greatest significance to local communities, and fishers are struggling to cope with this reduced resource availability.

1.5.3 Enhance security

Lesson 1: Within communities, poorer and richer households require secure access to different environmental assets

An important lesson is that communities are seldom homogenous and therefore relate to ecosystem services in different ways. The case study on common pool resources and income poverty in rural India (Chapter 16) concludes that secure access to ecosystem services is equally important for the poorer as well as the richer households. Nevertheless, the poor and the rich tend to rely on different types of natural resources to supplement income and sustain livelihoods. In the Jhabua district, for instance, the poor require secure access to flowers and seeds, while richer households rely to a larger extent on an uninterrupted provision of fodder and construction wood. Especially the wildlife-related chapters in this book demonstrate that where some suffer from wildlife protection others might benefit, such as those engaged in ecotourism. This influences community-based conservation or co-management initiatives, as depending on the representation of the different stakeholder interests, different decisions will be reached. For instance, in the example of rhino conservation in Nepal (Chapter 2) the landless would prefer more days of access to the protected area and farmers would prefer more compensation for the crops lost, whereas tour operators would prefer stricter enforcement of the rules.

Acknowledging the different interests, understanding the severity of the underlying livelihood issues and making sure all interests are represented in the decision-making is crucial to avoid local trade-offs. The analysis of livelihood–poverty–protected area linkages (Chapter 4), however, indicates that the poor especially often do not feel properly represented in the decision-making: to avoid negative trade-offs wildlife protection and protected area management should specifically consider how their interests can be met. Similar patterns are seen in the Vietnam study (Chapter 6) where fishermen feel underrepresented and therefore excluded from the benefits generated by the MPA.

Lesson 2: In absence of secure ownership, individuals fail to internalize environmental externalities and ecosystems become degraded
Government action is necessary to restrict access and hence prevent a 'tragedy of the commons' from occurring. This lesson is prevalent in almost all of the studies presented in this book. The case study on pasture-land management and poverty among herders in Mongolia (Chapter 18) suggests that herders in rural Mongolia fail to grasp the consequences of simultaneously pursuing increases in their livestock. Data from 60 house-holds in Ugtaal and Gurvansaikhan reveal that herders treat local pasture-land as a common resource and fail to coordinate their actions to prevent ecosystem degradation and overgrazing. Our case study on poverty and wetland/forest degradation in Uganda (Chapter 19) suggests that this result also extends to the case of local wetlands and forests in sub-Saharan Africa – impoverished households resort to excessive extraction of local resources with little understanding of negative side effects to other resource users in the present or in the future. Similar lessons can be learned from the deforestation studies in Tanzania (Chapter 8) and Zambia (Chapter 10), where public forest is disappearing at alarming rates due to charcoal production as a result of which a number of regulating and cultural ecosystem services are declining rapidly. In both cases, national and local government agencies try hard to impose command and control policies, which prove to be extremely inefficient. Changing the tenure systems seems to be one of the crucial ingredients of a more sustainable management regime.

Lesson 3: The government needs to work together with local communities in order to ensure that ownership of natural resources is mutually respected
The case study on poverty and land-use changes (wetlands/forest cover) in rural Uganda (Chapter 19) concludes that any top-down enforcement of

secure ownership is likely to be counterproductive. There are often tensions and mistrust between government enforcement agents and local communities, with the latter viewing the protection of ecosystems as a direct threat to their most immediate sources of income. In such cases, the government can stimulate community-level management of forests and wetlands and provide technical assistance to local councils when necessary. In particular the locally managed marine areas (LMMAs) in Fiji (Chapter 5) are a classic example of how cautious collaboration between local communities, the national government and knowledge institutes can lead to a sustainable and self-propelling movement of improved management of the marine environment. In a period of less than 10 years, the majority of the coastal communities in Fiji adopted the concept of LMMAs. The main role of the government was to provide the legal conditions for this sustainable concept to flourish. However, an important lesson learned from the analysis of community co-management in protected areas (Chapter 4) is that decentralizing protected area management to local communities should not automatically be expected to work. When conservation has little impact on local livelihoods, when the drivers of biodiversity depletion are non-local and when the perceived legitimacy of protected area establishment is limited then the potential for community co-management is small.

Lesson 4: In order for the government to enforce user rights, economic instruments are needed to collect sufficient revenue from, among others, ecosystem-related activities

Expecting the government to assign user rights and enforce these at the same time requires agencies to have sufficient means to fulfil their role as public authorities. Lack of government funds is one of the causes of poor government interventions. A typical example of ineffective government policies in ecosystem management is described in the case study in Tanzania (Chapter 8). The government recognizes the role of charcoal as the main driver of deforestation in Tanzania, yet partly due to insufficient funds, their response of simply banning charcoal production is very short-sighted and therefore ineffective. By improving fiscal instruments in the charcoal chain, revenues can be collected for enforcement while simultaneously providing the incentives for more sustainable production and consumption patterns. This certainly also holds for the situation in Zambia (Chapter 10) where copper production could easily provide the means for enforcing sustainable policies in the related forestry sector. The best example of how fiscal reform could lead to 'double dividend' is given by the case study on the water sector in South Africa (Chapter 15). This study analyses the potential outcomes of introducing a water charge that, combined with a tax break on food, would be the most effective policy to achieve the joint goals of diminished water consumption, reduced poverty and growth of the South African economy.

Lesson 5: Security issues in human–wildlife conflicts call for special attention for the negative impacts of healthy ecosystems

The two chapters addressing people–park conflicts (Chapters 2 and 3) show that when wildlife conservation increases local vulnerability conflicts between people and wildlife increase. In the case of rhinoceros poaching in Nepal (Chapter 2), farmers suffer from crop damage and the landless suffer because rhino conservation robs them from access to common land. When these people are not compensated, and if no efforts are made to improve their lot, they are the ones who start poaching rhinos, which is exactly what happened in the Nepalese case. In the case of human–elephant conflicts in Sri Lanka (Chapter 3), poor farmers not only lose part of their crops to the increasing number of elephants but in the ensuing conflicts many also get killed. Clearly, the security issue here becomes very pressing, and a highly innovative solution in the form of elephant insurance was found. Similar responses are observed in the Inner Niger Delta in Mali (Chapter 13) where farmers use toxic pesticides and nets to prevent migratory birds from feeding on their crops, thereby not only killing the birds but also heavily polluting the aquatic environment. Compensation measures and awareness raising could prevent these unsustainable practices from occurring.

References

Adhikari, B., Di Falco, S. and Lovett, J.C. (2004). Household characteristics and forest dependency: evidence from common property forest management in Nepal. *Ecological Economics*, **48**(2): 245–257.

Agrawal, A. (2001). Common property institutions and sustainable governance of resources. *World Development*, **29**(10): 1649–1672.

Agrawal, A. and Gibson, C. (1999). Enchantment and disenchantment: the role of community in natural resource conservation. *World Development*, **27**(4): 629–649.

Akter, S., Brouwer, R., Choudhury, S. and Aziz, S. (2009). Is there a commercially viable market for crop insurance in rural Bangladesh? *Mitigation and Adaptation Strategies of Global Change*, **14**: 215–229.

Akter, S., Brouwer, R., van Beukering, P.J.H. *et al.* (2011). Exploring the feasibility of private micro flood-insurance provision in Bangladesh. *Disasters*, **35**(2): 287–307.

Anand, S. and Sen, A. (1997). Concepts of human development and poverty: a multidimensional perspective. Human Development Papers, UNDP.

Andam, K. S., Ferraro, P. J., Sims, K. R., Healy, A. and Holland, M. (2010). Protected areas reduced poverty in Costa Rica and Thailand. *PNAS*, **107**(22): 9996–10001.

Andersson, C., Mekonnen, A. and Stage, J. (2011). Impacts of the Productive Safety Net Program in Ethiopia on livestock and tree holdings of rural households. *Journal of Development Economics*, **94**: 119–126.

Araujo, C., Araujo Bonjean, C., Combes, J.-L., Combes Motel, P. and Reis, E. J. (2009). Property rights and deforestation in the Brazilian Amazon. *Ecological Economics*, **68**: 2461–2468.

Armitage, D. M. Marschke, M. and R. Plummer (2008). Adaptive co-management and the paradox of learning, *Global Environmental Change*, **18**: 86–98.

Asche, F. (2008). Farming the sea. *Marine Resource Economics*, **23**: 527–547.

Barrett, C., Lee, D. and McPeak, J. (2005). Institutional arrangements for rural poverty reduction and resource conservation. *World Development*, **33**(2): 193–197.

Bawa, K., Joseph, G. and Setty, S. (2007). Poverty, biodiversity and institutions in forest-agriculture ecotones in the Western Ghats and Eastern Himalaya ranges of India. *Agriculture, Ecosystems and Environment*, **12**: 287–295.

Belcher, B, Ruiz-Perez, M. and Achdiawan, R. (2005). Global patterns and trends in the use and management of commercial NTFPs: implications for livelihoods and conservation. *World Development*, **33**(9): 1435–1452.

Berkes, F. (2006). From community-based resource management to complex systems. *Ecology and Society*, **11**(1): 45.

Botzen, W. J. W. and van den Bergh, J. C. J. M. (2008). Insurance against climate change and flooding in the Netherlands: present, future and comparison with other countries. *Risk Analysis*, **28**(2): 413–426.

Bouma, J. A., Bulte, E. H. and van Soest, D. P. (2008). Trust and cooperation: social capital and community resource management. *Journal of Environmental Economics and Management*, **56**: 155–166.

Boyce, J. K. (2002). *The Political Economy of the Environment*. Cheltenham, UK: Edward Elgar Publishing Ltd.

Brouwer, R. and Akter, S. (2010). Informing micro insurance contract design to mitigate climate change catastrophe risks using choice experiments. *Environmental Hazards*, **9**: 74–88.

Brouwer, R., Tesfaye, A. and Pauw, P. (2011). Meta-analysis of institutional-economic factors explaining the environmental performance of payments for watershed services. *Environmental Conservation*, **38**(4): 1–13.

Bulte, E. H., Lipper, L., Stringer, R. and Zilberman, D. (2008) Payments for ecosystem services and poverty reduction: concepts, issues and empirical perspective. *Environment and Development Economics*, **13**: 245–254.

Carlsson, L. and Berkes, F. (2005). Co-management: concepts and methodological implications. *Journal of Environmental Management*, **75**(1): 65–76.

Cernea, M. M. and K. Schmidt-Soltau (2006). Poverty risks and national parks: policy issues in conservation and resettlement. *World Development*, **34**(10): 1808–1830.

Chen, S. and Ravallion, M. (2007). Absolute poverty measures for the developing world: 1981–2004. World Bank Policy Research Working Paper 4211.

Chopra, K. R. (2005). *Ecosystems and Human Well-being: Volume 3: Policy Responses.* Millennium Ecosystem Assessment Series Vol. 3. Washington DC: Island Press.

Comim, F., Kumar, P. and Sirven, N. (2009). Poverty and environment links: an illustration from Africa. *Journal of International Development*, **21**: 447–469.

Danielsen, F., Burgess, N., Balmford, A. *et al.* (2008). Local participation in natural resource monitoring: a characterization of approaches. *Conservation Biology*, **23** (1): 31–42.

Dellink, R. B. and Ruijs, A. (eds.) (2008). *Economics of Poverty, Environment and Natural-Resource Use.* Frontis Series, 25(VI). Wageningen, the Netherlands: Springer.

Duraiappah, A. K. (1998). Poverty and environmental degradation: a review and analysis of the nexus. *World Development*, **26**: 2169–2179.

Ehrlich, P. R. and Ehrlich, A. (1981). *Extinction: The Causes and Consequences of the Disappearance of Species.* New York: Random House.

Engel, S., Pagiola, S. and Wunder, S. (2008). Designing payments for environmental services in theory and practice: an overview of issues. *Ecological Economics*, **65**: 663–674.

Ferraro, P. J. and Kiss, A. (2002). Direct payments to conserve biodiversity. *Science*, **298**(28): 1718.

Fisher, B. and Cristoph, T. (2007). Poverty and biodiversity: measuring the overlap of human poverty and the biodiversity hotspots. *Ecological Economics*, **62**: 93–101.

Fisher, B., Turner, K. and Morling, P. (2009). Defining and classifying ecosystem services for decision-making. *Ecological Economics*, **68**: 643–653.

Hazell, P. B. R. (1992) The appropriate role of agricultural insurance in developing countries. *Journal of International Development*, **4**: 567–581.

Hobley, M. and Shields, D. (2000). *The Reality of Trying to Transform Structures and Processes: Forestry in Rural Livelihoods.* London: Overseas Development Institute.

Kerr, J. (2002). Watershed development, environmental services and poverty alleviation in India. *World Development*, **30**: 1387–1400.

Kubo, H. and Supriyanto, B. (2010). From fence-and-fine to participatory conservation: mechanisms of transformation in conservation governance at the Gunung Halimun-Salak National Park, Indonesia. *Biodiversity and Conservation*, **19**(6): 1785–1803.

Maffi, L. and Woodley, E. (2010). *Biocultural Diversity Conservation.* London: Earthscan.

Martinez-Alier, J. (2002). *The Environmentalism of the Poor: A Study of Ecological Conflicts and Valuation.* Cheltenham, UK: Edward Elgar Publishing Ltd.

Millennium Ecosystem Assessment (MEA) (2005). *Ecosystems and Human Well-being: Biodiversity Synthesis.* Washington DC: World Resources Institute.

Murphee, M. (2009). The strategic pillars of communal natural resource management: benefit, empowerment and conservation. *Biodiversity Conservation*, **18**: 2551–2562.

Niesten, E. and Milne, S. (2009). Direct payments for biodiversity conservation in developing countries: practical insights for design and implementation. *Oryx*, **43** (4): 530–541.

Ostrom, E. (1990). *Governing the Commons: The Evolution of Institutions for Collective Action.* Cambridge: Cambridge University Press.

Ostrom, E. (2009). A general framework for analyzing sustainability of social-ecological systems. *Science*, **325**: 419–422

Ostrom, E. and Hess, E. (2010). Private and common property rights. In B. Bouckaert (ed.), *Property Law and Economics*. Cheltenham, UK: Edward Elgar Publishing Ltd., pp. 53–106.

Pearce, D. (2007). Do we really care about biodiversity? *Environmental and Resource Economics*, **37**: 313–333.

Plummer, R. and Fitzgibbon, J. (2004). Co-management of natural resources: a proposed framework. *Environmental Management*, **33** (6): 876–885.

Salafsky, N. and Wollenberg, E. (2000). Linking livelihoods and conservation: a conceptual framework and scale for assessing the integration of human needs and biodiversity. *World Development*, **28**(8): 1421–1438.

Sen, A. (1983). Poverty, relatively speaking. *Oxford Economic Papers*, **35**(2): 153–169.

Sen, A. (1995). Rationality and social choice. *American Economic Review*, **85**(1): 1–24.

Sims, K. (2010). Conservation and development: evidence from Thai protected areas. *Journal of Environmental Economics and Management*, **60**: 94–114.

Somanathan, E., Prabhakar, R. and Mehta, B. S. (2009). Decentralization for cost-effective conservation. *Proceedings of the National Academy of Sciences*, **106**(11): 4143–4147.

Sunderlin, W, Angelsen, A., Belcher, B. *et al.* (2005). Livelihoods, forests and conservation in developing countries: an overview. *World Development*, **33**(9): 1383–1402.

TEEB (2009). The economics of ecosystems and biodiversity for national and international policy makers. Summary: responding to the value of nature. Available at: http://www.teebweb.org/.

UN (2010). High-level meeting of the General Assembly as a contribution to the international year of biodiversity, A/64/865. United Nations General Assembly.

WHO (2007). *Country Profiles of Environmental Burden of Disease*. Geneva: World Health Organization.

Wittmyer, G., Elsen, P., Bean, W. T. *et al.*(2008). Accelerated human population growth at protected area edges. *Science*, **321**: 123–126.

World Bank (2001). *Attacking Poverty. World Development Report 2000/2001*. Washington DC: World Bank.

World Bank (2009). *Moving Out of Poverty*. Washington DC: World Bank.

Wunder, S. (2008). Payments for environmental services and the poor: concepts and preliminary evidence. *Environment and Development Economics*, **13**: 279–297.

Part I
Biodiversity–related ecosystem services

JETSKE BOUMA

Biodiversity is being depleted at an alarming rate (UN 2010). This is a problem, not only because biodiversity is intrinsically valuable, but also because biodiversity is the underlying asset from which all ecosystem services are produced. Biodiversity is best conserved in protected areas and national parks (Bruner *et al.* 2001). Establishing protected areas often has significant consequences for local communities. On the one hand, negative impacts occur when local communities are denied access to natural resources or are relocated to more marginal parts (Cernea and Schmidt-Soltau 2006). But even when communities are allowed to continue living near or inside protected areas the impacts of protected area establishment can be great: increased populations of wildlife increase crop damages and sometimes even cause personal injuries (Woodroffe *et al.* 2005) and the resource-use restrictions inside protected areas limit resource harvests for subsistence use (Wilkie *et al.* 2006). On the other hand, recent studies indicate that protected area establishment can also have positive livelihood effects: protected areas attract ecotourism, which can generate significant revenues at regional scale (Andam *et al.* 2010, Sims 2010). If these revenues trickle down to the local level, then protected area establishment can also have positive livelihood impacts at local scale.

This part of the book focuses on the potential trade-offs and synergies occurring when protected areas are established in areas where people are living as well. The rationale behind protected area establishment is often the meeting of international biodiversity targets or the protection of national

Nature's Wealth: The Economics of Ecosystem Services and Poverty, ed. P. J. H. van Beukering, E. Papyrakis, J. Bouma and R. Brouwer. Published by Cambridge University Press, © Cambridge University Press 2013.

interests such as the safeguarding of ecosystem provisioning services, including drinking water, timber or fish. Locally, people are often not consulted in the process of protected area establishment and their interests are hardly reflected in nature conservation plans. Biodiversity is a truly global public good and in principle the whole world profits from the conservation of genetic, species and ecosystem variety, both for its existence value and for its role in ecosystem resilience and regulating and preventing disease (MEA 2005). Local communities benefit from these services also, as well as from some of the more regional provisioning and regulating of ecosystem services, but since they are most affected by the resource use restrictions implied in protected area establishment they also tend to bear most of the costs. Given that poverty is high in most biodiversity-rich regions (Fisher and Christoph 2007), it is crucial to pay attention to local interests and avoid negative livelihood effects.

Acknowledging the crucial link between nature protection and local livelihoods, more inclusive approaches to protected area management have been developed over the past decades, including approaches that invest in alternative livelihoods (i.e. the marketing of non-timber forest products, community forestry, etc.) and approaches that directly compensate local users for restricted resource use (i.e. payments for ecosystem services) (Ferraro and Kiss 2002, Salafsky and Wollenberg 2000). Experiences with the different approaches vary. The literature indicates that approaches that directly link livelihood improvements to conservational outcomes are more effective than approaches with only an indirect link (Ferraro and Kiss 2002). Overall, the literature shows that creating synergies between biodiversity protection and local livelihood development is difficult and requires a good understanding of local dependencies on the natural resource base. The following chapters will illustrate some of these interdependencies and address how biodiversity can be protected while avoiding negative livelihood effects.

Organization of Part I

The first two chapters of Part I focus on park–people conflicts in Sri Lanka and Nepal. The Nepalese case study presented in Chapter 2 analyses the conflicts in the Royal Chitwan National Park. This park harbours a rich variety of ecosystems and wildlife, including 25% of the world's population of one-horned rhinos, of which only 400 remain. These rhinos are creating substantial damage to surrounding communities as they trample and eat agricultural crops. Many of the indigenous communities that were displaced because of the establishment of the park have become active in poaching rhinos. This is threatening the very survival of the rhino population and, in its wake, the increasing revenues from rhino ecotourism as well. Using stakeholder analysis, discrete choice experiments and simulation, the analysis indicates that conservation–development synergies are feasible but difficult to realize because of the diverging interests of the different stakeholders: landholding households seem most interested in crop damage compensation and job creation in ecotourism, whereas landless households are more interested in alternative livelihood investments at household and community scales and increased access to the protected resource base. The authors conclude that a policy mix of strong anti-poaching interventions combined with additional income generation investment (job creation in ecotourism, alternative livelihood development) will be the most optimal policy response and acceptable to all stakeholders.

Chapter 3 analyses the human–elephant conflict in Sri Lanka. This conflict has become very pressing: in the conflict an average of 50 people and 150 elephants die annually. The main reasons for the conflict are habitat destruction caused by illegal human encroachments, interruption of elephant corridors and scarcity of food and water in the national parks. Apart from increased mortality, elephants are causing havoc by destroying crops and property, which is causing substantial economic losses as well. Using spatial analysis and GIS mapping, the chapter first assesses which regions are prone to human–elephant conflicts, as well as the specific factors causing these conflicts. The analysis then zooms in on four human–elephant conflict prone areas, where household survey data are collected to analyse the relationship between rural poverty and the human–elephant conflict. The authors use multi-criteria analysis to rank the possible management responses, ranging from electric fences to compensation schemes. The analysis indicates that there is no general best practice for dealing with human–elephant conflicts, and that in some cases electric fences and in other cases compensation works best. The authors conclude that insurance can be a feasible mechanism for compensation as well.

Chapter 4 takes a broader perspective on poverty, livelihood and conservation issues by assessing poverty–nature linkages in four biodiversity hotpots around the world. Using household survey and village meeting data, the study analyses the potential synergies for nature conservation and local livelihood improvement in India, South Africa, Vietnam and Costa Rica. In all four regions biodiversity is conserved in protected areas, but the type of management, the type of ecosystem pressures and the type of livelihood strategies differ between the sites. The different approaches are compared and the interdependencies between nature conservation and local livelihoods are systematically assessed. The findings suggest that whereas biodiversity protection is currently creating livelihood trade-offs in most of the sites, synergies can be accomplished by more directly linking local livelihoods and nature conservation. Crucial for such a direct linking is that local livelihoods significantly depends on the protected resource base, that protected area management is open to direct community involvement and that local communities can influence the factors driving ecosystem degradation and biodiversity depletion at local and regional scales. To avoid adverse impacts on poor households, specific attention has to be paid to incorporating the interests of the poor in protected area management. In all four sites, income poor and food insecure households perceive the direct collection of products from nature as more important for their livelihoods than more wealthy households. However, they also feel less able to influence protected area management and decision-making at village scale. Specifically, minority ethnicity households feel less well represented in protected area decision-making, and they generally perceive the legitimacy of protected area establishment as low.

References

Andam, K. S., Ferraro, P. J., Sims, K. R., Healy, A. and Holland, M. (2010). Protected areas reduced poverty in Costa Rica and Thailand. *PNAS*, **107**(22): 9996–10001.

Bruner, A. G., Gullison, R. E., Rice, R. E. and da Fonseca, G. A. B. (2001). Effectiveness of parks in protecting tropical biodiversity. *Science*, **291**: 125–128.

Cernea, M. M. and Schmidt-Soltau, K. (2006). Poverty risks and national parks: policy issues in conservation and resettlement. *World Development*, **34**(10): 1808–1830.

Ferraro, P. J. and Kiss, A. (2002). Direct payments to conserve biodiversity. *Science*, **298**(28): 1718.

Fisher, B. and Cristoph, T. (2007). Poverty and biodiversity: measuring the overlap of human poverty and the biodiversity hotspots. *Ecological Economics*, **62**: 93–101.

Millennium Ecosystem Assessment (MEA) (2005). *Ecosystems and Human Well-Being: Biodiversity Synthesis*. Washington DC: World Resources Institute.

Salafsky, N. and Wollenberg, E. (2000). Linking livelihoods and conservation: a conceptual framework and scale for assessing the integration of human needs and biodiversity. *World Development*, **28**(8): 1421–1438.

Sims, K. (2010). Conservation and development: evidence from Thai protected areas. *Journal of Environmental Economics and Management*, **60**, 94–114.

UN (2010). High-level meeting of the General Assembly as a contribution to the international year of biodiversity, A/64/865. United Nations General Assembly.

Wilkie, D., Morelli, G., Demmer, J., Starkey, M., Telfer, P. and Steil, M. (2006). Parks and people: assessing the human welfare effects of establishing protected areas for biodiversity conservation. *Conservation Biology*, **20**(1), 247–249.

Woodroffe, R., Thirgood, S. J. and Rabinowitz, A. (eds.) (2005). *People and Wildlife: Conflict or Co-existence?* Conservation Biology series, Vol. 9. Cambridge: Cambridge University Press.

2 · Park–people conflicts, rhino conservation and poverty alleviation in Nepal

BHIM ADHIKARI, DUNCAN KNOWLER,
MAHESH POUDYAL AND
WOLFGANG HAIDER

2.1 Introduction

The conflict between protected area management and the local use of natural resources from those areas – often termed the 'park–people conflict' – has been at the forefront of management and policy debates during the last two decades. However, this conflict goes far beyond the use of natural resources, and as numerous studies have highlighted, includes problems such as 'human–wildlife conflict',[1] and the 'poaching problem' to name a few. The starting point for most, if not all, of the park–people conflicts seems to be the restriction of the use of resources inside the protected areas, especially of the traditional user rights (Maikhuri *et al.* 2000, Nepal, 2002). Heinen (1996) argues that most of the conflicts in developing countries which relate to the access and use of resources within protected areas have arisen where countries adopted a Western approach to protected area management. For the most part, this approach prohibits the use and extraction of natural resources inside the protected area. Heinen further points out that any such conflicts in developed countries can be easily resolved due to lower dependence of people on resource extraction. However, it is much harder to resolve these conflicts in countries where users are likely to depend on the protected area resources for their very

---

[1] Human–wildlife conflict usually has three sources: (1) livestock depredation by the protected wildlife; (2) crop damage by the protected wildlife; and (3) injuries to or loss of human life to the protected wildlife.

Nature's Wealth: The Economics of Ecosystem Services and Poverty, ed. P. J. H. van Beukering, E. Papyrakis, J. Bouma and R. Brouwer. Published by Cambridge University Press, © Cambridge University Press 2013.

survival. For instance, people living around the park buffer zone depend on forest products such as food, meat, fruits, medicines, fuels, animal food and building materials for their subsistence. Forests, trees and wildlife are also important components of the natural capital of the poor in the buffer zones of CNP, the central Terai zone of Nepal. Specifically non-farmer and landless households belonging to marginalized ethnic groups, such as *Chepang*, *Bote* and *Majhi*, depend on the forest and aquatic ecosystems for their livelihood. These communities have been living in the forestlands and on the riversides for generations but due to park establishment they were displaced.

The establishment of national park and wildlife reserves normally excludes communities from using the resources on which they traditionally depend. Quite often, establishment of protected areas has severely impacted indigenous customary right-, value-, belief- and livelihood-support systems (Nepal 2002). Studies have shown that huge efforts have been directed towards resolving these conflicts, especially regarding the access and use of protected areas over the last decade or so (Straede and Helles 2000, Mehta and Heinen 2001, Nepal 2002). However, as mentioned earlier, these conflicts do not end with a granting of access or user rights. As numerous studies have shown, other forms of conflicts, including human–wildlife conflicts, which affect people living close to protected areas, and the poaching problem affecting the population of the protected species of concern, remain a problem in park–people relations around the world. Although in Nepal protected areas for biodiversity conservation have been established since the late 1960s, park–people conflicts are still among the most important conservation trade-offs. For instance, a number of studies highlighted the nature of park–people conflicts in CNP, which is a crucial conservation area in terms of abundance of rare and endangered fauna, flora and rich cultural heritage (Mishra and Jefferies 1991, Sharma 1990). Although conservation and protection of wildlife in CNP is well recognized internationally, the conflict between park management and the local community is a serious matter of concern. Local communities have not seen the park favourably since its inception; they feel they are 'worse off' due to crop and livestock losses and restricted access to park resources. About 7% of the rhino population lives outside the national park, disturbing the people and their livelihood (Martin and Vigne 1996). Nepal and Weber (1994) estimated that wild animals, especially rhinos, boars and spotted deer destroy 13.2% of the crops around CNP. Furthemore, 60% of the paddy lost to wild animals is caused by rhinos, which often trample the paddy at night (Martin and Vinge 1996).

Despite considerable loss of crops and livestock from wild animals, local farmers do not receive any compensation for this damage. The revised

National Parks and Wildlife Conservation Act has provision for disbursing 30 to 50% of the income of a park to the corresponding Buffer Zone Management Committee (BZMC). Based on this legal provision, CNP has provided about NRs 183.8 million (Nepalese rupees, equivalent to US$ 2.2 million) to its BZMC over the last 6 years. However, the BZMC has so far spent only NRs 77.4 million (US$ 1 million dollars), which is less than half of the revenue it has received from CNP. This is largely due to the lack of a clear policy on the part of the BZMC, especially on how to distribute the money to different buffer zone user committees and for what purpose. So far, support has been concentrated on income-generating and conservation activities rather than on enhancing alternate livelihood opportunities for the poorest sections of the society.

Local farmers generally see wildlife as a threat and a nuisance, whereas local indigenous people often poach rhinos due to lack of alternate livelihoods and high and increasing poverty levels. Since rhinos are one of the most attractive animals in the park, poaching these animals may create a significant economic impact by reducing revenues from ecotourism. Considering this impact, and the status of the one-horned rhino as an endangered species, an all-out effort is needed to understand the reasons behind the recent increase in poaching and to identify new policy directions to protect the future of rhinos in Nepal. In addition, it is important to analyse the options for reducing poverty and promoting economic development in the Terai region, to reduce the vulnerability of local livelihoods.

This chapter presents the findings of an economic analysis of rhino poaching in Nepal and assesses the policy scenarios under which conservation and poverty reduction could be achieved. The study addresses three major research questions: (1) What external and policy-related factors have influenced poaching historically and what could be the effectiveness of conventional poaching control? (2) Who are the main stakeholders in the management of the one-horned rhino population and what are the gains and losses for these stakeholders under current rhino management policies? (3) What potential management options could provide incentives for communities to become involved in eradicating poaching but also help reduce poverty in and around the buffer zone of CNP?

The chapter is organized as follows. Section 2.2 deals with survey, data collection and analytical approaches. Results and discussions are provided in Section 2.3. The chapter concludes with a set of policy implications (see Section 2.4).

2.2 Data and methodology

In the study sites, agriculture and livestock are an integral part of the livelihood strategy of households, as in other parts of Nepal. The main impacts of park establishment for these households are the damage caused by rhinos to their crops. The damage caused by rhinos is particularly evident in Sauraha, Bagmara, Ratnanagar, Pithauli, Dibyapuri and Patihani Village Development Committees (VDCs). In these VDCs, rhinos damage paddy, wheat, maize and lentil plants, and eat cauliflower, potato, radish, banana and other vegetables and plants. Many local farmers in these VDCs have abandoned the cultivation of wheat, as rhinos like to graze in wheat fields during winter.

Economic losses in terms of crop damage and loss of livestock from rhinos have made villagers even poorer. Hence, it should come as no surprise that many are not happy with park management. Further, due to widespread poverty and lack of alternative income opportunities, some local people, particularly from ethnic groups, are involved in poaching rhinos and other rare and endangered species, and rhino poaching has been steadily increasing each year in the park. Although rhino poaching dropped to almost zero from 1976 to 1983, it became a serious problem in 1984. Poaching reached a peak in 1992 with 17 animals killed from a population of less than 500 (Maskey 1998). More recently, the Department of National Parks and Wildlife Conservation (DNPWC) office in Sauraha reported that 46 rhinos (31 in Chitwan and 15 in Nawalparasi) were killed in the single year 2002. Local indigenous people, particularly *Chepang, Tharu, Tamang, Magar* and *Darai* have been involved in poaching rhinos and their involvement is believed to be triggered by economic poverty. If this situation continues, there will be a vicious 'downward spiral' between poverty and environment, in which poor people, facing immediate survival needs, mine the very resources that could underpin their livelihoods and generate ecotourism returns.

In 1994 the government initiated a 'Park and People Project' around the buffer zone with financial support from the United Nations Development Programme (UNDP). The main objective was to empower local communities living in and around the park buffer zone to initiate various community development works with a focus on biodiversity conservation. However, the buffer zone programme could not provide real incentives to local households, as the projects did not target their needs. One of the more frequent failures in buffer zone management occurs when the link between stakeholders and authorities is weak. Though the buffer zone programme was viewed as a successful example of community-based conservation, there is still doubt as to whether community welfare has actually improved and resulted in an equitable distribution of costs and benefits among the different stakeholder groups.

In an attempt to address these pertinent policy issues, field data were collected by surveying six VDCs in the buffer zone of CNP (i.e. Sauraha, Bagmara, Ratnanagar, Pithauli, Dibyapuri and Patihani). To capture a wide range of stakeholders, their interests and perceptions of park-related issues, the survey sites were selected based on two criteria: (1) proximity to CNP and (2) proximity to the main tourism centre.

A combined sample of approximately 450 households was interviewed in the six VDCs listed above. The household survey gathered information on socio-economic status of households, production systems, household use of park resources and households' perceptions of rhino conservation and park management. In addition, qualitative and quantitative village-level data were collected with a village survey regarding infrastructure development, area and utilization pattern of park resources and related issues at village scale. Finally, a discrete choice experiment (DCE) was administered during the household survey as well.

A variety of economic and biological modelling approaches were used to examine historical poaching losses (see Poudyal and Knowler 2005 for detail), to simulate poaching behaviour and assess alternative policy scenarios (see Knowler et al. 2005). In order to understand the historical influences on variables of policy interest (e.g. poaching of rhinoceros), econometric techniques were used to explore the significance of the different variables. We then used simulation modelling to analyse the impact of various socio-economic and policy variables on the level of poaching of the one-horned rhinoceros in CNP for the period 1973–2003. Table 2.1 describes the variables incorporated into the econometric estimation.

In the discrete choice experiment (DCE), respondents were asked to evaluate hypothetical scenarios. The survey instrument for this study was developed through a series of discussions with experts, focus groups and pre-tests in the field (see Adhikari et al. 2005 for detail). The main purpose of these stages was to identify and describe the most relevant attributes and their associated levels. The first attribute reflects the need to describe changes in park benefits (number of rhinos protected), an attribute which also doubles as a key ecological indicator. The second attribute describes the gains to the community from rhino protection in terms of tourism

Table 2.1 *Variables used in the econometric modelling of poaching of the one-horned rhinoceros in CNP from 1973–2003*

Variable code	Definition
Dependent variable:	
POACH_NP	Number of rhinoceros poached during the year inside CNP
Independent variables:	
POPN	The population of rhinoceros inside CNP at the end of the year
REAL_PEN	The penalty imposed for convicted poachers in real terms
APU	The number of anti-poaching units active during the year
GDPC_NEP	Per capita GDP of Nepal in constant 1990 prices
GDPC_HK	Per capita GDP of Hong Kong in constant 1990 prices
MAOIST	Dummy variable that equals to 0 up to year 1996 and 1 for the year 1997 onwards (to account for the effect of Maoist insurgency in poaching)

development and employment. The third attribute describes an income-generation programme for the benefit of the entire community. Two other attributes describe compensatory policies, such as compensation paid to farmers for crop damage and household access to park resources (specifically for collecting roof thatching grasses). Each of these attributes was described on four levels, which were derived from field-testing.

The scenarios were created by combining the five attributes in different combinations resembling future outcomes (results) of possible management actions. For the choice experiment, two scenarios were combined into one choice set (Figure 2.1). Each choice set contained two hypothetical alternatives 1 and 2, and one additional scenario describing the status quo situation. Respondents were asked to choose among one of these three alternatives, or to state that none of these three alternatives was acceptable.

Respondents were residents of communities within the buffer zone of CNP. Given the level of illiteracy at 41% in the buffer zone (DNPWC/PPP 2001), the choice sets were designed with pictograms representing the respective concepts, and vertical bars representing the respective levels (Table 2.2) so that even illiterate respondents could recognize the attributes and their respective level.

For the simulation exercise, a model was constructed drawing on the historical poaching analysis described earlier. The first step involved construction of a population dynamics model for the rhino (see Rothley *et al.* 2004). Subsequent steps in constructing the simulation model involved linking the poaching model with the population dynamics model to form

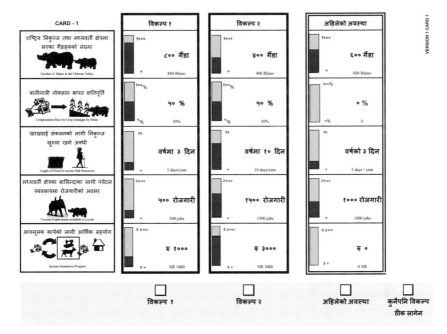

Figure 2.1 Example of a choice set from the discrete choice experiment

a single sub-model. Then the simulation model was extended to include an additional sub-model that compiled components of household income in the buffer zone (regional income) and a further one that computed the revenue–expenditure balance in the community share of national park revenues. The simulations differed in terms of policy scenarios, including variations in the level of anti-poaching enforcement, compensation paid to farmers for crop losses due to rhino and local employment initiatives. Table 2.2 describes the policy scenarios used for the simulation model.

2.3 Results and discussion

2.3.1 Description of the stakeholders

Stakeholder analysis identified five major stakeholder groups with respect to rhino conservation: non-farmers, farmers, hotel and lodge owners, government and non-governmental conservation organizations, and tourists and visitors. Generally, the group of non-farmers and landless households is composed of households belonging to marginalized ethnic groups such as the *Chepang*, *Bote* and *Majhi* communities. Non-farmer households can

Table 2.2 *Assumptions used in policy scenarios analysed*

Policy variables★	Policy scenarios			
	1. Baseline	2. Conventional conservation strategy	3. Incentives-based conservation strategy (farmer emphasis)	4. Incentives-based conservation strategy (non-farmer emphasis)
Community patrols (APUs)	0	15	4	4
Crop damage compensation	0	0	25%	10%
Tourism jobs subsidized for locals (% of total jobs)	0%	2%	5%	15%
NRs collection days in CNP	3	3	5	7
Income generation/ micro-credit	0	1500★★	1000	2000

★ Cost assumptions: APUs: NRs 500 000 per APU per year; job subsidy: NRs 23 000 per job per year
★★ Only from excess budget after allocating to APUs

further be divided into two different groups i.e. non-farmers landless/ marginal households and non-farmer households with at least one member employed in the private or public sector, including CNP. The non-farm, landless households, often from the indigenous groups, are households that used to collect wild fruits, vegetables, mushrooms and other forest products and a variety of medicinal plants from the forests. After the establishment of the national park, these communities lost their traditional rights over their lands, territories, forests, rivers and other natural resources, leaving them without access to land. Some people belonging to this group are now involved in rhino poaching. Interviews with poachers revealed that economic gains from rhino poaching were one of the main reasons why people get involved (see detail in Adhikari *et al.* 2005).

Local farmers around the park buffer zone are the second type of stakeholders who were further divided into three sub-stakeholder groups, namely poor, middle wealth and rich farmers. Farmers of large and medium-sized holdings near the buffer zone reported heavy crop damage from the rhinos, with average losses of respectively NRs 3913 and 2727 per year. Furthermore, they had to spend about NRs 1000 in constructing and maintaining defensive measures in their fields. The production of small farms is decreasing each year, mainly for two reasons. First, like the rich and medium farmers they suffer from crop damage worth about NRs 2200 every year, plus the costs associated with the defensive measures. Second, the national park has affected the number of livestock that households keep, which has had a negative impact on agricultural production. Despite these adverse impacts, most of the stakeholders mentioned that if provided with lands or a job, they would support the conservation of rhinos and other wildlife.

Hotel and lodge owners are another important stakeholder group. They include business people in Sauraha and the surrounding areas of the CNP. This is the group that benefited most from the park. They generally consider rhinos to be an important source of income as every year the CNP attracts a number of tourists from around the world. These tourists would not come if the rhino ceased to exist. Development of tourist facilities around the CNP buffer zone has been rapid in response to the increasing number of visitors.

Government and non-governmental conservation organizations represent national and international stakes in rhino conservation. While the national interest consists of a combination of economic and social concerns, the global interest in the conservation of rhinos has to do with the wish to conserve a unique genetic resource and the wish to maintain ecosystem resilience in and around the park. The park authorities indicate that park establishment has not only contributed to national and international interests but to local community development as well. Of the total income generated from the park, 50% goes to local community development. The official records of the national park show that to date NRs 11 million have been spent on local communities. Of the funds, 30% were used for community development and 20% for conservation programmes. There are also income-generating programmes, such as mushroom farming, knitting and weaving, biogas plant projects, supported by the King Mahendra Trust for Nature Conservation (KMTNC) which is a well-known non-governmental conservation organization in Nepal. These income-generating programmes are targeted particularly towards the park-affected communities.

Finally, the last stakeholder group consists of the global stakeholders in rhino conservation and ecotourism, the (international) tourists and visitors. Preservation of biodiversity in the CNP provides opportunity to observe the greatest diversity of landscapes, plants and endangered wild animals for tourists and visitors. Environmentally concerned visitors prefer to go for ecotourism to enjoy natural wilderness. As tourism promotion is one of the key strategies in CNP, visitors are important stakeholders who want to have easy access to rhinos and other wild animals. At present, over 70% of the 96 000 tourists visit CNP mainly to see rhino and Bengal tigers. The number of tourists to CNP has increased from 836 in 1974 to over 64 000 in 1994/95 (KMTNC 1996).

2.3.2 Analysis of stakeholder preferences

Results from the discrete choice experiment show that respondents prefer a steady increase in the number of rhinos. This is a bit surprising given that people treat rhinos as a threat to their livelihoods because of the damage they cause: this came out strongly during the group discussion. One important point to note here is that farmers dislike the nuisance created by rhino but are not necessarily against rhino per se. It appears that most stakeholders are quite happy to accept a larger rhino population for various forms of other benefits. The simulation modelling based on the results of the DCE makes these trade-offs very apparent.

At present, compensation for crop damage is not in place in the CNP buffer zone. It was expected that farming households would prefer increasing amounts of compensation against agricultural crops damaged by rhinos. The level of support rose sharply for the initial rises in the level of compensation, however, as the compensation rate approached 60%, the changes in the level of support were not so clear and farming households were rather indifferent to any further rise in the compensation rate. Various stakeholder groups reacted rather differently to the varying amounts of compensation, with the high-income farmers being much less sensitive to increasing amounts of compensation at a lower compensation level as opposed to mid- and low-income farmers. On the other hand, non-farm households, especially the landless marginalized segment, supported crop compensation only at the lower level; as the level of compensation rose above 40%, their support for the compensation policy started to decrease again.

All local stakeholder groups, except the landless marginalized households, clearly preferred more tourism-related jobs, and for all groups the linear estimates were significant. Landless, marginalized households probably did not prefer more tourism-related jobs because their education level is very low and they feel that they would have little chance for any of these jobs even if the employment opportunities increased. Furthermore, obtaining a job either in CNP or related sectors depends on the individual's networking capacity, connection to authority and power relationship with other individuals within the community, connections and relations that marginalized households tend not to have.

Access to the park was an important variable for many groups of the local population because it allowed them to satisfy certain subsistence needs. All groups desired some access to the park's resources, but were satisfied with 6 or 7 days per year. Only the landless, marginalized group had strong preference for additional days of access, i.e. the maximum number of days offered in this model.

The income-generation programme for the community was also considered very important by all groups. The high-income farmers show a clear linear preference pattern, while the middle-income group were happy with smaller loan amounts. The landless marginalized had much stronger support for this programme. Direct financial support in the form of micro credit schemes was the most popular policy option for non-farming communities, who felt that NRs 2000 per recipient household significantly raised their socio-economic status.

Finally, we tested whether the results would hold for specific sub-groups. Female respondents were significantly less in favour of the alternative management options compared to males, they were also less supportive of higher numbers of rhinos and had a higher preference for income-generation programmes. This may be due to the fact that women members of a household have more interaction with park management, and the rhino population, given their higher engagement in collection and gathering activities inside and outside the CNP. Respondents living closer to tourism centres preferred more rhinos, considered an increase in tourism-related job opportunities as more important, and were also more in favour of income-generation plans. As to be expected, respondents who currently spend three or more days per year for subsistence purposes in the park were in favour of additional days of access, while the ones who currently do not use the park were much less in favour. On the other hand, the group who currently does not use the access was more in favour of income-generation programmes.

2.3.3 Historic analysis of poaching in CNP

The historic analysis considered the factors that were expected to influence the level of poaching, analysing the likely impact of these factors on the historic level of poaching of one-horned rhinoceros in CNP. The main factors influencing poaching levels considered in the literature are: (1) effectiveness of anti-poaching efforts; (2) penalties when caught poaching; (3) available economic alternatives (i.e., the opportunity cost of poaching); (4) direct costs of poaching; and (5) the price of rhino horn on international (black) markets (see for example, Jachmann and Billiouw 1997; Leader-Williams *et al.* 1990; Milner-Gulland 1993; Milner-Gulland and Leader-Williams 1992). Count data models (Poisson and Negative Binomial) were used to estimate the poaching model. The variables used in the poaching model are described in Table 2.1.

Of all the factors considered of influence on the level of poaching only the penalty (in real terms) was insignificant in all estimations. It is worth noting that the level of penalty over the years has been fixed at two (nominal) levels. Thus, the penalty in real terms demonstrated a decreasing trend over the study period, which could have caused a reduced influence of this variable on the level of poaching. Yet, as the coefficient on penalty in the estimation remained insignificant it might also be argued that the maximum fine established under the Wildlife Act has not been effective in deterring poaching in CNP.

Real per capita gross domestic product (GDP) for East Asia (lagged by a year), which was used as a proxy for the price of rhino horn on international markets, remained insignificant in all the models. The sign on the coefficient of this variable stayed positive over all the estimations, showing the consistency of its effect on the dependent variable. Furthermore, the significance level (*p*-value) for the coefficient on this variable was close to 10%, when estimated using the negative binomial model. Given the small sample size (31 observations), this can be considered relatively significant. An alternative model using the real per capita GDP for Hong Kong lagged by a year (GDPC_HK) was estimated as well. The rationale for using per capita GDP for Hong Kong follows from its location as the first international port of trade for rhino horn from Nepal, as well as being a significant consumer of rhino horn itself. However, the results from the estimation of this alternative model did not show any improvement over the original model.

Since the rhino population in CNP increased in most of the study period, poachers had an increasing stock from which to harvest. This is reflected in the highly significant and positive coefficient for this variable. It also suggests that a larger exploitable population might have increased the incentives to poach by lowering the search costs of poaching. Offsetting this effect was the influence of anti-poaching effort. Although the number of anti-poaching units (APU) active during the year represented only one element in anti-poaching effort (the other part being the involvement of the Royal Nepalese Army), it showed a consistently negative and highly significant impact on the level of poaching in CNP, which indicates their importance in rhino conservation in the park.

The real per capita GDP of Nepal also showed a consistently negative and highly significant effect on the level of poaching in CNP, indicating the importance of alternative economic opportunities in reducing the level of poaching. This is especially important to deter local poachers from being involved in poaching as they come from very poor and landless groups. Thus, consistent with earlier findings (e.g. Martin 1998, Milner-Gulland 1993, Milner-Gulland and Leader-Williams 1992), this analysis provides evidence for the creation of local alternative economic opportunities to deter poachers.

The ongoing Maoist insurgency in Nepal has been considered a major influence on poaching in CNP in recent years (Martin 1998, Yonzon 2002). The significant rise in the level of poaching during the insurgency years compared to the years before, as shown by the estimated model, provides support for this conjecture. However, it must be clarified that this study did not consider whether the Maoist rebels were involved in poaching. The only conclusion that can be drawn from the analysis is that the Maoist uprising has helped poachers indirectly by making the anti-poaching efforts less effective, and by creating a climate of instability. This reflects the importance of political stability in biodiversity conservation, especially in the conservation of highly valuable species like the one-horned rhinoceros.

2.3.4 Simulation model of alternative policies to combat rhino poaching

A key objective of our study was to test the usefulness/applicability of an ecological–economic simulation model in assessing policy scenarios at the aggregate level (e.g. the buffer zone). The results from these runs provide important insights into the effects of various policy variables on the rhino population, level of poaching, tourist visitation and local community revenues and household income (the latter through tourism jobs, rice production or micro-credit schemes). Changes in some specific policy variables, such as crop damage compensation or micro-credit schemes, only contribute to the income of certain stakeholder groups (as per our assumptions). However, as components of regional income, they contribute to overall community income in the buffer zone.

Three programme scenarios were simulated, in addition to a current or 'baseline' scenario. The scenarios varied according to a number of factors that influence the rhino population and the income of the buffer zone households (Table 2.2). In the baseline scenario (Scenario 1), all the policy variables are kept at their 2003 level. For example, community patrols, crop damage compensation and funding under the micro-credit scheme are set at zero. In addition, no local tourism jobs are subsidized and the park is open for 3 days per year for natural resources collection. The other scenarios are described in Table 2.2 and comprise the following:

Scenario 2 – a conventional conservation strategy with emphasis on APUs;

Scenario 3 – an incentives-based conservation strategy with emphasis on farmers; and

Scenario 4 – an incentives-based conservation strategy with emphasis on non-farmers.

One policy variable that affected all stakeholders, either negatively or positively, was the level of community patrolling (APUs). Although community patrols do not directly affect rice production, they do so indirectly by decreasing poaching and increasing the population of rhinos. On the other hand, an increase in community patrols has a positive effect on household incomes through the dual effects of paid employment while patrolling and increased job opportunities created by higher visitor numbers; this, in turn, is partly the result of higher numbers of rhino (due to increased patrols). With regard to the rhino population, Scenario 2 (with 15 APUs) was the best scenario, as it gave the highest population at the end of the simulation period and a higher rhino population in each year of the simulation, compared to other scenarios. Although buffer-zone communities under Scenario 2 suffer higher losses of rice (and hence lower production), the increased share of revenue they receive from the CNP and the increased income they receive from the growing tourism sector more than compensate for this loss at the community level. In fact, the aggregate community income with this policy option is over 12% higher at the end of the simulation period compared to that of the baseline scenario.

In contrast to Scenario 2, Scenarios 3 and 4 assume fewer APUs and more community income devoted to economic development activities. As a result, Scenarios 3 and 4 provide greater benefits to specific household groups (such as farmers, through higher compensation in Scenario 3, and non-farmers through increased jobs subsidies, park open days, and micro-credit facilities in Scenario 4). However, the aggregate community income under these policy scenarios is lower than that in Scenario 2.

This result can be explained by examining the calculus of agricultural and other income gains and losses under the various policy scenarios. For example, the income from rice production is higher under Scenarios 3 and 4 than under Scenario 2. But this increase is not as significant as the increase in tourism-related employment income and the communities' share of national park revenues in Scenario 2. In the latter case, these additional funds can be used to generate additional household income via community-based and funded economic activities. Hence, aggregate income suffers in Scenarios 3 and 4 compared to that in Scenario 2.

We can analyse the impact of each policy scenario on specific groups of households. For example, Scenario 2 has a negative impact on the income of farming households. This is because it substantially increases their losses of rice due to rhino compared to the baseline scenario. Scenarios 3 and 4 also have a negative impact on the income of farming households, though these would be less severe than under Scenario 2. However, by compensating 25% of the rice loss in Scenario 3, farm household income (after this compensation) is actually slightly higher, compared to the baseline; hence, farming households are more likely to prefer this policy option than Scenario 2. Although Scenario 4 provides 10% crop compensation, rice income is still lower than in the baseline scenario (but significantly higher than in Scenario 2), making this scenario less desirable to farmers. In terms of non-farmer households, all of the alternative policy scenarios (Scenarios 2, 3 and 4) represent an improvement from the baseline. However, since Scenario 4 provides a higher level of job subsidy, greater micro-credit per recipient and a longer collection period at CNP, it is more likely to be preferred by non-farmer households compared to other scenarios.

In summary, we find that communities may well have self-interest in pursuing anti-poaching activities from an income generating perspective but that this would not be universal across all households. To demonstrate the magnitude of these effects we calculated the present value of income under each scenario for each of the stakeholder groups and for changes at the household versus community revenue levels, and compared these to the baseline values from Scenario 1. The results support the discussion above and argue for a policy mix that emphasizes strong anti-poaching interventions combined with the redistribution of community revenues from tourism to affected stakeholders.

2.4 Conclusions and policy implications

Stakeholder analysis in the buffer zone of CNP revealed that there are five major stakeholders: landless/marginalized households, farmers, tourism and related sectors, visitors and non-users and government/NGOs – who represent different interests with regard to park management and rhino conservation. Compensation of crop and livestock losses are most important for local farmers while non-farmers are more interested in direct incentives in the forms of micro-credit, employment and direct access to park resources.

Analysis of the historic level of poaching provided valuable insights into the factors that have affected the level of poaching in CNP over the years. Although factors such as the international price of rhino horn cannot be influenced by policies at national level, a number of factors, such as the effectiveness of the anti-poaching enforcement, can be influenced by national policy initiatives and may help reduce the level of poaching in CNP in the years to come. The most important of these, given the current situation, seem to be the APU structure, and creation of local alternative economic opportunities at the local level. In addition, resolution of the Maoist uprising is very important, as this has had a stark impact on the level of poaching in CNP since the beginning of the uprising in 1996.

The discrete choice experiment confirmed that most of the respondents are in favour of compensation for damages as well as for a community development programme funded by parks revenues. Interestingly, the general preference for these compensatory measures peaks at about 50%, indicating either that they do not require full compensation for the damage, or that most respondents simply do not believe that higher amounts might be feasible. Respondents also considered increases in the number of tourism and employment opportunities as very important. Most importantly, the majority of respondents indicated that with the above-mentioned compensations in place they would have a clear linear preference for more rhinos. The various stakeholder groups behaved as to be expected; the highest income farmers regard the compensatory measures as more important, while the landless marginalized group considers more opportunities for park access as more important, together with the income generation programme.

The simulations that we ran on the basis of our hypothetical policy scenarios provided some insight into the possible effects of the different scenarios on the rhino population. Outputs included the expected level of poaching, tourist visits and income for local households through tourism jobs, through rice production or through the micro-credit schemes. In general, all the alternative policy scenarios predicted a positive outcome with regard to rhino conservation and the overall welfare of the buffer zone communities. However, in terms of maintaining a high level of rhino population and generating higher park revenue (and subsequently higher share of the revenue for communities), the policy scenario emulating the conventional conservation approach seemed to be more successful than the other approaches. Nevertheless, the two alternative policy scenarios that focused on incentive provisioning to households in the buffer zone (in terms of crop compensation, job subsidies in tourism and conservation, micro-credit facilities and park access) also predicted significant improvements from the status quo (in terms of preserved rhino population, controlled poaching, park revenues and economic performance of the buffer zone communities). Our key recommendation is that park management focuses on non-farm, income-generating activities in order to increase the opportunity cost of poaching. This would serve to diminish the need to seek out this illegal activity as a result of the limited availability of alternative income sources. Indeed, we believe that our model might have underestimated the potential impact of such policies on the rhinoceros population: fostering goodwill and providing alternative opportunities could actually shift the parameters of the poaching function, leading to lower rates of poaching, all other things remaining equal. Moreover, the increasing consensus that crop damage compensation schemes cause substantial problems in practice argues further for an emphasis on a Scenario 4 type policy mix.

Clearly, emphasizing an income generating policy (while not completely ignoring losses incurred by farmers) is also the more consistent policy option, and our analysis indicates that this is a policy option that is acceptable to all stakeholders, regardless of their socio-economic conditions. Providing information, education and training to people from marginalized groups will help them build confidence and find employment as park guides, caterers, safari organizers, etc. Similarly, pro-active efforts are needed to ensure a fair distribution of tourism benefits and impacts on local communities and ecosystems. The challenge for a pro-poor conservation policy is to integrate the voice and needs of poor people into efforts to conserve an internationally significant resource (rhinos), and to ensure that poor farming households are compensated for the costs they incur in supplying this unique public good.

References

Adhikari, B., Haider, W., Gurung, O. *et al.* (2005). Economic incentives and poaching of one-horned Indian rhinoceros in Nepal: stakeholder perspectives in biodiversity conservation. PREM Working Paper 05/12, IVM Institute for Environmental Studies, VU University Amsterdam, the Netherlands.

DNPWC/PPP (2001). *Royal Chitwan National Park: Bufferzone Profile.* Kathmandu, Nepal: Department of National Parks and Wildlife Conservation (DNPWC) and Park People Programme (PPP).

Heinen, J. T. (1996). Human behaviour, incentives, and protected area management. *Conservation Biology,* **10**(2): 681–684.

Jachmann, H. and Billiouw, M. (1997). Elephant poaching and law enforcement in the Central Luangwa Valley, Zambia. *Journal of Applied Ecology,* **34**(1): 233–244.

KMTNC (1996). *Royal Chitwan National Park: An Assessment of Values, Threats and Opportunities.* Lalitpur, Nepal: King Mahendra Trust for Nature Conservation (KMTNC).

Knowler, D., Poudyal, M. and Adhikari, B. (2005). Combating poaching of rhinoceros in Nepal: an ecological and economic simulation of alternative policies. PREM Working Paper, Poverty Reduction and Environmental Management (PREM) Programme. IVM Institute for Environmental Studies, VU University Amsterdam, the Netherlands.

Leader-Williams, N., Albon, S. D. and Berry, P. S. M. (1990). Illegal exploitation of black rhinoceros and elephant populations: patterns of decline, law enforcement and patrol effort in Luangwa Valley, Zambia. *Journal of Applied Ecology,* **27**(3): 1055–1087.

Maikhuri, R. K., Nautiyal, S., Rao, K. S. *et al.* (2000). Analysis and resolution of protected area-people conflicts in Nanda Devi Biosphere Reserve, India. *Environmental Conservation,* **27**(1): 43–53.

Martin, E. (1998). Will new community development projects help in rhino conservation in Nepal? *Pachyderm,* **26**: 88–99.

Martin, E. B. and Vigne, L. (1996). Nepal's rhinos: one of the greatest conservation success stories. *Pachyderm,* **21**: 10–26.

Maskey, T. M. (1998). *Sustaining Anti-Poaching Operations and Illegal Trade Control.* Kathmandu, Nepal: WWF Nepal Program.

Mehta, J. N. and Heinen, J. T. (2001). Does community-based conservation shape favourable attitudes among locals? An empirical study from Nepal. *Environmental Management,* **28**(2): 165–177.

Milner-Gulland, E. J. (1993). An econometric analysis of consumer demand for ivory and rhino horn. *Environmental and Resource Economics,* **3**: 73–95.

Milner-Gulland, E. J. and Leader-Williams, N. (1992). Illegal exploitation of wildlife. In T. Swanson, and E. Barbier, (eds.), *Economics for the Wilds.* London: Earthscan, pp. 195–213.

Mishra, H. R. and Jefferies, M. (1991). *Royal Chitwan National Park: Wildlife Heritage of Nepal.* Kathmandu, Nepal: King Mahendra Trust for Nature Conservation.

Nepal, S. K. (2002). Involving indigenous peoples in protected area management: comparative perspectives from Nepal, Thailand, and China. *Environmental Management,* **30**(6): 748–763.

Nepal, S. K. and Weber, K. E. (1994). A buffer zone for biodiversity conservation: viability of the concept in Nepal's Royal Chitwan National Park. *Environmental Conservation*, **21**(4): 333–341.

Poudyal, M. and Knowler, D. (2005). Poaching of the one-horned Indian rhinoceros in the Chitwan Valley Nepal: a retrospective econometric analysis. PREM Working Paper 05/07. IVM Institute for Environmental Studies, VU University Amsterdam, the Netherlands.

Rothley, K., Knowler, D. and Poudyal, M. (2004). Population model of the greater one-horned rhinoceros (*Rhinoceros unicornis*) in Royal Chitwan National Park, Nepal. *Pachyderm*, **37**: 19–27.

Sharma, U. R. (1990). An overview of park–people interactions in Royal Chitwan National Park, Nepal. *Landscape and Urban Planning*, **19**, 133–144.

Straede, S. and Helles, F. (2000). Park–people conflict resolution in Royal Chitwan National Park, Nepal: buying time at high cost? *Environmental Conservation*, **27**(4): 368–381.

Yonzon, P. (2002). The wounds of neglect. *Habitat Himalaya*, **9**(1).

3 · Rural poverty and human–elephant conflicts in Sri Lanka

RON JANSSEN, L. H. P. GUNARATNE,
ROY BROUWER, VITHANARACHCHIGE
D. N. AYONI, PRIYANGA
K. PREMARATHNE AND
H. P. L. K. NANAYAKKARA

3.1 Introduction

The conflict between wild elephants and rural farmers is of great concern for wildlife management and rural development in most parts of Asia and Africa. In Sri Lanka, the conflict is at a very acute stage, as revealed by the fact that between 1992 and 2001 a total of 536 people were killed by wild elephants and that elephant mortality has reached a level of more than three elephant losses per week. Habitat destruction caused by illegal human encroachments and settlements, interruption of elephant corridors and scarcity of food and water in the national parks are some of the causes that have led to the increase of the human–elephant conflicts (HEC)[1] in all wildlife regions in Sri Lanka. Rapid economic development in the dry zone of the country led to large-scale deforestation. In addition, the civil war further increased the conflict by forcing elephants to shift from forests and national parks in the north and east to the south and the west. As their habitats are increasingly lost, wild elephants often have no other choice than invading nearby farm fields causing damage to crops, trees, storage, houses and other properties, sometimes also resulting in the loss of human lives. In

---

[1] The IUCN-SSC African elephant specialist group define HEC as 'any human intervention, which results in a negative effect on human social economic or cultural life, on elephant conservation or on the environment'.

Nature's Wealth: The Economics of Ecosystem Services and Poverty, ed. P. J. H. van Beukering, E. Papyrakis, J. Bouma and R. Brouwer. Published by Cambridge University Press, © Cambridge University Press 2013.

response, farm households are compelled to retaliate in various ways to protect their crops, including the killing of elephants. As stated by Heffernan (2005), the factors driving HECs are complex and case specific to environmental and social variables. The intensity of the HEC depends on elephant-related parameters such as aggressiveness to post-traumatic stress disorder or propensity to feed on field crops instead of the more difficult to find forest plants. Thus, HECs are created in the interaction between the forest environment and people's actions such as encroachment and land claims.

Elephant are the largest terrestrial animal on earth and they require approximately 130–300 kg of feed and 100–200 l of water per day (WWF 2002). As their natural habitats become fragmented, elephants move from one forest patch to the other for food and water, travelling about 20–25 km per day. In Asian countries, including Bangladesh, Bhutan, China, Cambodia, India, Indonesia, Laos, Myanmar, Malaysia, Nepal, Sri Lanka, Thailand and Vietnam, elephants are restricted to a few isolated and fragmented populations, a total number of approximately 35 000–50 000 elephants thus being affected by rapid population growth (Santiapillai and Jackson 1990, Sukumar 1989). In fact, 70% of elephant habitats in Asia were lost between 1960–2000 (Wickramasinghe and Santiapillai 2000).

The conservation of elephants is imperative for a number of reasons. First of all, the Asian elephant (*Elephas maximus*) is considered an endangered species (Fernando 1999). The Asian elephant is on the 'red list' of the World Conservation Union (IUCN) and is mentioned in the Conservation in International Trade in Endangered Species (CITES). Ecologically, elephants are known as 'flagship species' and the animal is biologically important to the ecosystems as a 'keynote species' as well (Desai 1998). Second, there is public concern about the survival of the Asian elephant, much greater than, for example, for the African elephant. This can be attributed to the long-term coexistence of humans and elephants in Asia and the religious and cultural values of elephants in the Asian context. The Sri Lankan elephant (*E. m. maximus*) is different from the rest of the Asian elephant species (i.e. *E. m. indicus and E. m. sumatranas*) (Fernando *et al.* 2003). The incessant conflict with humans caused the decline of the elephant population in Sri Lanka to 3500–4500 (De Silva 1998) and more than 50% of the wild elephants live outside park areas.

During the past two decades, the Department of Wildlife Conservation (DWLC) has tried to mitigate the impact of the HEC by investing in measures such as elephant drives, translocation of aggressive elephants and electric fencing. Fernando (1998) points out that these interventions have rarely been monitored for their effectiveness. De Silva (1998) states that most of the actions implemented are transient in nature and implemented on an ad hoc basis. Actions are effective only if they are implemented as part of a broader management plan for elephant conservation. Fernando (1999) emphasizes the need for proper land use zoning for developmental and conservational needs.

Bandara (2005) argues that the technologies implemented by the DWLC are inadequate and most of the time inappropriate as the views and needs of the resource poor farmers who are the main stakeholders in the conflict have not been taken into account into the planning and implementation of elephant conservation measures. Negligence of the preferences, concerns and economic status of the rural households is likely to result in non-effective policies for elephant conservation. In other words, due to high rural poverty, effective conservation of elephants and protection of their habitats can only be achieved when the needs and demands of the local population are also addressed. In fact, the main argument presented in this study is that poverty and the HEC are related so that successful conservation policy should understand and incorporate the economic reality and concerns of the main stakeholders, i.e. the rural population. Against this background, this study attempts to assess the relationships between rural poverty and the HEC in Sri Lanka in order to identify solutions that both help the rural community and improve the conservation of the Sri Lankan elephant.

One way of supporting both rural communities and elephant conservation is through the introduction of compensation schemes for farmers that suffer elephant damage. WWF (2002) summarizes the case for compensation schemes: 'one of the simplest ways to mitigate conflict without affecting elephant behaviour or population size is to compensate people for the damage they have suffered or would have suffered had they not protected their crops' (WWF 2002). Similar beliefs apply to the conservation of other vertebrate species such as tigers, leopards and monkeys (Bulte and Rondeau 2000). The available literature suggests, however, that the success of existing compensation schemes is mixed due to biased assessments, prolonged delays, low payments and overall inefficiency of the bureaucracy. After lengthy studies carried out on compensation schemes operated in Kenya, Ghana, South Africa, Malawi, Zimbabwe and Bostwana, Osborn and Anstey (2002) concluded that compensation schemes for HECs suffer from a number of deficiencies such as administrative inefficiencies, corruption (bogus claims, inflated claims, deliberate cultivation in HEC-prone zones), insufficient funds and inability to address the social opportunity costs. This implies that market-based alternatives such as insurance mechanisms need to be set up to help governments and NGOs to achieve their objectives whilst encouraging wildlife conservation. Ray (1998) indicates that it is highly unlikely that farmers in developing countries can purchase insurance against wildlife damage from the market unless premium rates are low. This study will further explore the possibilities of setting up a compensation-cum-insurance scheme, transferring financial resources from the urban rich to the rural poor.

In the following, Section 3.2 describes the methodology, data collection and analysis. Section 3.3 presents the findings, Section 3.4 deals with the development of an insurance-based compensation scheme, while the last section summarizes the findings and concludes.

3.2 Data and methodology

Several methods were used to collect and process information. In the data collection process, both experts and local farmers were consulted to provide details about the costs of elephant damage, stakeholder attitudes and perceptions with regard to elephant conservation, prevailing and potential mitigation options and stakeholder willingness to accept compensation (WTAC) for elephant damages. In particular, the following three methods were used for data collection and analysis: (1) spatial analysis; (2) survey of 480 households; (3) multi-criteria analysis (MCA). They will be further explained in the following.

3.2.1 Spatial analysis using Geographical Information Systems (GIS)

Given that the HEC and the factors influencing the conflict such as the distribution of the elephant population and their habitats, human population density, land-use patterns in elephant areas and the proximity of human settlements to elephant habitats are spatially determined, GIS is used as an appropriate analytical tool for understanding the present status and intensity of the HEC. First, we identified the scale and intensity of the HEC at national level. Second, we focused the analysis on selected study sites where high risks prevailed. In this analysis we also tested the spatial relationship between poverty and the occurrence of HECs in Sri Lanka. Both primary as well as secondary data sources and information were used in the GIS analysis. The sources of secondary information include the DWLC, Department of Census and Statistics of Sri Lanka (DCS), Water Resource Management Institute and Divisional Secretary offices. Primary data were collected through the field survey. The software ArcView was used for the analysis.

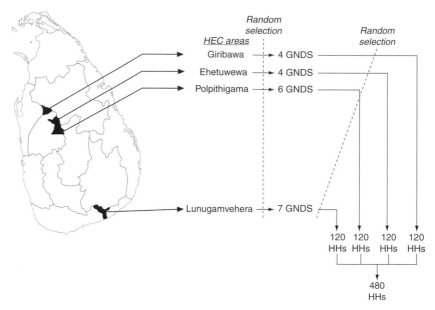

Figure 3.1 Sampling of households for the survey

3.2.2 The household survey

Primary data were collected through a field survey that covered 480 randomly selected households located in four major HEC prone areas throughout Sri Lanka, namely Giribawa, Ehetuwewa, Polpithigama and Lunugamvehera (see Figure 3.1). These four study sites were chosen based on the spatial analysis exploring the intensity of the HEC, the severity of poverty and the prevailing and potential mitigation measures. The survey included questions regarding the socio-economic condition in the affected communities, details regarding elephant-related incidences such as the numbers and types of damage, the intensity of the damage and the costs incurred because of the damage. In addition, information was gathered about the attitudes and perceptions of households with respect to elephant-related activities, about the existing and potential options for conflict mitigation and WTAC.

3.2.3 Multi-criteria analysis

We used MCA to analyse the information collected and select the optimal management response. MCA is especially suitable for decision-making problems that have multiple dimensions (ecological, social and economic) and for which a number of alternative management options are available.

To rank the alternative management options, different MCA methods can be used. In this study we used the weighted summation method because it is simple and transparent. An appraisal score is calculated for each alternative by first standardizing the criterion scores. Next, the standardized scores are multiplied by their appropriate weight. This is followed by summing the weighted scores for all criteria (Janssen 1992). The software package DEFINITE was used for the application (Janssen and Herwijnen 2011). Before weighted summation could be applied, qualitative weights were transformed into quantitative weights using the expected value method. Monte Carlo analysis was used to test the sensitivity of the ranking for uncertainties in the criterion scores. An uncertainty percentage was attached to each score and the ranking was calculated a large number of times. In each run the score could at most be the percentage fixed to be higher or lower than the score in the table assuming a normal distribution with the original score as its mean. The necessary data for the weighting process were obtained by interviewing the respective officers in each of the study sites about their expert views, supplemented by the household survey data and secondary data from DWLC.

3.3 Results and discussion

3.3.1 Spatial analysis

It is important to understand how poverty and the HEC are linked. Poverty is a rural phenomenon in Sri Lanka as approximately 28% of the rural people live below the poverty line. A large part of the dry zone of Sri Lanka is prone to HECs, and this zone is also very poor. Using GIS, the spatial distribution of HECs and poverty were first mapped for Sri Lanka as a whole and then for the four study sites. The level of conflict was expressed as the number of deaths/1000 inhabitants that resulted from HECs in the year 2004. These numbers are scaled between 0 (no deaths) and 100 (DS division with the highest deaths/1000 inhabitants and presented in three classes: low (0–33), medium (33–66) and high (66–100)). Poverty scores were calculated as the percentage of households below the poverty line in the year 2002 and also divided in three similar classes: low–medium–high. The map in Figure 3.2 indicates that poverty is medium or high in almost all districts with HEC. For example, the district Giribawa shows high conflict and medium poverty. Vanathavillu, Maho, Kahatagasdigiliya, Koralaipattu-North, Palugaswewa, Elahera and Dimbulagala districts show medium-level HECs and medium poverty.

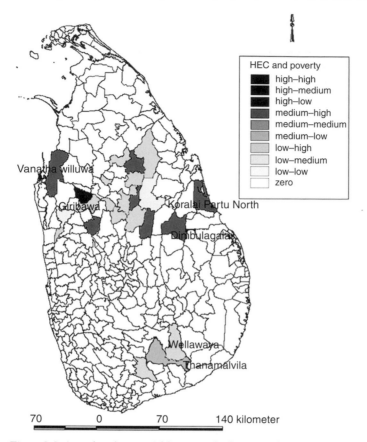

Figure 3.2 Actual and potential human–elephant conflict (HEC) at national level

Mapping of potential HEC areas reveals that the north-western region, which includes three of the four study sites, had the highest incidences. Specifically, the analysis revealed that the districts Giribawa, Ehetuwewa, Nawagaththegama, Galanewa, Ipalogama, Karuwalagaswewa, Mihinthale, Vanathavillu, Lankapura, Padiyathalawa, Sevanagala, Ampara and Lunugamvehera are most prone to HECs.

Further analysis revealed that elephant death intensity is especially high when farm fields are surrounded by protected areas. This can be explained by the fact that elephants tend to migrate from one protected area to the other. Therefore, identification of elephant migratory patterns is crucial for developing effective HEC mitigation strategies. Fencing can control the damage of elephant migration, and is thus expected to be an effective way of conflict mitigation if properly planned and managed. The mapping of potential elephant habitats shows that there is an area of about 9652 km^2 of potential elephant habitats in the entire country. If the minimum land area required for a single elephant is 2 km^2, this means that a maximum of about 4800 elephants could be accommodated in the country, about 25% more than the present population of elephants.

The association between poverty and HECs is largely determined by elephant migratory patterns, household proximity to protected areas and availability of water. To avoid adverse impacts of elephant conservation on poor rural households the introduction of suitable land-use policies based on the identification of HEC-prone zones and resettlement of high-risk villages is recommended. In addition, the impact of the HEC should not be ignored in poverty-alleviation programmes that are already being implemented in high-risk areas. The approach can be extended to develop a realistic overall HEC mitigation plan for all the human–elephant risk areas in order to identify alternative locations for humans and elephants, risk zoning and linking or reconnecting park areas.

3.3.2 Analysis of the farm household survey data

The household survey was implemented in the four study sites characterized by different degrees of HEC intensity (in increasing order of intensity: Giribawa, Ehetuwewa, Polpithigama and Lunugamvehera). The main objective of the survey was to improve insight into the relationship between HECs and poverty. The key results are discussed here.

Poverty is substantial in the different HEC areas, well above the national average rural poverty level. Of the HEC intensity variables, value of paddy crop damage is significantly and positively correlated with the poverty head-count index. At the national level, the number of human deaths and the poverty head count index also show a significant positive relationship. Household characteristics such as family composition, household income, education level, ownership of livestock and self-reliance were statistically different across the four sites.

Analysis of the household survey data furthermore indicated that crop-raiding elephants are a serious problem for poor farmers. HECs were ranked as the most important problem by a majority of the respondents compared to other poverty-related issues and problems such as health care, electricity, clean drinking water and paved roads. More than 50% of the respondents indicated that the primary responsibility of solving HECs lies with the government. Only a limited fraction of the sample admitted shared responsibility. On average, revenue from crop farming is the most

Table 3.1 *Percentage of households per region losing 100% of the crop yield*

	Giribawa	Ehetuwewa	Polpithigama	Lunugamvehera
Rice	14.1	33.6	–	50.0
Maize	4.5	16.0	47.0	69.1
Coconut	17.8	75.0	72.9	82.9
Banana	12.4	100.0	47.5	63.9
Average	12.2	56.2	41.8	66.5

important source of income for nearly half of the households. Poverty is exacerbated through frequent crop depredation and loss of productive working days through injuries and death from elephant attacks. Some sites depend more on crop revenue than others. There have been many occasions where households have lost all their crops. Farmers in Lunugamvehera suffer, as expected, the greatest relative crop losses (Table 3.1). Farmers in these villages are the most vulnerable to the conflict, since in addition to major crop losses, many do not have the safety net of private and government sector wages contributing to household income that, for instance, many Giribawan families have.

The importance of preventive measures was acknowledged by the survey respondents, including the use of firecrackers, elephant drives, changing cropping patterns and patrolling at night. Protecting crops is expensive, requiring investments in labour, firecrackers and watchtowers. It is also stressful and disruptive to family life. Almost 75% of the households experienced sleepless nights patrolling their fields during the crop season. Elephants are large animals and can cause extensive damage to property (including houses, sheds and stored paddy). Damage to property and loss of income were significantly higher in Ehetuwewa and Lunugamwehera than in Giribawa and Polpitigama.

The analysis revealed that the annual value of crop damage due to elephant intrusion is substantial and varies across the four sites, from SLR 14 072 in Giribawa to SLR 27 411 in Polpithigama. A remarkable finding was that respondent WTAC for crop damage is lower than the total reported losses due to crop damage. Possible explanations for this may be the lack of trust in existing compensation mechanisms or the belief that households will not receive full compensation for the damage they experienced. Multivariate regression analysis revealed that income damage costs, number of elephants in the area and gender of respondent significantly influenced the stated WTAC values. Compared to results of Bandara and Tisdell (2004) with respect to urban willingness to pay

(WTP) to compensate farmers for crop damage, the crop losses due to HEC could be mitigated by compensation and increase the welfare of both the urban rich and the rural poor.

3.3.3 Multi-criteria analysis

Solutions are needed which help both the conservation of wild elephants and at the same time (directly or indirectly) alleviate rural poverty. Several management alternatives were identified as options to deal with the problem. One commonly used technical measure is electric fencing. This reduces human damage since elephants can no longer enter the paddy fields, but it does not support elephant conservation because it reduces the size of the elephant habitat. Other popular management alternatives are translocation, live fencing and financial compensation. In this section, the different management alternatives available according to their ecological, social and economic performance will be evaluated, using multi-criteria decision analysis to rank the different alternatives.

The analysis is presented as an illustration for the Giribawa site only; see Gunaratne *et al.* (2006) for the other results. The effects presented in Table 3.2 are used to compare the alternatives. In addition to the alternatives (the columns), the effects table includes effects/criteria (the rows) and their scores (the cells). The criteria are divided according to the three main policy objectives: ecological, social and economic.

Table 3.2 shows a mixture of quantitative scores (e.g. number of elephants), qualitative scores (e.g. quality of family life), and monetary scores (e.g. property and crop damage). Quantitative scores are obtained from secondary DWLC data or are calculated from the survey results. The qualitative scores are assessed through interviews with local experts. Experts included local wildlife managers, district managers and researchers. A −−−/+++ scale was used for the assessments.

This is a convenient approach if not all effects can be scored quantitatively but competent experts are available and capable of providing qualitative expert judgements. This approach is acceptable as long as the meaning of the pluses and minuses is clear to the experts. Table 3.3 lists the descriptions used to identify the number of pluses and minuses together with the experts.

In order to be able to rank the alternatives based on their relative importance, the weights of the criteria in relation to the objectives are needed. Expert judgement was also used to assess the criterion weights. Within each objective the experts were asked to rank the criteria from

Table 3.2 *Effects table of management alternatives for the study site Giribawa*

	C/B	Unit	Current policies	Do nothing	Live fence	Electric fence	Habitat enrichment	Fenced chena inside the park	Compensation
Ecological									
Number of elephants	B	Number/DS	170	160	180	180	190	190	190
Quality of elephant habitat		0/+++	+	+	+	+	+++	++	+++
Land fragmentation	C	Percentage	15	20	10	0	15	30	20
Social									
Human injuries	C	Number/DS/year	1	3	0.75	0.5	0.5	0.75	1
Human deaths	C	Number/DS/year	1	3	0.75	0.5	0.5	0.75	1
Social cohesion		0/+++	0	0	++	++	++	+	++
Freedom of movement		0/+++	+	0	++	+++	++	++	0
Acceptance		0/+++	+++	0	+	++	+	++	+++
Quality of family life		---/0	---	---	-	0	--	-	0
Economic									
Poverty	C	Percentage	75	75	70	50	65	50	25
Property/crop damage	C	Million SLR/year	124	186	62	0	62	62	0
Implementation cost	C	Million SLR/year	0.270	0	0.076	0.075	0.7	1	125
Implementation time	C	Years	0	0	5	1	4	5	0.5

Table 3.3 *Assessment classes used for the qualitative criteria*

Elephant habitat	Social cohesion	Freedom of movement	Acceptance	Family life
Water and food are insufficient due to encroachments; disturbance is high (0)	Individual measures (crop guarding by themselves to protect their own fields) (0)	There is always a threat (0)	People disagree with the measures (0)	No disruption (0)
Occasional disturbance (+)	Measures taken together with neighbours (+)	It is safe to move with precautions (torches/teams) (+)	People support if they are assigned by the farmer organizations (+)	Disruption in the absence of alarm system (−)
Disturbance only in non-rainy periods for slash-and-burn (chena) cultivation, etc. (++)	Activities done by the farmer organizations (++)	It is safe to move most of the time (even in harvesting period) (++)	People support with satisfactory awareness programmes by DWLC (++)	Family life is seriously disrupted for 3 months of the year (vegetable cultivation) (−−)
Sufficient food and water within the habitat; no disturbance (+++)	Protection projects are undertaken by the whole community, costs are shared (+++)	It is safe to go out at all times (+++)	Strong support, intuitively (+++)	Family life is seriously disrupted for 6 months of the year (paddy and other crops) (−−−)

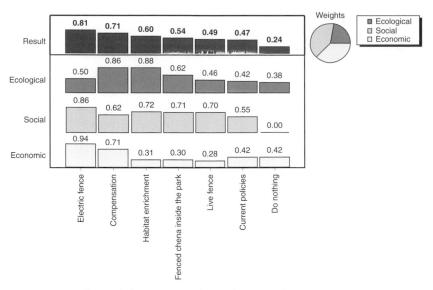

Figure 3.3 Ranking of alternatives in the study site Giribawa

most important to least important. This is referred to here as 'expert weights'. In the rural household survey, farmers were furthermore also asked to rank the three objectives. The number of times farmers ranked an objective as most, second or least important was used to determine the 'stakeholder weights'.

Because the effects table differed for each location, it was expected that both expert and stakeholder weights depend on the status quo at each study site. In practice, the stakeholder weights appeared to be similar across sites, while the expert weights differed substantially.

The ranking of the alternatives for Giribawa is shown in Figure 3.3. The top row shows the overall result. Expert weights were used to calculate the values for the three objectives and stakeholder weights to generate the overall ranking. The stakeholder weights used to produce this result were derived from the household survey and are shown in the pie diagram to the right of the histogram (ecological 0.22; social 0.41; economic 0.37). The top row of Figure 3.3 shows that for Giribawa the electric fence is considered the best solution. Compensation comes second, but scores highest on the ecological objective. Do nothing is the worst alternative due to its bad performance in both ecology and society terms. Figure 3.3 also shows the ranking for the three objectives separately. Habitat enrichment and compensation are best for ecology since both these alternatives increase the food supply for elephants. The electric fence performs best on

Table 3.4 *Rankings for all four study sites*

	Giribawa	Megaleewa	Polpithigama	Lunugamvehera
1	Electric fence	Electric fence	Compensation	Elephant corridor
2	Compensation	Compensation	Habitat enrichment	Compensation
3	Habitat enrichment	Habitat enrichment	Electric fence	Habitat enrichment
4	Fenced chena	Elephant drive	Current policies	Current policies
5	Live fence	Current policies	Elephant drive	Electric fence
6	Current policies	Do nothing	Do nothing	Translocation
7	Do nothing			Live fence
				Do nothing

the social objective, because it provides maximum security. It also performs best on the economic objective due to the absence of crop and property damages despite the high cost of construction.

Finally, Monte Carlo analysis was used to test the sensitivity of the ranking for uncertainties in the criterion scores. An overall uncertainty of 25% was attached to the qualitative scores and an overall uncertainty of 10% to the quantitative scores. Using these uncertainties, the positions of the top three and last alternatives appeared to be very stable. The ranking of intermediate alternatives could not be determined with sufficient certainty.

Table 3.4 shows the rankings for all four study sites. As the effects tables and expert weights differ substantially for the four sites the rankings also differ substantially (see Gunaratne *et al.* 2006 for more details). The electric fence ranks best for Giribawa and Megaleewa. Compensation ranks best for Polpithigama and second for the other three sites.

There is a lot of experience with electric fences. Electric fences are often built as community projects around villages. In these cases the material is supplied by the DWLC and the labour by the villagers. The experiences with compensation are less favourable, but since it was considered a promising option a follow-up study was conducted to further explore this option.

3.4 Insurance-based compensation scheme

A solution is needed that helps both the elephants and the rural poor. An electric fence solves the problem for the people since the elephants can no longer enter the paddy fields but does not support the elephants because it reduces the size of elephant habitat. Compensation for crop damage could potentially support elephants and people: by allowing the elephant to eat farmers' crops, additional food and habitat is created for the elephants. If farmers know that they are guaranteed financial compensation for the loss of their crops, they will not chase or try to frighten off the elephants by using firecrackers. Thus, the chances of being killed by a crop-raiding elephant are much reduced. It is also no longer necessary to staff the watchtowers at night, which would greatly improve the quality of life for farmers and their families.

In principle there is enough money available for compensation. However, due to administration costs and local corruption in handling these types of transactions, collecting and distributing the money is currently problematic. It would need to be handled by a different government institution since existing compensation schemes do not provide enough compensation to all those who are entitled to it. But the outlook is positive. The rural community has accepted the right to survival for elephants for generations, while forests in Sri Lanka can accommodate more wild elephants. Eventually, this conflict may be transformed into co-existence.

With this background, the Poverty Reduction and Environmental Management Programme (PREM) extension study attempted to develop an insurance scheme with the help of the private sector. The process involved a series of workshops with all the relevant stakeholders. At first, a number of workshops were held at national level (with the private and public sector decision makers) and local level (with village level government officers) with the aim of assessing the perspectives and obtaining their suggestions for a new financial scheme. Then, a set of focus group discussions and participatory rural appraisals were held in the conflict areas to get their overall support for an insurance scheme, identify their compensation needs (i.e. a matrix of components and levels of an alternative scheme) and will-ingness to contribute. Then this information was presented to the top-level decision makers representing all the leading insurance firms in the country. Since there is huge gap between farmers' WTP premium and the required premium to run the insurance scheme, most of the firms were reluctant to intervene. However, Dr Lalith Kotelawela, Chairman of Ceylinco Insurance, welcomed the idea and agreed to intervene from the perspective of corporate social responsibility. The annual premium computed using actuarial calculations was SLR 650 per annum to compensate life and disabilities, property and crop damage in the affected farm lands based on their initial requirement. Simultaneously, a trust fund was established to meet the balance of SLR 500 per policy for 227 000 households who live under the HEC risk. This example clearly demonstrates how a comprehensive economic assessment can lead to creative policy solutions that may be replicated in other contexts in which both conservation and livelihoods are at stake.

Local farmers were invited to a presentation of the insurance scheme in Giribawa. All present indicated that they would participate in the scheme as soon as it started. The scheme was introduced at a press conference attended by local television stations and newspapers. The insurance without the subsidy was introduced a few months after completion of the project. Complaints by other insurance companies unfortunately created political problems, which resulted in a delay of the full implementation of the scheme.

3.5 Conclusions and policy implications

Since the majority of wild elephants live outside the park areas, conflicts between rural farmers and elephants are inevitable and occur frequently in Sri Lanka. With an increasing rural population and development activities in the dry zone of Sri Lanka, these conflicts will become more severe in the future. Therefore, the identification of short- and longer-term solutions to conserve wild elephants and their habitats while alleviating poverty is of paramount importance. Otherwise, the intended benefits of elephant conservation as well as poverty alleviation cannot be achieved.

In the study areas, no general best-practice alternative could be identified for avoiding HECs. Electric fencing is considered the best alternative for two of the study sites. Compensation is best in one of the sites and second in three others. As a result of bureaucracy and corruption, however, current compensation schemes do not function adequately. Thus, an extension of the project was conducted to develop an insurance-based compensation scheme. This scheme shows great potential and can be used where electric fences are not feasible.

For the short term, therefore, promotion of both electric fences and insurance-based compensation is recommended. For the long-term success of the electric fences, it is important that construction and maintenance is community based. To avoid bureaucracy, the insurance scheme should be run by private companies in combination with a trust fund with an independent control mechanism. It must be kept in mind, however, that if these management strategies are successful the number of elephants will grow in the long term.

The analysis of countrywide issues and site-specific analyses indicates that more attention should be paid to longer-term strategies for wildlife conservation and economic development. Solutions are only sustainable if the underlying causes of HECs are addressed, including issues pertaining to current resource-use patterns (i.e. land and water use), and when the social and biological requirements of both elephants and human beings are being systematically considered . Besides, further infrastructure development, especially provision of paved roads and electricity, awareness, education and agricultural extension should be an integral part of the proposed plans. Spatial analysis can be further extended to identify elephant migratory routes, risk zones and interconnect park areas.

References

Bandara, R. (2005). The economics of human-elephant conflict. Area Management and Wildlife Conservation Project, Ministry of Environmental and Natural Resources, Sri Lanka.

Bandara, R. and Tisdell, C. (2004). The net benefit of saving the Asian elephant: a policy and contingent valuation study. *Ecological Economics*, **48**: 93–107.

Bulte, E. and Rondeau, D. (2000). A bioeconomic model of community incentives for wildlife management before and after CAMPFIRE. WWF-International, Gland, Switzerland.

Desai, A. (1998). Conservation of elephants and human-elephant conflict. Technical report. Department of Wildlife and Conservation, Colombo, Sri Lanka.

De Silva, M. (1998). Status and conservation of the elephant and the alleviation of man–elephant conflict in Sri Lanka. *Gajah*, **19**: 1–25.

Fernando, P. (1998). Genetics, ecology, and conservation of the Asian elephant. PhD thesis. University of Oregon, OR, USA.

Fernando, P. (1999). Elephants in Sri Lanka: past, present and future, *Loris*, **22**(2): 38–44.

Fernando, P., Vidya, T. N. C., Linda Ng, S. G., Schikler, P. and Melnick, D. J. (2003). Conservation genetic analysis of the Asian elephant: a range-wide study. *Proceedings of the Symposium for Human–Elephant Relationships and Conflicts*, Colombo, August 2003.

Gunaratne, L. H. P., Ayoni, V. D. N., Premaratne, P. K. *et al.* (2006). Final report of the Poverty Reduction and Environmental Management # 14, Sri Lanka. IVM Institute for Environmental Studies, VU University Amsterdam, the Netherlands.

Heffernan, J. (2005). Elephants of Cabinda: Mission Report, Angola, April 2005. Fauna and Flora International and United Nations Development Programme in Cooperation with the Department of Urban Affairs and Environment, Canada.

Janssen, R. (1992). *Multiobjective Decision Support for Environmental Management*. Dordrecht, the Netherlands: Kluwer Academic Publishers.

Janssen, R. and van Herwijnen, M. (2011). DEFINITE 3.1 A system to support decisions on a finite set of alternatives (Software package and user manual). IVM Institute for Environmental Studies, VU University Amsterdam, the Netherlands.

Osborn, F. V. and Anstey, S. (2002). Human–elephant conflict and community development around the Niassa Reserve, Mozambique. Report for WWF and SARPO.

Ray, D. (1998). *Development Economics*. Princeton, NJ: Princeton University Press.

Santiapillai, C. and Jackson, P. (1990). The Asian elephant: an action plan for its conservation. IUCN SSC Action Plan.

Sukumar, R. (1989). *The Asian Elephant: Ecology and Management*. Cambridge Studies in Applied Ecology and Resource Management. Cambridge: Cambridge University Press.

Wickramasinghe, C. S. and Santipillai, C. (2000). Can wildlife pay for its conservation? *Loris*, **22**: 18–22.

WWF (2002). WWF Fact Sheet, 12th Meeting of the Conference of the Parties to CITES Santiago, 3–15 November 2002, WWF-International, Gland, Switzerland.

4 · Poverty, livelihoods and the conservation of nature in biodiversity hotspots around the world

JETSKE BOUMA, K. J. JOY AND
MARONEL STEYN

4.1 Introduction

Globally, biodiversity is being depleted at an alarming rate (UN 2010). This is a problem, not only because biodiversity is intrinsically valuable, but also because it provides the basis for many ecosystem services that are crucial for human well-being. Biodiversity is best conserved in protected areas (Bruner *et al.* 2001) but the establishment of protected areas often has negative livelihood effects (Cernea and Schmidt-Soltau 2006). Particularly, since biodiversity is highest in the poorest parts of the world (Fisher and Cristoph 2007), there is an ongoing debate about how to protect biodiversity while improving local livelihoods and alleviating poverty at the same time. Recently, literature has started emerging that suggests that linking conservation outcomes directly to local livelihood improvement is the best way of protecting biodiversity and alleviating poverty at the same time (Ferraro and Kiss 2002, Niesten and Milne 2009). In this chapter we assess local dependence on the surrounding ecosystem, but focus the analysis on the potential for community co-management of protected areas as a means of improving conservation and avoiding negative livelihood effects.

The study analyses the potential synergies for nature conservation and local livelihood improvement in four biodiversity hotspots around the world (India, South Africa, Vietnam, Costa Rica). This makes it possible to compare approaches and assess the interdependencies between nature conservation and local livelihoods. In all four regions biodiversity is

Nature's Wealth: The Economics of Ecosystem Services and Poverty, ed. P. J. H. van Beukering, E. Papyrakis, J. Bouma and R. Brouwer. Published by Cambridge University Press, © Cambridge University Press 2013.

conserved in protected areas, but the type of management, the type of ecosystem pressures and the type of livelihood strategies differ between the sites. The main aim of the analysis is get an understanding of whether synergies between nature conservation and local livelihoods are indeed possible, with a specific focus on the potential for community co-management in the different sites.

4.1.1 Community co-management

Decentralizing protected area management to local communities creates non-monetary incentives for conservation by partly transferring user rights: indigenous protected area or co-management approaches are examples of this approach where communities define, monitor and enforce resource-use restrictions themselves (Maffi and Woolley 2010, Murphee 2009, Plummer and Fitzgibbon 2004). This has another advantage as local communities usually have more knowledge about the ecosystem and can better monitor and enforce sustainable resource use (Danielsen et al. 2008). This not only increases the effectiveness of conservation, but also lowers protected area monitoring and enforcement costs (Somanathan et al. 2009).

The literature on common-pool resource management has convincingly shown that communities can sustainably manage forests, pastures or other types of collective natural resources (Ostrom 2009). Effective community management requires that communities self-enforce resource-use restrictions through informal rules and social control (Agrawal 2001, Ostrom 1990). Communities may also collaborate with governmental or non-governmental organizations (NGOs) in co-management arrangements (Berkes 2006). In the case of protected area co-management, for example, communities collaborate with park management in monitoring and enforcing restricted resource use (Kubo and Supriyanto 2010). Finally, the external context plays a crucial role in determining the potential for effective community management as it determines the profitability of resource extraction and the institutional context in which community management takes place (Berkes 2006).

With regard to this last factor, community co-management is most effective when the threats to nature are local in nature as well. When threats are regional, or even international (hydropower dam, climate change), community management may not be able to contribute much. In fact, effective community management is seldom only local and representation at higher decision-making levels is required to influence decision-making affecting the ecosystem and local livelihoods (Armitage *et al.* 2008). Co-management arrangements are by their nature cross-level (Berkes 2006), local communities cooperating with regional government or the management of the park. This requires not only that the role of the community in nature conservation be formally acknowledged, but also that the community acknowledges the legitimacy of the park. Several studies have indicated that for communities to collaborate with park management they need to trust park management and perceive the park as legitimate (Carlsson and Berkes 2005, Kubo and Supriyanto 2010, Plummer and Fitzgibbon 2004, Stern 2008).

4.1.2 Poverty, livelihoods and decision-making

MEA (2005) and TEEB (2009) suggest that biodiversity protection will alleviate poverty by improving the quality of the ecosystem on which poor people depend. The idea behind this is that poor people tend to rely more on the ecosystem than non-poor people, since they have fewer assets and resources of their own (Bawa *et al.* 2007, Deland 2006). For example, poor people are more likely to depend on common pastures and forests for grazing their livestock (Kerr 2002) and improving the quality of the common pastures is expected to benefit the poor especially. Often, this is implicitly assuming that most locals are poor: this might be the case by Western standards, but in local terms not. In fact, there is quite some evidence that poor people usually do not benefit from improved resource management, but instead bear most of the costs (see for example Adhikari *et al.* 2004, Kerr 2002). This has to do with the fact that poor people are usually not well represented in the decision-making so that resource-use restrictions are not defined in their interests but in the interests of the better-off. Scholars such as Amartya Sen have pointed out that this is actually an important determinant of poverty, that people have no voice in decision-making and lack the capacity and capability to pursue their needs (Sen 1983, 1995). It also means that ensuring participation of local communities in park-related decision-making is important for better representing the needs of poor people and avoiding adverse livelihood effects.

Using data from 670 households in 22 villages, we assess local livelihood strategies, their relation to nature and respondent perceptions of protected area management. Unfortunately, we cannot evaluate the impacts of protected area establishment on local livelihoods since we have no baseline or time series data and the heterogeneity of our sample is such that we cannot meaningfully compare impacts between sites. The analysis does, however, illustrate the possible linkages between poverty, livelihoods and nature conservation and addresses whether it seems realistic to expect conservation–development synergies to evolve. In this, we pay specific attention to the interests of poor people and the potential for protected area co-management.

We expect our findings to be useful for policymakers and practitioners, who tend to forget that livelihood improvements differ from poverty alleviation impacts and that investments in livelihoods do not necessarily improve conservational outcomes too. Given the recent statements in leading policy documents that biodiversity protection contributes to poverty alleviation (TEEB 2009, UN 2010) it is important to assess whether this is wishful thinking or likely to be true. Clearly, policymakers want to create a 'can do' mentality and win–win solutions make ideas easier to sell, but if this implies that the poor become more vulnerable the message should be corrected and interventions should be changed. Avoiding trade-offs is important, especially since we are considering fragile ecosystems in the world's poorest regions where vulnerabilities are high: sometimes 'do no harm' might be more sensible than wanting to conserve biodiversity and alleviate poverty, whereas in other cases synergies might be possible and create win–win outcomes in the long term. Understanding the linkages between poverty–livelihood and conservation is crucial for creating such synergies, which is what this chapter is all about.

The chapter is structured as follows. In Section 4.2 we introduce the study sites. Section 4.3 elaborates on the empirical approach and Section 4.4 presents the findings. In the last section we discuss our findings and conclude.

4.2 Study sites

We selected our study sites in countries well-known for their biodiversity and national interest in biodiversity protection: Costa Rica and South Africa are probably the two countries in the world that are best known for their biodiversity and India and Vietnam are good examples of biodiversity management and protection as well. Clearly, these countries are not representative of biodiversity conservation or protected area management in general, but in terms of geographical coverage, socio-economic context and other issues they give a good spread.

In each country we selected a riparian ecosystem that included one or two protected areas. All the areas selected are considered biodiversity hotspots and were selected as part of the EU funded FP7 LiveDiverse

project. For more information about the LiveDiverse project, see www.
livediverse.eu.

In South Africa we selected the Mutale basin (including Makuya Park –
bordering the Kruger Park – and the indigenous protected area of Lake
Fundudzi), in India the Warna Basin (including Chandoli National Park)
and in Costa Rica the Terraba Basin (including the Terraba–Sierpe wetland and the Boruca indigenous reserve). In Vietnam, the study region did
not include a riparian ecosystem, but was confined to Ba Be Lake National
Park and Na Hang Nature Reserve.

The Makuya Park in South Africa is located in the downstream part of
the Mutale Basin and was originally laid out in the mid-1980s as part of
the Kruger 'National' Wildlife Park. The park is one of the successful
land restitution claims in the province, where land was given back to
the rightful owners, the tribal authorities of Makuya, Mutele and
Mphaphuli (Lahiff 1997). The park covers an area of 18 000 ha and is
co-managed by two government departments and the three tribal
authorities, although no formal co-management agreement has been
signed by the community due to political conflict (Medvey 2010). The
Makuya Park is an arid savannah ecosystem that is rich in wildlife,
including lions, elephants, hyenas and others. Makuya Park is managed
by full-time staff mostly appointed from the three communities who
own the land. Apart from the money received from government for
rental of the land on an annual basis, the communities also have restricted
access to the conservation area (via permission from their traditional
authority) to collect firewood, plants for medicinal purposes and to fish.
Communities from time-to-time receive meat from animals killed during culling or hunting concessions. Illegal hunting permits and poaching
of wildlife is the major cause for concern, although it is important to note
that poaching is mostly international (China) for rhinoceros horn.

The sacred Lake Fundudzi is located at the top of the Mutale Basin, and
has traditionally been protected by local communities. Outsiders are not
allowed near Lake Fundudzi and it plays an important cultural–spiritual
role. The lake is surrounded by the Thathe Vondo forest – so full of spirits
that few locals venture into it for fear of hauntings. The respect for the lake
and the taboos that prevent visits have meant that the lake has survived in
quite good condition, although long-held traditions are not sustained
with the same authority as before.

The Chandoli National Park is situated at the top of the Warna Basin, in
the northern part of the Western Ghats, India. It covers an area of 31 900
ha and was initially declared a Wild Life Sanctuary in 1985 under section 18

of Wildlife (Protection) Act 1972. In 2004 the sanctuary was upgraded to that of a National Park and in 2007 it was declared part of Sahyadri Tiger Reserve, which also comprises the adjoining Koyana and Radhnagari Sanctuaries. The park is rich in biodiversity and includes various endemic animal and plant species. Wild animals such as the Indian gaur, wild boar, sambar, leopard and tiger have been reported in Chandoli. The forest types include western tropical hill forests, semi-evergreen forests and southern moist mixed deciduous forest. The area is also known for the plant *Narkya* (*Mappia foetida*) a plant believed to be an important curative agent for breast cancer. Smuggling has been going on for more than 5 years, and *Narkya* is mostly exported to Japan where there is a huge demand. Because of the strict action by the forest department in the last couple of years, this illegal extraction and trade have been to some extent controlled. Out of a total 33 villages/hamlets within the park area, 29 villages have been relocated and 4 villages/hamlets are still inside the park.

Ba Be National Park is situated in north-east Vietnam, 254 km from Ha Noi. The total area of Ba Be Park is 7610 ha, including 3226 ha of strictly protected forest, 4084 ha of buffer zone and 300 ha surface water. Ba Be has been recognized as the national history cultural heritage in 1986 and ASIAN's Heritage in 2003. There are 524 households (3200 people) living in the strict core zone and 6000 living in the buffer zone. Ethnic minorities are Tay, H'mong, Dao, Nung and Kinh. Tay people live in wooden houses and are traditionally fishers and weavers, H'mong traditionally depend on hunting and shifting cultivation. The establishment of the park constrained local livelihoods, but also resulted in additional income sources: households are paid for forest protection, local tourism has been developed and agricultural extension services are offered to park inhabitants. The main threats to biodiversity protection in the park are continuing forest clearance for upland agriculture, illegal burning, collection of timber and other plant products, and hunting for meat and the wild animal trade.

Na Hang Nature Reserve is located near Ba Be National Park and comprises 41 930 ha, including a strict protection area of 27 520 ha, 12 910 ha buffer zone and an administrative area of 1500 ha. Ethnic minority groups include the Tay, Dao, Kinh and H'mong. The nature reserve was designated as a reversed forest area to protect biodiversity. The principal attraction of the nature reserve is the endemic population of about 200 Tonkin snub-nosed monkeys, a globally endangered species. The Thac Mo waterfall is the other attraction of the nature reserve and annually some 15 000 tourists visit. The construction of the Gam river dam near the park

Box 4.1   *Biodiversity in Ba Be National Park and Na Hang Nature Reserve, Vietnam*

According to Hill *et al.* (1997) the flora in Ba Be National Park and Na Hang Reserve is composed of 603 vascular plant species, 10 of which have been listed in the *Vietnam Red Data Book*. Vegetation coverage mainly includes two types of forests: the limestone and evergreen forests. Limestone forests are distributed on steep mountainsides, with thick plant cover and take up most of the park area. Evergreen forests are distributed on low earthen hills covered with a thicker soil layer. Species of lowland forests are more diverse than those found on limestone mountains (Hill *et al.* 1997). Rare mammals include Francois' langur (*Semnopithecus francoisi*) and Ownston's banded civet (*Hemigulus ownstoni*). Francois' langur was found in 1995 along the second bank of the Ba Be Lake and in 2001 was rediscovered nearby Da Dang Guard Post. In the park, the snub-nosed monkey (*Rhinopithecus avunculus*) can be found in core zone. The animal was reported to have disappeared until it was rediscovered in Na Hang Nature Reserve in 1992. Interviews with hunters confirmed the existence of three groups in the south of the park. The Vietnamese salamander (*Paramesotriton deloustali*) was recently discovered in the southern sector of the park. Finally, there are 54 species of fish listed for Ba Be Lake, covering 25% of the freshwater fish fauna of North Vietnam, among them, 10 species listed in *Red Data Book*.

affected the reserve. The main threats to biodiversity come from the people living in and around Na Hang Nature Reserve. There is no definite buffer zone so far but effectively the reserve comprises nine communes with an estimated population of 6215 households and 35 302 people. The District Forest Protection Division is responsible for activities in the buffer areas. In addition 2159 households with a total population of 11 233 people live inside the boundary of the nature reserve. Shifting cultivation, livestock grazing and overexploitation of forest products are specifically threatening biodiversity conservation in the park.

The Terraba Sierpe National Wetland is a protected area located in the Osa Conservation Area on the Southern Pacific coast of Costa Rica representing the biggest mangrove zone in Costa Rica. Due to its ecological characteristics and the ecosystem services that it provides, it was declared a Ramsar Site in 1951. The wetland is characterized by a great variety of mangrove forest. Different fish species, migratory birds and reptiles such as boas, crocodiles, caimans and turtles, are the most representative fauna found in the wetland. Despite its importance, the legal and institutional framework in the wetland has not been clear and the laws regulating land use and fishing inside the wetland are not being enforced (i.e. one guard for 16 700 ha of wetland; Uribe 2010). Inside the wetland there are several fishing communities, but the main pressures on the wetland are intensive agriculture and the possible construction of a hydropower dam upstream.

The Boruca indigenous territory located in the more upstream part of the Terraba Basin, was among the first indigenous reserves established in Costa Rica in 1956. The lands currently on the reservations were named *baldíos* (common lands) by the General Law of Common Lands, passed by the National Government in 1939, making them the inalienable and exclusive property of the indigenous people. The subsequent law of the Instituto de Tierras y la Colonización de Costa Rica (ITCO or Institute of Lands and Colonization) passed in 1961, transferred the baldíos to state ownership. The Indigenous Law of Costa Rica (La Ley Indígena de Costa Rica) passed in 1977 laid out the fundamental rights of the indigenous peoples. This law defined 'indigenous', established that the reserves would be self-governing and set limitations on land use within the reserves. The population of the Reserva Indígena Boruca subsists mainly on small-scale agriculture and the sale of indigenous crafts. Agriculture and overexploitation of natural resources put some pressure on the ecosystem, but the main pressure is the possible construction of a large hydropower dam upstream.

4.3 Data and methodology

To collect information about the poverty–livelihood–nature linkages in the study sites we developed a household survey. The main goal of the survey was to collect data about household livelihood strategies and ecosystem dependence, and to gain insight into the distribution of poverty at village scale. Also, we collected information about people's perceptions of protected area management, perceived capability to influence decision-making at different levels and the household's willingness to contribute to protected area management and participation in voluntary organizations at village scale. For the survey a selection of study villages was made within the study sites. The criteria for village selection were (1) location near or inside a protected area and (2) location in the basin (up, down). Per site 4–10 villages were selected. Within these villages a random selection of approximately 10–20% of the households was made. Due to differences in village size the total number of completed household surveys varies: in Costa Rica, 123 surveys were conducted in 4 villages; in South Africa, 116 surveys in 5 villages; in Vietnam 292 surveys in 10 villages; and in India 509 surveys in 9 villages. For this chapter, a selection of villages from the total sample was made based on the distance of the study villages from the protected area. Only villages located less than 15 km from a protected area were included, meaning that one village was dropped from the South African sample and five villages were dropped from the Indian database.

To assess poverty levels we asked respondents to indicate their monthly average household income level from a predefined range of income categories. We related these figures to regional poverty levels, acknowledging that this is an imperfect poverty measurement since we could only include monetary income (whereas livelihoods might be partly subsistence based). To correct for this, we also collected information about land and livestock ownership and inquired whether the household usually had sufficient to eat. Information about livelihood strategies was collected by asking about the income earning strategies and other activities each household member was involved in. To get insight into the household's dependence on the surrounding ecosystem questions were asked about the main source of drinking water, fuel and fodder and the products that the household collected in the surrounding ecosystem (and whether this was inside a protected area or not). Finally, to get insight into the household's perception of protected area management, voice in decision-making and willingness to contribute to improved management of the park, we asked some questions about the respondents' trust in others, perceived influence in local decision-making and the perceived impact of the park. A full version of the household survey is available online at www.livediverse.eu.

For the analysis we combined data from all the study sites but conducted separate analyses for the different countries. We did, however, estimate the same statistical models to allow for a comparison of the results. Data from different villages were pooled in the analysis, but we controlled for village characteristics such as distance to the protected area and location in or outside the park. We did not control for park characteristics in the statistical analysis, but we address park characteristics in the text. In the first part of the analysis we focus on the differences in poverty–livelihood–nature linkages between the study sites. We compare averages for important poverty, livelihood and ecosystem dependence indicators and discuss the differences between the sites. In addition, we estimate a bivariate probit model using Stata, a statistical software package, to assess the factors determining the probability that respondents consider the collection of non-timber forest products (NTFP) important for the household's livelihood. In this, we specifically consider the impact of households being poor.

The second part of the analysis focuses on protected area management and the extent to which respondents feel they can influence decision-making at different scales. First, we consider respondent perceptions of park management and the extent to which respondents feel they can influence park management or not. Again, we estimate a probit model, assessing the factors that determine whether respondents feel they can influence the management of the park. In this, we account for respondent and village characteristics. Second, we estimate a probit model for explaining the probability that a respondent is willing to report people who break the rules. Using the same explanatory variables as in the preceding models, we show that this willingness depends on household and village characteristics and on the perceived influence in the management of the park. Qualitative findings of village meetings and interviews are used to interpret the findings and illustrate the results. Using the same model specifications for each study site resulted in a number of dropped variables since optimal model specification differed between the sites. We tried to find the specification that best fitted the different study sites, but in some cases trade-offs between the model specifications for the different countries did arise. In the explanation of the analytical outcomes we will address such trade-offs and explain the impact of different model specifications on the results.

4.4 Results

4.4.1 Livelihood strategies and ecosystem dependence

The survey indicates that there are significant differences in livelihood strategies between the sites (see Table 4.1). In Vietnam and India agriculture is the main livelihood strategy, whereas in Costa Rica and South Africa small commerce plays a more important role.

In the Terraba Basin, Costa Rica, only a quarter of the households still depends on agriculture, as most households sold their land to large commercial enterprises (pineapple and palm oil plantations) and earn their livelihood with jobs in commerce, the public sector or as agricultural labourers on the plantations. Between villages, differences are large: the

Table 4.1 *Summary statistics on household composition and livelihood strategies*

	Mutale Basin, South Africa	Terraba Basin, Costa Rica	Ba Be–Na Hang, Vietnam	Warna Basin, India
Number of observations	96	123	292	159
Number of villages	4	4	10	4
Average household size	4.8 (2.1)	3.9 (1.7)	4.8 (1.6)	5.3 (2.4)
Households (self-) employed in:				
– agriculture	8%	24%	96%	75%
– industry/commerce	22%	33%	20%	28%
– tourism	7%	17%	15%	12%
– fish/forestry	7%	4%	8%	0
– public sector	4%	21%	24%	8%
Households with multiple strategies	32%	47%	53%	34%
Households receiving remittances	8%	14%	5%	40%

livelihoods of people living inside the two protected areas are still rather traditional (i.e. fishermen in the protected wetland and indigenous people depending on craft-making for tourism and subsistence agriculture in the protected forest) but the households in villages outside the protected area have more urban lifestyles and a much weaker connection with the surrounding land. Tourism is relatively important in the basin, with 17% of the households gaining income from tourism.

In South Africa, apartheid fundamentally changed livelihoods from agriculture-based to migratory labour dependent and currently households in the Mutale Basin receive most of their income through government pensions and grants. Surprisingly, the percentage of households receiving remittances is quite low: this might be because of underreporting, but the surveyors indicated that respondents forcefully argued that household members who had migrated keep the income they earn for themselves. This finding is in line with Posel and Casale (2006) who found that the proportion of households receiving remittances had declined since 1999 due to declining average real wages, and a reduced 'perceived' need to pay remittances due to an increase in the value and coverage of social grants.

Tourism hardly plays a role in local livelihoods, which is surprising given that the study sites are located near the Kruger Park. Between villages, differences are minor, except that in the upstream village, located near the sacred Fundudzi Lake, rainfall is much higher than in the arid downstream villages and most households in the upstream village grow crops on their land. There are no villages located inside the protected areas.

In the Vietnamese study site, almost all households have one or two members engaged in agriculture but more than half of the households are engaged in non-agricultural activities as well. This reduces household vulnerability since multiple livelihood strategies increase the household's capacity to deal with a poor harvest, job loss or something else. In this sense, Vietnamese and Costa Rican households are less vulnerable than Indian and, especially, South African households who tend to depend solely on government grants. Differences in livelihood strategies between the Vietnamese villages are relatively small, although in more remote villages dependence on (subsistence) agriculture is higher and livelihood diversification less. Interestingly, almost a quarter of the households have a public sector job, which seems to reflect Vietnam's political system. Villages located inside the protected area face strong resource-use restrictions but especially those located in Ba Be gain from tourism as well.

In the Warna Basin, India, most households depend on agriculture for their livelihood, but a significant share of households also depends on small commerce or has other non-agricultural jobs. Almost half of the households report receiving remittances; an indication that migration is an important livelihood strategy as well. Between villages, the main difference is between villages located up- and downstream of the hydropower dam: downstream households have access to irrigation and as a result agriculture is more intensive and commercial, whereas upstream agriculture is extensive and subsistence based. Of the two upstream villages one is located inside the park and the other on the border. Both villages are indigenous; one is a nomadic tribe.

Table 4.2 summarizes the results of the household assets measured in terms of livestock and land holdings. Given that only 8% of the households in South Africa indicated that they depended on agriculture for their livelihood, the share of households with access to land is surprisingly high. Most of the year this land is left fallow, however, since irrigation is lacking and aridity is high. In Costa Rica, households with land are mostly located inside the indigenous territory; the remainder are few large landowners with pastureland. In the Vietnamese and Indian sites most households are landowning and land use is distributed quite equally, although average landholding is larger in India than in Vietnam. In Vietnam more households have access to irrigation, but landholding size is very small. In India, landholdings near and inside the park are relatively large compared to

Table 4.2 *Summary statistics on household assets: livestock and land holdings*

	Mutale Basin, South Africa	Terraba Basin, Costa Rica	Ba Be–Na Hang, Vietnam	Warna Basin, India
Households with access to land	58%	24%	98%	92%
Average landholding (acres)	2.4 (4.7)	14.4 (74.2)	1.3 (1.4)	3.4 (6)
Access to irrigation (of the landowners)	0	0	71%	25%
Households with livestock	59%	9%	91%	65%
Average no. of chickens per household	4.8 (5.7)	4.5 (11.2)	12.9 (21.4)	2.7 (7.6)

more downstream farmers, but without access to irrigation. Agricultural productivity is low in Vietnam and India, mostly because traditional production technologies are used.

With regard to livestock holdings, in Vietnam most households are livestock owning: cattle for ploughing and pigs for household consumption. In India, livestock holdings are mostly related to agricultural production (bullocks) or reflect the household's identity, i.e. that the household is part of the traditional sheep or cattle rearers' caste. In South Africa, households hold free-grazing cattle and goats for meat and donkeys for transportation. In Costa Rica few households own livestock, which reflects the urban lifestyle most households have. In all sites, chickens were an important livestock asset as well.

Clearly, livestock holdings might have been affected by park establishment, which holds for access to land and landholding size as well. Given that we have no baseline data, we cannot assess the impacts of park establishment on livestock holdings and landholding size. However, we do know that the establishment of Chandoli Park, India, caused displacement and loss of livestock and land. In South Africa, people also lost their land due to park establishment but this was more than 30 years ago. Still, even without displacement, protected area establishment affects land and livestock ownership: for example, fishermen communities in the Terraba-Sierpe wetland in Costa Rica are not allowed to own land in the wetland and land-use restrictions in Vietnam are such that households cannot increase the size of their agricultural land. Unfortunately, for lack of baseline data, we cannot evaluate the size of these effects.

Box 4.2 *The dam, park and conflicts in the Warna Basin, India*

The Warna Dam and the Chandoli National Park have caused a multi-layered conflict in the area especially on the issue of displacement and resettlement. The Warna Dam became an issue in 1985. The catchment of the dam was first declared a wildlife sanctuary and then elevated to the status of a national park in 2004 with an area of 317.67 km^2. The process of resettlement began in 1995–97. Twenty-nine villages were relocated to areas outside of the park, four villages remain inside the park as they have been resisting eviction (but the process to move them out of the park is still ongoing). The construction of the dam and later the establishment of the park have caused violent disruption of the livelihoods and socio-cultural practices of the people because of displacement and also severe restrictions on the collection of biomass, including wood and other forest resources from the park. Civil society organizations in the area have tried to organize the people affected by the dam and the park, and during the last two decades there have been many intense struggles and protests to get the demands of the displaced people accepted by the government. There have also been conflicts between the ousted people and the host communities as the oustees are rehabilitated by taking over land from larger land-holders by applying land-ceiling rules. There have also been conflicts within the affected villages on issues such as whether to move out or not, where to move to and so on. Though initially all the villagers were very united in resisting eviction, gradually fissures appeared.

Like land and livestock ownership, park establishment also tends to affect direct ecosystem use. Table 4.3 gives an overview of the current collection of forest products and the share collected inside protected areas and parks. In Costa Rica around a third of the households do not collect any products from nature at all: these are mostly urban households who visit nature only occasionally for recreation. With regard to the type of resources collected, most households reported collecting fruits, vegetables, firewood and construction materials, but in South Africa and Costa Rica a significant proportion of the households also collects medicinal plants and dyes (for traditional craft-making). In Costa Rica these are indigenous households located inside the Boruca indigenous territory (see also Box 4.4). Hunting is not often reported, but this might also be because hunting

Table 4.3 *Summary statistics on ecosystem dependence: water, fuel and other products*

	Mutale Basin, South Africa	Terraba Basin, Costa Rica	Ba Be–Na Hang, Vietnam	Warna Basin, India
Households collecting:				
– meat	9%	4%	0%	1%
– fish	36%	10%	10%	3%
– fruits	62%	64%	8%	25%
– vegetables, mushrooms	34%	24%	31%	6%
– medicinal plants, dyes	26%	22%	3%	2%
– timber, construction material	43%	7%	41%	33%
– firewood	88%	2%	91%	82%
– flowers	1%	3%	1%	13%
Average no. of products collected	3.7 (1.3)	1.41 (1.30)	1.94 (1.14)	1.7 (1.2)
Considers collection important	95%	41%	76%	85%
Collected in protected area	17%	17%	74%	22% (47%)
Does not collect any products	2%	29%	3%	13%
With individual/collective tap	40%	99%	80%	77%
With electricity	82%	100%	94%	78%
With wood as main fuel source	100%	15%	98%	84%

is usually not allowed. It is important to note that, except for Vietnam, most of these products are not collected in a protected area. For India, the figure is likely to be biased since in the one village located inside the protected area all households indicated that they did not collect any products in the park. Hence, between brackets we added a figure that includes the households inside the park.

Considering the direct use of other ecosystem services, it is interesting to note that in South Africa almost 50% of the households still depend on surface water for drinking and washing and that, except for Costa Rica, firewood is the main source of fuel in all the sites. This is surprising given that most households have access to electricity, and often have to pay for firewood as well. As firewood collection is a major driver of environmental

Table 4.4 *Summary statistics on income, poverty and food security*

	Mutale Basin, South Africa	Terraba Basin, Costa Rica	Ba Be–Na Hang, Vietnam	Warna Basin, India
Monetary income per capita (US$/month)	91 (96)	214 (253)	13 (14)	9.9 (7.7)
Below the regional poverty line	49%	25%	62%	50%
Households with sufficient to eat	30%	90%	60%	85%

degradation, it is important to acknowledge the important role that firewood plays. With regard to water, better management of the ecosystem can improve the quality of surface water, and thus contribute directly to poverty alleviation by reducing health risks. However, in the villages considered, the main protected area is located downstream so that better protection of the ecosystem is unlikely to have any health-related impacts on the villages upstream.

Turning to the distribution of income and poverty in the project sites, Table 4.4 indicates that monetary incomes are on average low: compared to local poverty levels between 25–60% of the households in the study sites should be considered poor. Note that for each country we used the lower bound of the regional poverty level to assess whether the household should be classified as poor. Given that this only includes monetary income, the estimates are likely to be an overestimate: most households produce (some) food for their own consumption and this is not included in the monetary income measurement.

A better indicator might be the number of households that indicates sometimes having insufficient to eat. The data show that in India and Costa Rica 10–15% of the households indicate not having sufficient to eat, whereas in Vietnam and South Africa these numbers are much higher, i.e. 40 and 70%, respectively.

Comparing food security and income poverty at village level suggests that food producing households are often income poor but usually food secure, and that non-food producing households are less often income poor, but might be food insecure. This last result especially holds for South Africa where the three villages in the downstream arid part of the Mutale Basin are highly food insecure. Half of the households in the Mutale Basin are income poor, which can be explained by the fact that many

Box 4.3   *Products sourced from nature in the Mutale River Basin, South Africa*

Trees, plants and wild fruits make for an important ecosystem service in the Mutale Basin, South Africa. Food products that are mainly sourced from nature include wild grapes, groundnuts (*dovhi*) and *mbuyu* (Tshihwanambi 2007). *Mbuyu* is the fruit of the baobab tree. The VhaVenda people make a porridge where the acidic flesh of this fruit is crushed and when mixed with milk it causes the milk to thicken and mildly ferment, giving a distinct taste. Mopani worms or what they call *mashonzha*, although not necessarily recognized by the younger generations as such, form an integral part of the protein intake of these people two seasons of the year. These worms can be eaten dried, deep-fried or cooked in stews. Other sources of protein directly sourced from nature come from hunting wild rabbits and hyraxes more commonly known as *dassies*. Wild leaves from the field are cooked like spinach and are collectively known as *morogo* or *imifino* (Faber *et al.* 2010). *Dovhi* groundnuts are then used to thicken sources or add flavour to wild spinach dishes such as *tshidzimba*, *tshimbundwa* and *tshigume*. Various plants and shrubs are sometimes used for medicinal purposes. *Delele* (herb better known as wild jute) and *mushidzhi* (also known as black Jacks) are important for health reasons (Tshihwanambi 2007). *Matumba* or wood is a particularly important product sourced from nature. Wood is used for cooking, serves as fuel to provide heat and also for construction of houses, building storage space for maize (the staple food) as well as the *kraal*, where livestock are kept. It is said that the VhaVenda, although some 82% of the respondents interviewed have electricity, prefer to make their porridge on the open fire since it tastes better (Makhado *et al.* 2009). When asked about collection of firewood, the quantities they consume on a daily basis and conservation of nature, one of the respondents said: 'The plants and trees will always be there for our use, because we are not cutting the living trees, only the dead ones.'

households depend for their livelihood on a government pension or one or two child grants. In India, landholding households are generally not food insecure. Income poverty is high in India because most households only grow a few cash crops and are largely subsistence based. In Costa Rica, most households are food secure. The income poor are mostly located in Boruca village, the indigenous village located inside the indigenous protected area where livelihoods are based on subsistence agriculture and traditional craft-making.

In Vietnam the story is different. Most households are landholding and produce food, but still 40% of the households indicate being food insecure. Village interviews and field visits indicate that food insecurity is indeed a problem, caused by low agricultural productivity and scarcity of agricultural land. This last factor is partly caused by protected area establishment, which limits agricultural use of land: given that households still use traditional, extensive land use agricultural practices, the harvest they reap from their land is simply too small. The high share of income poor households in Vietnam is also caused by the subsistence nature of local agriculture. Households that diversified their livelihood strategy are generally not income poor, nor food insecure, a result that holds for the other sites as well.

Considering the importance of natural resource collection for different households, Table 4.5 presents the results of a probit analysis explaining

Table 4.5 *Probit model explaining stated importance of natural product collection*

	Mutale Basin, South Africa	Terraba Basin, Costa Rica	Ba Be–Na Hang, Vietnam	Warna Basin, India
Village located inside PA	Dropped	−0.22	0.12	−.015
Female respondent	−0.01	0.05	0.015	0.16
No. of household members	0.01	0.009	−0.02	−0.02
Household's per capita income (US$/month)	−0.000	−0.00	0.00	0.00
Households below poverty line	−0.05	0.23	0.07	0.14
Household has sufficient to eat	0.008	−0.18	−0.13	0.07
Livestock owning household	0.005	0.007	0.28	0.01
No. of products collected	0.003	0.26	0.12	−0.001
Wood main source of fuel	Dropped	0.34	0.30	0.56
Households with diversified income	−0.003	0.03	0.18	0.45
No. of observations	92	118	266	136
Wald chi^2 (8)	14.3	40.16	34.19	37.13
Pseudo R^2	0.21	0.30	0.20	0.31

Reported variables are marginal effects (i.e. no constant). Shaded boxes are significant factors (1–10%).

the perceived importance of NFTPs. As expected, the results indicate that the more products a household collects, the more important the activity is considered to be. More interesting is the finding that in Vietnam and India it is the poor households that consider the collection of NFTPs important. In Vietnam, the finding that food insecure households are more likely to consider collection of NTFPs important indicates that natural resource harvesting may play an important nutritional role. This is confirmed in village meetings and field visits, where respondents indicated that forest products form an important part of the local diet. A similar finding holds for South Africa, where people explained in the village meetings that the *mopani* worms, vegetables and fruits they collect are important elements in their food (also see Box 4.3). In India this is much less the case as people mostly use the forest for the collection of (fire) wood. In Costa Rica people collect fruits and plants (for dye). Land size has no significant impact on the findings, and different model specifications generate similar results.

Summarizing, we find different poverty–livelihood–nature linkages between the study sites. In the Vietnamese and Indian site, livelihoods are mostly agriculture based, whereas in the South African and Costa Rican sites livelihoods are much less linked to the land. In the Terraba Basin in Costa Rica there are considerable differences between the villages. Villages located outside the protected areas have rather urban lifestyles, whereas the indigenous communities living inside the Boruca indigenous territory and the fisher communities living inside the Terraba-Sierpe wetland depend on nature directly for their livelihood. In the Mutale Basin in South Africa, dependence on the ecosystem for non-productive uses is high (water, fuel), but for income people migrate to the city or depend on government pensions and child grants. Given that people have few assets and are generally very poor, the safety net function of the ecosystem seems especially important in the South African context whereas in Costa Rica the ecosystem hardly plays a safety net role.

In the Vietnamese and Indian study sites, most households own a plot of land on which they cultivate (subsistence) crops. Livestock is important for agricultural production and consumption, and most households depend on the ecosystem for fodder and grazing land. Although in both countries more than half of the households are income poor, in Vietnam food security is an issue as well. This is worrisome, since most households own land and livestock, and sufficient water seems to be available as well. However, agriculture is still rather extensive and productivity is low. Before the establishment of the protected areas, households could expand their agricultural production by expanding the size of their land (i.e. slash and burn cultivation), but now land-use restrictions are constraining households in their traditional practices and few investments in agricultural intensification are made. In India protected area establishment strongly affected local households through total bans on resource use and displacement. Displaced households had to transform their livelihood from subsistence-based agriculture to wage labourer, and remaining villages lost access to common land. Still, households in the study villages indicate being food secure, households with small land sizes having access to irrigation and households without access to irrigation having rather large plots of land.

With regard to the relationship between poverty and nature, the analysis indicates that the collection of NTFPs plays an important role in the local diets of Vietnamese and South African households and that in all

---

Box 4.4   *The use of natural resources in the Boruca indigenous community*

The Boruca Indigenous Territory is located in the province of Puntarenas, in the county of Buenos Aires, and forms part of the lower basin of the Grande de Térraba River catchment area. Few opportunities for employment are available in Boruca communities, but due to the introduction of tourism the local economy has begun to transform itself over the past decades. Today, the production of handicrafts is one of the principal productive activities in the Boruca community. Boruca handicrafts are famous for their craftsmanship and use of natural pigments and dyes. Traditionally, the extraction of natural pigments and dye-making took place in two areas: in the indigenous settlement itself, where dyes were extracted from plants and flowers, and in the nearby coastal area, where the indigenous communities migrated to in the summer months to extract dyes from shells and molluscs native to the littoral zone. Both areas are now protected, e.g. the Boruca Indigenous territory and the Ballena Marine national park, which has complicated and constrained the making of handicrafts. In addition, the scarcity of shellfish and sea snails from which many dyes are extracted has increased, both because of the increase in harvesting for textile production, as well as because of over fishing and habitat destruction. In response, the Boruca artisan community has reduced its use by not completely dyeing their handicrafts, producing textiles with less-intense coloration.

sites, except Costa Rica, poor people consider the collection of NFTPs especially important for their livelihoods. The analysis also indicates that, except for Costa Rica, the majority of households in the study regions are income poor. Partly, this has to do with our use of an estimator that does not account for non-monetary sources of income, such as subsistence agriculture. On the other hand, it is also an indication that compared to the regional averages the study regions are relatively poor. Food insecurity, an indicator of absolute poverty, is a considerable problem in South Africa and Vietnam: according to the findings of the survey, respectively 70 and 40% of the households are food insecure.

Before discussing the implications of these findings for nature conservation, poverty alleviation and livelihood development in the study sites, we need to get some further insight into local representation in the decision-making, specifically of the poor. Hence, in the next section we present our findings on people's perceived influence in protected area management and willingness to contribute to rule enforcement and control.

4.4.2 Influence in protected area management and decision-making

Table 4.6 summarizes the findings of the survey with regard to respondent perceptions of the household's influence in local decision-making, the management of the protected area and the village's influence in decision-making at higher scales. Also, it gives an indication of local activity in

Table 4.6 *Summary statistics on local decision-making and voluntary organization*

	Mutale Basin, South Africa	Terraba Basin, Costa Rica	Ba Be–Na Hang, Vietnam	Warna Basin, India
Households that believe they can influence protected area management	48%	70%	48%	24%
Households that believe they can influence local decision-making	77%	41%	64%	64%
Households that believe their village influences decision-making	78%	44%	54%	60%
Households active in voluntary groups (excl. church)	82% (62%)	73% (37%)	80% (80%)	51% (44%)

voluntary organizations (micro-credit, women, burial, union), including membership of religious organizations (temples, churches).

The high percentage of Costa Rican households stating that they can influence protected area management is surprising, given that these are mostly urban households, and suggests that respondents refer to a general notion of protected area management rather than to the specific influence of the household itself. Interestingly, in Vietnam and South Africa, both examples of co-managed protected areas, fewer than half of the households feel capable of influencing protected area management. In India, with top-down park management, only a quarter feel able to influence management of the park. In all sites, households are active in voluntary organizations, although less so when excluding church and temple membership. In Vietnam, a staggering 80% of the population is socially active, but this also reflects the political organization of Vietnamese society (see Box 4.5). Relevant organizations for protected area and natural resource management are the forest fire and/or inspection teams in Vietnam and the farmer and fisher cooperatives in Costa Rica.

To analyse the factors explaining the perceived influence in protected area management we again estimated a probit model to predict the probability that a respondent feels able to influence management of the park (see Table 4.7). Interestingly, in all study sites the perceived ability to

---

Box 4.5   *Grassroot democracy and community consultation in Vietnam*

In Vietnam, the government has recognized the importance of involving households in decisions that affect their lives and of making local government more transparent and accountable in the announcement of Decree 29 on Grassroot Democracy. Normally the ways of carrying out community consultation are through (1) large meetings, (2) direct contact and (3) by soliciting written comments. Specific tasks are identified for commune officials. They include discussing the draft manual socio-economic plan with villagers and seeking their feedback; disseminating government policies to villagers; providing information to villagers on projects and programmes being implemented at the local level; holding biannual meetings between electorate and elected members of the People's Council; holding meetings to review their work in the presence of villagers, listening to their criticism; and planning and implementing village infrastructure works. Thus, the village chief has a key role to play in the implementation of grassroot democracy.

Table 4.7 *Probit model explaining perceived influence in PA management*

	Mutale Basin, South Africa	Terraba Basin, Costa Rica	Ba Be–Na Hang, Vietnam	Warna Basin, India
Distance to PA	−0.03	0.08	−0.005	−0.03
Location inside PA	Dropped	0.29	0.02	0.07
Female respondent	0.11	−0.14	−0.17	0.06
Agriculture	0.15	0.20	−0.28	0.19
Small commerce/Industry	0.006	−0.01	0.16	0.03
Tourism	−0.02	0.13	0.26	0.27
Public sector	−0.04	0.03	0.28	0.004
Household per capita income (US$/month)	0.0008	0.000	−0.000	0.002
Household below poverty line	0.21	−0.47	0.11	0.008
Household has sufficient to eat	−0.12	0.12	0.05	−0.22
Landholding size (acres)	−0.02	−0.001	0.08	−0.002
Livestock-owning household	−0.04	−0.03	0.10	−0.29
Household can influence DS	0.29	0.10	0.14	0.17
Village can influence DS	0.49	0.40	0.18	0.05
Household participates in voluntary organization (incl. church)	0.28	0.11	−0.11	0.24
Household majority ethnic group	0.13	0.20	0.03	0.10
Household majority religious group	−0.13	0.15	0.32	0.12
No. of observations	87	109	246	147
Wald chi^2 (17)	29.56	31.84	54.12	45.80
Pseudo R^2	0.24	0.29	0.20	0.28

Reported variables are marginal effects (i.e. no constant). Shaded boxes are significant factors (1–10%).

influence village decision-making is an important explanatory factor, suggesting that influence in protected area management is only partly determined by the organization of protected area management itself. Also, the perception that respondents have of the village's influence in regional decision-making is a significant determinant, underlining the importance of village representation at higher governance scales.

Another important finding is that, as expected, poor people feel less able to influence decision-making concerning the management of the park. In Vietnam, household income status does not matter, but larger landholders and households with jobs in the public sector (i.e. the better-off) feel better able to influence park decision-making. In India, surprisingly, food secure households feel less able to influence park management, but the significant negative sign for livestock-owning households is an indication that the indigenous (and poor) *dhangar* households feel less capable of influencing management of the park. Interestingly, in Vietnam and India households involved in tourism feel better able to influence park management, whereas in Costa Rica it is the households that are agriculture based. Finally, in Vietnam the significant positive sign for majority religion (note that in Vietnam this is atheism) indicates that the indigenous and Protestant *H'mong* people might not feel represented in decision-making.

Turning to the perceived effectiveness of protected area management, we asked respondents whether they believed the rules regarding the use of natural resources in the PA were well enforced. As the results presented in Table 4.8 indicate, rule enforcement is perceived to be relatively strong in Vietnam, and relatively weak in India. The very low percentage of households knowing of rule-breakers in India seems to have to do with fear. In anonymous interviews people said that they knew many examples of rule-breaking (Trepp 2010), but this is not reflected in the survey results. The potential willingness to report rule-breakers is in all sites relatively high, which could be an indication that there is a potential for protected area co-management, with communities playing a larger role in enforcement and control. Danielsen *et al.* (2008) indicate that community involvement in monitoring and enforcement can greatly improve the effectiveness of protected area management, local communities having more information about the (illegal) use of resources, both by their own community members and by outsiders.

Table 4.8 *Summary statistics of rule enforcement and willingness to report*

	Mutale Basin, South Africa	Terraba Basin, Costa Rica	Ba Be–Na Hang, Vietnam	Warna Basin, India
Households that believe that rules are well-enforced	64%	61%	78%	49%
Households knowing of rule-breaking	21%	38%	57%	3%
Households willing to report	66%	86%	66%	58%

Considering the factors that determine the willingness to report rule-breakers, the results in Table 4.9 suggest that this differs per study site and depends on household and community characteristics such as ethnicity, landownership and type of livelihood activity in which the household is involved. In the Mutale Basin, households involved in agriculture are more likely to report rule-breakers, possibly because, compared to other households, they are more strongly connected to the land. Livestock-owning households, on the other hand, are less eager to report rule-breakers, possibly because they intensively use common lands for grazing

Table 4.9 *Probit model explaining the household's willingness to report rule-breakers*

	Mutale Basin, South Africa	Terraba Basin, Costa Rica	Ba Be–Na Hang, Vietnam	Warna Basin, India
Distance to PA	0.005	−0.01	0.009	−0.04
Location inside PA	Dropped	−0.22	0.045	−0.2
Female respondent	0.09	0.007	−0.19	−0.08
Agriculture	0.25	0.01	−0.12	−0.19
Small commerce/industry	0.05	0.025	0.12	0.07
Tourism	Dropped	0.03	0.04	−0.02
Public sector	Dropped	0.03	0.15	−0.24
Household per capita income (US$/month)	0.000	−.0000	0.000	0.005
Household below poverty line	0.009	−0.11	0.12	−0.05
Household has sufficient to eat	0.39	0.03	0.10	0.04
Landholding size (acres)	−0.02	0.000	0.04	−0.04
Livestock-owning household	−0.24	−0.71	0.07	0.06
Household can influence DS	0.29	0.018	0.10	−0.07
Village can influence DS	0.32	0.016	0.01	0.10
Household participates in voluntary organization	0.11	0.017	0.02	0.18
Household majority ethnic group	0.11	0.05	0.30	0.01
Household majority religious group	−0.11	0.24	0.28	0.26
Household knows of rule-breakers	−0.003	−0.02	0.17	−0.36
No. of observations	76	110	246	147
Wald chi^2 (18)	34.90	53.51	62.66	39.93
Pseudo R^2	0.24	0.40	0.23	0.20

and livestock feed. In Costa Rica, households involved in public sector jobs and tourism are more likely to report, whereas livestock-owning households (mostly urban households with landholdings outside the village) are not. This seems to reflect the household's costs and benefits associated with rule-enforcement: urban households having higher opportunity costs and tourism dependent households having direct benefits in ecosystem protection. In Vietnam, women are less willing to report rule-breakers, which could either reflect women being more engaged in NFTP collection or an intrinsic unwillingness to report. In India livelihood strategy as such has no significant impact, but households with large landholdings are less willing to report. This is an indication that households inside and near the park are unwilling to cooperate, since these are the households with large landholding size. The relationship between household income, poverty levels and willingness to report is ambiguous: in India and South Africa better-off households are more willing to report, in Costa Rica poorer households are more willing to report and in Vietnam income and poverty levels do not have a significant effect.

Interestingly, only in Vietnam does knowledge of rule-breaking influence willingness to report, which seems a bit awkward since one would expect to see such a relationship in the other sites too. Household ethnicity and religion play a significant role in Costa Rica and Vietnam. In Costa Rica the positive sign for majority religion (i.e. Catholic) indicates that the Protestant fishermen in the wetland might be less willing to report. Similarly, in Vietnam the positive sign for the majority ethnic group (Tay) and majority religion (atheism) indicate that minority groups might be less willing to cooperate with park management. In India ethnicity has no significant influence, but households active in voluntary organizations are more likely to cooperate in rule enforcement. In South Africa ethnicity and involvement in voluntary organizations has no significant influence, but households that feel capable of influencing village decision-making are more likely to collaborate in enforcement and control.

Summarizing, there are many factors influencing household willingness to report rule-breaking and ability to influence protected area management, but as expected the household's livelihood strategy, asset and income status (food security, landownership, etc.) and ethnicity play an important role. This is in line with the broader literature and suggests that communities may be involved in protected area management, but only when they see sufficient benefits in collaborating and when they feel they can influence the management of the park.

4.5 Discussion

This chapter analysed the potential synergies for nature conservation and local livelihood improvement in four biodiversity hotspots around the world. In all four regions biodiversity is conserved in protected areas, but the type of management, the type of ecosystem pressures and the type of livelihood strategies differ between the sites. By assessing the differences in livelihood strategies and park management, and the influence of these differences on the household's willingness to participate in rule enforcement, we addressed the potential for community co-management as a means of reaching conservation–development synergies in the sites.

Our findings suggest that, whereas biodiversity protection is currently creating livelihood trade-offs in most of the sites, synergies would be possible when local livelihoods and nature conservation could be more directly linked. In Ba Be National Park and Na Hang Nature Reserve, Vietnam, communities officially participate in park management, but with little conservational responsibilities and no influence on how resource-use restrictions are defined. Community members suffer from stringent resource-use restrictions and although they participate in forest fire and other committees, biodiversity monitoring and rule enforcement is controlled top-down. Making communities more directly responsible for conservational outcomes, and rewarding them for their efforts through payment mechanisms or a granting of more flexible user rights is expected to help improve biodiversity protection and reduce negative livelihood trade-offs.

In the Terraba-Sierpe wetland, Costa Rica, monitoring and enforcement of sustainable resource use is missing and local communities suffer from a lack of clear user rights and a continuing degradation of the surrounding ecosystem on which their livelihoods depend. Park officials do not acknowledge the role of local communities in wetland conservation, and although the wetland is threatened by external factors (construction of a hydropower dam, commercial agriculture upstream) acknowledgement of the potential role of communities could help induce improved wetland conservation and better representation of local interests at higher governance levels. In the Boruca Indigenous Territory communities are already responsible for conservation, but, like the communities in Terraba-Sierpe wetland, their livelihoods suffer from developmental pressures at other scales. Effective representation of local interests in development-related decision-making seems crucial in the Terraba Basin to avoid conservation–development trade-offs.

In Chandoli National Park, India, displacement of local communities for park establishment increased socio-economic vulnerability and thus created large livelihood trade-offs. Regranting local communities access to the protected area could help reduce the adverse impacts on local livelihoods, but effective co-management seems unlikely since the perceived legitimacy of the park is low. A more structural change of park management seems required to create synergies, since currently biodiversity protection is creating serious livelihood trade-offs.

In Makuya Park, South Africa, park establishment created trade-offs during apartheid, but at present the dependence of local livelihoods on the protected area is low. Hence, better management of the park is unlikely to generate clear livelihood benefits, although the non-protected part of the ecosystem does play an important safety net role. Formally, community co-management has already been established, communities being represented by tribal authorities that directly participate in and benefit from the management of the park. Benefits do not seem to trickle down to community members, however, and given that benefits are mostly generated through ecotourism, alternative livelihood investments in ecotourism might have more substantial livelihood improvement effects. Similarly, at Lake Fundudzi, communities formally already manage the ecosystem, but given the high dependence on government grants and pensions the incentive for community members to become actively engaged in ecosystem protection is relatively low. Again, investments in community-level ecotourism might have more substantial livelihood impacts, but it is outside the scope of this chapter to elaborate what these interventions might be.

Whether involving the poor in decision-making is sufficient for alleviating poverty seems unlikely since in all sites income poor and/or food insecure households feel less able to influence park management and decision-making at village scale. Income poor and/or food insecure households perceive the direct collection of products from nature as more important for their livelihoods than non-poor and/or food secure households, but since we could not evaluate the impact of park establishment on local livelihoods we could not determine whether park establishment especially affected the poor. The likelihood of conservation contributing to poverty alleviation seems, however, small. Conservation implies resource-use restrictions and given that the poor have fewer assets to fall back on they tend to be more strongly affected by restrictions than the better off. Finally, the findings indicate that the potential for co-management and direct linking of local livelihoods to conservation outcomes does not only depend on livelihood–nature linkages and the type of park management but on household ethnicity as well: minority ethnicity households are less willing to cooperate with park management, probably because household (and village) ethnicity influences trust in park officials, representation of household interests in decision-making and the perceived legitimacy of the park.

Further research is required to elaborate the potential for co-management in the sites mentioned and the factors driving ecosystem degradation and biodiversity depletion at local and regional scale. Also, the analysis did not elaborate the potential impact of alternative livelihood investments on nature conservation and local livelihoods: possibly, such investments could make a difference especially in cases where direct linkage approaches do not work. Finally, given the heterogeneity of our sample and lack of time series data we could not evaluate how people's livelihoods were affected by protected area establishment and what the causalities between poverty, local livelihood and the use and management of the ecosystem are. Still, we believe that the analysis has succeeded in giving a broad overview of the possible poverty–livelihood–nature linkages and an understanding of what it requires to create conservation–development synergies in biodiversity hotspots around the world.

Acknowledgements

The authors would like to acknowledge the support of the European Commission through the FP7 programme ENV.2007.2.1.4.3 Biodiversity values, sustainable use and livelihoods, LiveDiverse Project 211392. They would especially like to thank Vu Cong Lan (NIAPP-Vietnam) and Alexander Lopez Ramirez (UNA-Costa Rica) for their contributions. Also, they would like to acknowledge the efforts of the household survey data collection teams and of Elisa Trepp, Paulina Gonzalez Pichardo, Aine Nirain, Jelena Perunicic and Lisette van Marrewijk for their help with data collection.

References

Adhikari, B., Di Falco, S. and Lovett, J. C. (2004). Household characteristics and forest dependency: evidence from common property forest management in Nepal. *Ecological Economics*, **48**(2): 245–257.

Agrawal, A. (2001). Common property institutions and sustainable governance of resources. *World Development*, **29**: 1649–1672.

Armitage, D., Marschke, M. and Plummer, R. (2008). Adaptive co-management and the paradox of learning. *Global Environmental Change*, **18**: 86–98.

Bawa, K., Joseph, G. and Setty, S. (2007). Poverty, biodiversity and institutions in forest-agriculture ecotones in the Western Ghats and Eastern Himalaya ranges of India. *Agriculture, Ecosystems and Environment*, **12**: 287–295.

Berkes, F. (2006). From community-based resource management to complex systems. *Ecology and Society*, **11**(1): 45.

Bruner, A. G., Gullison, R. E., Rice, R. E. and da Fonseca, G. A. B. (2001). Effectiveness of parks in protecting tropical biodiversity. *Science*, **291**: 125–128.

Carlsson, L. and Berkes, F. (2005). Co-management: concepts and methodological implications. *Journal of Environmental Management*, **75**(1): 65–76.

Cernea, M. M. and Schmidt-Soltau, K. (2006). Poverty risks and national parks: policy issues in conservation and resettlement. *World Development*, **34**(10): 1808–1830.

Danielsen, F., Burgess, N., Balmford, A. *et al.* (2008). Local participation in natural resource monitoring: a characterization of approaches. *Conservation Biology*, **23**(1): 31–42.

Deland, C. (2006). Not just minor forest products: the economic rationale for the consumption of wild food plants by subsistence farmers. *Ecological Economics*, **59**: 64–73.

Faber, M, Oelofse, A., Van Jaarsveld, P. J., Wenhold, F. A. M. and Jansen van Rensburg, W. S. (2010). African leafy vegetables consumed by households in the Limpopo and KwaZulu-Natal provinces in South Africa. *South African Journal of Clinical Nutrition*, **23**(1): 30–38.

Ferraro, P. J. and Kiss, A. (2002). Direct payments to conserve biodiversity. *Science*, **298**(28): 1718–1719.

Fisher, B. and Cristoph, T. (2007). Poverty and biodiversity: measuring the overlap of human poverty and the biodiversity hotspots. *Ecological Economics*, **62**: 93–101.

Hill, M., Hallam, D. and Bradley, J. (1997). *Site Study; Ba Be National Park, Cao Bang Province, Vietnam. SEE-Vietnam Research Report 3*. London: Society for Environmental Exploration.

Kerr, J. (2002). Watershed development, environmental services and poverty alleviation in India. *World Development*, **30**: 1387–1400.

Kubo, H. and Supriyanto, B. (2010). From fence-and-fine to participatory conservation: mechanisms of transformation in conservation governance at the Gunung Halimun-Salak National Park, Indonesia. *Biodiversity and Conservation*, **19**(6): 1785–1803.

Lahiff, E. (1997). Rural land, water and local governance in South Africa: a case study of the Mutale River Valley. Resources Rural Livelihoods Working Paper Series. Paper No. 7.

Maffi, L and Woodley, E. (2010). *Biocultural Diversity Conservation*. London: Earthscan.

Makhado, R. A., Potgieter, M. J. and Wessels, D. C. J. (2009). *Colophospermum mopane* wood utilisation in the northeast of the Limpopo Province, South Africa. *Ethnobotanical Leaflets*, **13**: 921–945.

Medvey, J. (2010). Benefits or burden? Community participation in natural resource management in the greater Kruger Park area. MSc thesis. IVM Institute for Environmental Studies, VU University Amsterdam, the Netherlands, available at: www.livediverse.eu.

Millennium Ecosystem Assessment (MEA) (2005). *Ecosystems and Human Well-being: Biodiversity Synthesis*. Washington DC: World Resources Institute.

Murphee, M. (2009). The strategic pillars of communal natural resource management: benefit, empowerment and conservation. *Biodiversity Conservation*, **18**: 2551–2562.

Niesten, E. and Milne, S. (2009). Direct payments for biodiversity conservation in developing countries: practical insights for design and implementation. *Oryx*, **43**(4): 530–541.

Ostrom, E. (1990). *Governing the Commons: The Evolution of Institutions for Collective Action*. Cambridge: Cambridge University Press.

Ostrom, E. (2009). A general framework for analyzing sustainability of social-ecological systems. *Science*, **325**: 419–422.

Plummer, R. and Fitzgibbon, J. (2004). Co-management of natural resources: a proposed framework. *Environmental Management*, **33**(6): 876–885.

Posel, D. and Casale, D. (2006). Migration and remittances in South Africa. Background document on migration and first set of draft questions for inclusion in the National Income Dynamics Study. University of KwaZulu-Natal, Department of Economics, p. 58.

Sen, A. (1983). Poverty, relatively speaking. *Oxford Economic Papers*, **35**(2): 153–169.

Sen, A. (1995). Rationality and social choice. *American Economic Review*, **85**(1): 1–24.

Somanathan, E., Prabhakar, R. and Mehta, B. S. (2009). Decentralization for cost-effective conservation. *Proceedings of the National Academy of Sciences*, **106**(11): 4143–4147.

Stern, M. (2008). Coercion, voluntary compliance and protest: the role of trust and legitimacy in combating local opposition to protected areas. *Environmental Conservation*, **35**(3): 200–210.

TEEB (2009). The economics of ecosystems and biodiversity for national and international policy makers. Summary: responding to the value of nature. Available at: http://www.teebweb.org/LinkClick.aspx?fileticket=dYhOxrQWffs%3d&tabid= 1019&mid=1931.

Trepp, E. (2010). Chandoli National Park and Resettlement: impacts on local communities in Maharashtra, India. MSc thesis, IVM Institute for Environmental Studies, VU University Amsterdam, the Netherlands, available at: www.livediverse.eu.

Tshihwanambi, T. P. (2007). Consumption patterns of vitamin-A rich foods of 10–13 year old children living in a rural area in Venda. MSc thesis, Consumer Science, University of Pretoria

UN (2010). High-level meeting of the General Assembly as a contribution to the international year of biodiversity, A/64/865. United Nations General Assembly, New York.

Uribe, M. (2010). Terraba Sierpe Wetland's Management plan: struggling for policy change and its implementation. MSc thesis, IVM Institute for Environmental Studies, VU University Amsterdam, the Netherlands, available at: www.live diverse.eu.

Appendix

Table 4.10 *Overview of household survey variables used in analysis*

Variable	Type of variable and interpretation
Distance to PA	Km to the fence of the protected area (PA)
Location inside PA	Dummy variable, =1 when village is located inside PA
Female respondent	Dummy variable, =1 when respondent is female
No. of household members	Number of household members
Agriculture	Dummy variable, =1 when at least one member of household is self-employed in agriculture
Industry/commerce	Dummy variable, =1 when at least one member of household is self-employed in commerce/industry
Tourism	Dummy variable, =1 when at least one member of household is self-employed in tourism
Public sector	Dummy variable, =1 when at least one member of household is employed in the public sector
Household per capita income (US$/month)	Average monetary household income/per capita (US$/month)★
Household below poverty line	Dummy variable, =1 when household per capita income is below the underbound of the regional poverty line
Household has sufficient to eat	Dummy variable, =1 when household reports having always or usually sufficient to eat
Land-owning household	Dummy variable, =1 when household owns land (user right)
Size landholding (acres)	Size of the landholding in acres
Livestock-owning household	Dummy variable, =1 when household owns livestock
No. of products collected	No. of (non-)timber forest products collected in nature
Importance of (non-) timber forest product collection	Dummy variable, =1 when household considers collection very or somewhat important
Wood main source of fuel	Dummy variable, =1 when household reports using firewood/charcoal as main source of fuel
Household can influence PA	Dummy variable, =1 when households always, often or sometimes feels able to influence PA management
Household can influence village level decision-making	Dummy variable, =1 when household always, often or sometimes feels able to influence village decision-making

Table 4.10 (*cont.*)

Variable	Type of variable and interpretation
Village can influence decision-making	Dummy variable, =1 when household always, often or sometimes feels that the village influences decision-making at higher levels
Household participates in voluntary organization	Dummy variable, =1 when household participates in a voluntary organization (excluding church)
Household majority ethnic group	Dummy variable, =1 when household is part of the regions majority ethnic group (Vhembe in South Africa, Costarican in Costa Rica, Tay in Vietnam and Marathi in India)
Household majority religious group	Dummy variable, =1 when household is part of the region's majority religious group (i.e. Christian in South Africa, Catholic in Costa Rica, atheist in Vietnam and Hindu in India)

*1 US$ = 7 South African Rand = 500 Costarican colones = 20 000 Vietnamese Dong = 50 Indian RS

Part II
Marine-related ecosystem services

PIETER J. H. VAN BEUKERING

Coastal regions are special in many ways. First, coastal ecosystems are among the most productive systems in the world. These ecosystems produce disproportionately more services relating to human well-being than most other systems (MEA 2005). Second, 60% of the world's population lives in coastal zones. The total number of people living in coastal areas has doubled in the last 20 years (Goudarzi 2006). Coastal regions are home for more than 250 million poor people around the world (Brown *et al.* 2008). Therefore, coastal and marine resources are of increasing importance for human well-being in various ways (e.g. food, employment). Third, coastal ecosystems experience the heaviest impacts from human uses and environmental changes (Adger *et al.* 2005, Donner and Potere, 2007, Jackson *et al.* 2001). Future pressures from climate change, population increases in coastal areas, pollution, aquaculture development, greater human mobility and the spread of invasive species are likely to further exacerbate these trends (Brown *et al.* 2008). As a result, these characteristics of coastal ecosystems pose crucial challenges for the maintenance of ecosystem services and poverty alleviation.

The ecosystems covered in this part of the book cover coastal and marine resources as well as small islands. Coastal ecosystems form the interface between ocean and land, and include coastal lands, areas where fresh water and salt water mix, and nearshore marine areas. Coastal ecosystems cover the area between 50 m below mean sea level and 50 m above the high tide level or extending landward to a distance 100 km from shore (Nelson 2007). Typical ecosystems that can be found in the coastal zones include coral reefs, intertidal

Nature's Wealth: The Economics of Ecosystem Services and Poverty, ed. P. J. H. van Beukering, E. Papyrakis, J. Bouma and R. Brouwer. Published by Cambridge University Press, © Cambridge University Press 2013.

zones such as mangroves, estuaries and sea grass communities. Coastal and marine ecosystems provide a wide range of ecosystem services such as fish and other food sources, a sink for human wastes, coastal protection, building materials, tourism, recreation and cultural values. Marine areas are defined as those areas in the sea that are deeper than 50 m. Marine ecosystems are oceans where fishing is typically the major activity. Finally, small islands are lands surrounded by water. Island states cover 40% of the world's oceans (including their exclusive economic zones), and tend to have a higher proportion of coastal area to inland area. For some of the smaller islands, the entire land area is classified as 'coastal'. This is due to the geography (often comprising mangroves, wetlands, sea grass beds, coral reefs and sandy beaches) and the small size of the island (van Beukering *et al.* 2007).

The degradation of coastal and marine ecosystems can have serious implications for ecosystem services of importance to the poor. Unfortunately, data on the actual production of ecosystem services are rarely available and therefore the impact of ecosystem degradation on the poor is generally assumed rather than measured. A further complication is the fact that the link between coastal ecosystems and services provided often varies spatially due to geography and the social environment, and in some cases also is non-linear (Barbier *et al.* 2008). This makes it even more difficult to generalize the relationship between ecosystems and ecosystem services and emphasizes the need to conduct site-specific research on the link between coastal ecosystem and ecosystem services.

Another important question is whether coastal and marine ecosystems are a means of last resort or whether they form a basis for a mature and sustainable source of living. Although there is often limited precise information on the real contribution of fisheries to livelihoods of the poor, the scientific literature provides a mixed picture. Some studies associate coastal fisheries with poverty traps for the very poor (Béné 2003: 956). Other studies contradict the view of fishing as a last-resort activity and view coastal ecosystems as a source of average income levels (Allison and Ellis 2001, Allison and Horemans 2006). Knowledge of the role of other ecosystem services in supporting the livelihoods of the poor, besides provisioning services, is even more scant. Brown *et al.* (2008) conducted 15 focus groups among coastal communities in five different countries with the aim of quantifying the relative contribution of the four types of ecosystem services provided by coastal and marine ecosystems. Although significant differences were recorded between and within countries, on average, provisioning services were considered to be the main source of livelihood (i.e. 59% of the total benefits). Supporting, regulating and cultural services scored 22%, 10% and 9%, respectively.

Coastal and marine ecosystems only benefit the poor if they also have the ability to benefit from these ecosystem services. In general, numerous barriers exist in enabling the poor to reap the benefits provided by coastal and marine ecosystems. Brown *et al.* (2008) identify several categories of barriers. First, access to provisioning services in particular is an important factor in the ability of the poor to gain from ecosystem services. Generally, access to coastal and marine ecosystems is determined by a range of formal and informal property rights. Licensing is an example of formal property rights, which in theory can be used to protect access to resources for the poor. However, in practice, licenses and permits often have built-in biases against the poor, as will be demonstrated in the South African case study in Chapter 7. Although licenses may affect the poor in some developing countries, the majority of the coastal and marine ecosystem contexts are characterized as open access resources. The lack of regulations, or the enforcement thereof, may have both positive and negative impacts on the poor. On the one hand, the poor can access resources without formal rights or at relatively low cost. On the other hand, since other users cannot be excluded, the poor have no guarantees that sufficient benefits are left for themselves. Nor can they effectively manage these open access resources due to the presence of free-riders that are likely to violate any formal or informal rule. The second barrier is technology, which often constitutes an important constraint or enabler to access by the poor. Technology can advantage or disadvantage particular groups of people, and can shift the distribution of benefits from ecosystem services in important ways. Third, access to markets has a distinct impact on the extent to which the poor can generate an income from provisioning services. Limited market access means that middlemen can dictate the prices and so fishers have no choice but to sell their catch at prices that hardly cover the costs. Fourth, marine protected areas (MPAs) may have a positive or negative impact in terms of their ability to support poor people's access to ecosystem services. Although some MPAs provide various types of benefits, they may also impose restrictions to the poor through the displacement of local fishermen. Finally, competing uses such as industrial development, tourism and agriculture are also known to undermine the ability of poor people to benefit from coastal ecosystem services.

The reciprocal nature of the relationship between coastal and marine ecosystems and the poor implies that besides ecosystems affecting the livelihoods of local communities, activities by local communities equally affect coastal ecosystems. Unsustainable exploitation by the coastal poor can occur in various ways. Destructive fishing techniques such as dynamite fishing and small mesh sized nets result in overfishing and habitat loss. Overexploitation of mangroves and coral reefs for fuelwood and construction materials is another frequent cause for habitat loss. Agriculture in coastal areas leads to increased sediment loads and high nutrient levels in the coastal waters. Selected harvest of or easily accessible and high value species can cause sudden shifts in ecosystems leading to, for example, excessive algal growth. The cases described in this book also contain elements listed above in which poor people negatively affect coastal and marine ecosystems. However, the contribution of the wealthier stakeholders in society seems to outweigh that of the poor. In Vietnam, the tourist industry is a major contributor to coastal run-off and over-use of coral reefs. In South Africa, the richer fishers have larger boats and thereby claim a larger share of the fish catch. In conclusion, the poor are certainly responsible for part of the ecosystem degradation; however, whether their impact outweighs that of the non-poor is unlikely. Especially when seen from the perspective of the costs and benefits of ecosystem exploitation, these have not been equally distributed. In particular, the costs have been disproportionately borne by the world's poor, and the gap between the rich and the poor has increased (Turner and Fisher 2008).

There are various ways in which people can intervene to increase the provision of marine and coastal ecosystem services. The most straightforward manner is the enhancement of provisioning services through, for example, aquaculture for fish production. However, negative impacts of aquaculture on natural ecosystems may jeopardize the provision of other ecosystem services. Alternatively, the provision of ecosystem services can be developed, or at least stabilized at sustainable levels, by limiting fishing pressure or by the establishment of marine protected areas (MPAs). The MEA (2005) identified overexploitation of fisheries of most relevance to the coastal poor. For heavily overfished stocks, such as those involved in the live reef fish food trade (Sadovy *et al.* 2003), reductions in fishing pressure could result in increased production and more stable yields for local people. Both approaches are extensively addressed in the following chapters in which the system of fish quota in South Africa is explained and the costs and benefits of MPAs in various Asian countries is elaborated upon.

Organization of Part II

Part II of the book demonstrates how coastal and marine ecosystems and the lives of the poor are closely interrelated. Special attention is given to the ability of poor to benefit from coastal ecosystem services and ways to increase the livelihood opportunities derived from these ecosystem services. The three chapters presented in Part II each have a distinct methodological approach and also differ in terms of the focus of the study. Part II is organized as follows.

Chapter 5 describes the results of a comprehensive multi-country study, which specifically addresses the role of MPAs in alleviating poverty in the Asia–Pacific region. The study sites are in Fiji (Navakavu), the Solomon Islands (Arnavon Islands), Indonesia (Bunaken) and the Philippines (Apo Island). The sites are not a random sample but were deliberately chosen because local experts believe they have contributed to poverty reduction. The objective was to study potentially positive examples to see if there are common factors for success. There are in fact several shared reasons why these particular MPAs helped reduce local poverty. Replicating these success factors can help MPAs in general contribute more to reducing local poverty. The findings conclusively show that people in the community are now better off and this is because of the MPA. Across the four sites, there was clear evidence that poverty had been reduced by several factors: (1) improved fish catches; (2) new jobs, mostly in tourism; (3) stronger local governance; and (4) benefits to health and women. The researchers hope that governments will use these study findings to harness the full benefits of MPAs to improve the well-being of local people while conserving marine life. The study recommends several key strategies for strengthening the creation and management of MPAs. First, governments are recommended to commit to financial investment in protected areas, both in the initial set up and in subsequent years. If done correctly, it can be a worthwhile investment. 'The marine protected area is like a bank to the people', noted a Fijian community leader. Second, the team encourages the development of a network of smaller, ecologically connected MPA sites, each linked to a community, to increase local access to benefits. And finally, it will be crucial to empowering local communities in the decision-making and management of the MPA. The study convincingly shows that MPAs and local communities need each other. Without the support of the local community, MPAs will not succeed.

The second study in Part II also addresses the efficiency of an MPA but focuses on issues of distribution of the benefits of coastal and marine ecosystems and ways to set up a system of sustainable financing. Chapter 6 reports on the economics of conservation for the Hon Mun MPA in Vietnam. The study demonstrates that the overall benefits of the

Hon Mun MPA have indeed increased over time, yet these additional economic returns mainly benefitted the wealthy entrepreneurs in the tourist industry rather than the local fishermen who lost some of the fishing ground. Moreover, the study continues to recognize that despite the ecological and socio-economic benefits MPAs provide, the establishment and management of MPAs is often seriously constrained by both a lack of funding and a poor relationship with communities living around (or within) them. Although international financial support has ensured sound management of its natural resources in the short run, in the long term, the Hon Mun MPA needs to develop its own sustainable and autonomous financing regime. On the basis of a thorough analysis of the various (economic) instruments for generating funds for MPA management, the researchers conclude that the best way to 'appropriate' Hon Mun's potential economic benefits would be through a user-fee for eco-tourists. Subsequent revenues could be ploughed back into management of the park and its buffer zone, and could also support much-needed alternative livelihood schemes in the region. This latter measure is essential given that, notwithstanding the park's success, around one-third of local people in and around Hon Mun feel worse off since its creation. Similar to the conclusions of the multi-country MPA study presented in Chapter 5, this study emphasizes that without the support of the local people it will become increasingly difficult to enforce the park's no-take zones.

The last study presented in Part II addresses an alternative system to manage the marine ecosystem. Chapter 7 describes a multi-criteria approach to equitable fishing rights allocation in South Africa's Western Cape. The problems of the management of marine fishery resources, the allocation of rights to exploit those resources and of allocating associated quota (in terms of quantum or effort) are universal. This study has attempted to investigate and offer solutions to some of the problems in a context of the additional need for transformation. Due to a combination of insufficient access to fishing rights, overexploitation of marine resources and a lack of alternative livelihood opportunities, the coastal communities of South Africa's Western Cape have lost both social and economic stability. Significant changes in fisheries management and the fishing rights allocation system have occurred since the end of the apartheid era; however, complementary policies to support fishers in dealing with these developments have been lacking. In particular, total allowable catches (TACs) have been reduced in precisely the fisheries of greatest significance to local communities, and fishers are struggling to cope with this reduced resource availability. The objectives of the study were to investigate the skills and training needs precipitated by the introduction of new approaches to allocations in previous years and the empowerment of fishers for more successful application and use of fishing rights and, primarily, to develop a simple, transparent and legally defensible allocation system. To provide background material for the project, the development of rights allocation internationally and in South Africa was investigated. Three communities were chosen who contributed to the various parts of the study through workshops and other interactions. Four fisheries relevant to these communities were investigated in more depth in terms of the changes in rights allocation over time. The analyses and interactions undertaken to achieve the three objectives of the project led to a number of recommendations all of which have implications for environmental management, sustainable use, empowerment and poverty reduction. This study concludes that a just, transparent and broadly acceptable process for allocating fishing rights, and the empowerment of local people to make effective use of these rights, will be critical for both the protection of the fish stocks and the alleviation of poverty in South Africa's Western Cape.

References

Adger W. N., Hughes, T. P., Folke, C. *et al.* (2005). Social–ecological resilience to coastal disasters. *Science*, **309**(5373): 1036–1039.

Allison, E. H. and Ellis, F. (2001). The livelihoods approach and management of small-scale fisheries. *Marine Policy*, **25**: 377–388.

Allison, E. H. and Horemans, B. (2006). Putting the principles of the sustainable livelihoods approach into fisheries development policy and practice. *Marine Policy*, **30**: 757–766.

Barbier, E. B., Koch, E. W., Silliman, B. R. *et al.* (2008). Coastal ecosystem based management with non-linear ecological functions and values. *Science*, **319**(5861): 321–323.

Béné, C. (2003). When fishery rhymes with poverty: a first step beyond the old paradigm on poverty in small-scale fisheries. *World Development*, **31**(6): 949–975.

Brown, K., Daw, T., Rosendo, S., Bunce, M. and Cherrett, N. (2008). Ecosystem services for poverty alleviation: marine and coastal situational analysis. Synthesis Report, November 2008. University of East Anglia, Norwich.

Donner, S. D. and Potere, D. (2007). The inequity of the global threat to coral reefs. *BioScience*, **57**: 314–315.

Goudarzi, S. (2006). Flocking to the coast: world's population migrating into danger. *Live Science*, available at: http://www.livescience.com/environment/060718_map_settle.html.

Jackson, J., Kirby, M., Berger, W. *et al.* (2001). Historical overfishing and the recent collapse of coastal ecosystems. *Science*, **293**(5530): 629–637.

Millennium Ecosystem Assessment (MEA) (2005). *Ecosystems and Human Well-being: Biodiversity Synthesis.* Washington DC: World Resources Institute.

Nelson, S. A. (2007). Coastal zones. EENS 2040, Natural disasters. Tulane University, LA. Available at: http://www.tulane.edu/~sanelson/geol204/coastalzones.htm.

Sadovy, Y. J., Donaldson, T. J., Graham, T. R. *et al.* (2003). *The Live Reef Food Fish Trade while Stocks Last.* Manila, Philippines: Asian Development Bank.

Turner, R. K. and Fisher, B. (2008). Environmental economics: to the rich man the spoils. *Nature*, **451**(7182): 1067–1068.

van Beukering, P., Brander, L., Tompkins, E. and McKenzie, E. (2007). *Valuing the Environment in Small Islands: An Environmental Economics Toolkit.* Peterborough, UK: Joint Nature Conservation Committee (JNCC).

5 · The role of marine protected areas in alleviating poverty in the Asia-Pacific

PIETER J. H. VAN BEUKERING, LEA M. SCHERL AND CRAIG LEISHER

5.1 Introduction

Small-scale fishers in developing countries depend heavily on near-shore marine fish capture (Pauly 2006, SOFIA 2008). Yet marine fisheries in many developing countries are underregulated and overfished (Agnew et al. 2009, Le Gallic and Cox 2006, Varkey et al. 2009). Global marine capture fish production peaked in the mid-1980s, and one in three marine fisheries are now considered overfished (SOFIA 2008, Worm et al. 2009). The overlapping issues of local livelihoods and fisheries management are particularly apparent in coastal coral reefs.

Globally, the area called the 'Coral Triangle' is the epicentre for coral and marine fish diversity (Allen 2008, Allen and Erdmann 2009, Veron et al. 2009). This area of Indonesia, Malaysia, the Philippines, Papua New Guinea and the Solomon Islands is where the challenges of enhancing livelihoods, regulating fisheries and conserving coral reefs intersect. Of the 296 million people in the Coral Triangle area, more than 168 million people live below the international income poverty line (Ravallion et al. 2009) and there are an estimated 63 million people in the Coral Triangle living within 20 km of a coral reef (authors' GIS calculations). Marine ecosystems play an important role in the subsistence of many people in the Coral Triangle, and there is a growing emphasis in the region on the use of MPAs as fisheries management and conservation tools.

Most of the scientific literature on the effectiveness of MPAs concentrates on the biological effects of marine reserves. The effects of MPAs on poverty alleviation have been addressed by only a limited number of

Nature's Wealth: The Economics of Ecosystem Services and Poverty, ed. P. J. H. van Beukering, E. Papyrakis, J. Bouma and R. Brouwer. Published by Cambridge University Press, © Cambridge University Press 2013.

studies, and all these case studies are conducted in a single country only, thereby making it more difficult to draw generic conclusions on the role of marine protected areas in alleviating poverty (Aswani and Furusawa 2007, Baticados and Agbayani 2000, Gjertsen 2005, Katon *et al.* 1999, Maliao and Polohan 2008, Oracion *et al.* 2005, Pomeroy *et al.* 1997, Tobey and Torell 2006, Webb *et al.* 2004). This study investigates whether community-managed and co-managed MPAs can measurably improve livelihoods and contribute to poverty reduction.

Research questions such as these are hard to answer because there is a general lack of rigorous assessments of conservation initiative impacts on people (Ferraro and Pattanayak 2006), and much of the current thinking about the relationship between poverty and conservation is still primarily based on expert opinion rather than data from well-designed studies (Pullin *et al.* 2004, Scherl *et al.* 2004).

The evidence base in the Coral Triangle is no different. There is little empirical evidence for MPAs benefiting local people. Yet within the more than 675 formally designated MPAs in the Coral Triangle (WDMPA 2010), many experts believe there are at least a handful of standout successes. To highlight these successes, we used a social science approach from the health sector known as 'positive deviance' (Pascale 2010, Sternin and Choo 2000). Positive deviance is based on the premise that in every community there are outliers that are deviant in a positive way, and understanding *how* they achieve a measure of success where others have not is the first step to replicating the success elsewhere.

In this study, we assess the poverty reduction contributions of four MPAs. Understanding how these MPAs have contributed to poverty reduction can help poverty-focused organizations and conservation agencies do their jobs better. For government policymakers in particular, this study provides empirical evidence of how MPAs can contribute to poverty reduction and highlights the factors perceived to drive the MPAs' achievements.

Between November 2006 and May 2007, a study team conducted 991 interviews at households and with key informants, and held 18 group discussions with approximately 120 participants. In total, more than 1110 local people were consulted to determine whether four particular MPAs had contributed to poverty reduction, and if so, why. The four study sites do not represent a random sample but were deliberately chosen because local experts believe they have contributed to poverty reduction.

5.2 Data and methodology

5.2.1 Defining and measuring poverty

Income level is the most widely used indicator of poverty, and the Millennium Development Goals enshrined this concept in their target of halving the number of people living on less than US$ 1 a day by 2015. Yet defining poverty by income alone is widely recognized as too narrow an approach. To reduce poverty, greater income is important, but poverty reduction can also come from *increasing opportunities* for the poor through, for example, education and new livelihoods. It can come from *empowering the poor* in areas such as decision-making on public services and resource allocation. It can come from *enhancing the security* of poor people by reducing their risk from food shortages, natural disasters, health crises and other catastrophic events.

In recognition of the fact that poverty is multi-dimensional, this study uses the World Bank's definition of poverty, which comprises three elements: opportunity, empowerment and security (World Bank 2001). This was based on a previous literature review that identified this framework (amongst others) to be the most suitable to convey in a systematic way the links between protected areas and poverty reduction (Scherl *et al.* 2004). Using a single definition for all countries also helped the study team avoid analytical complications that could arise from different government definitions of poverty.

To make the definition of poverty measurable for an MPA, the three dimensions of poverty were subdivided into specific focal areas drawn largely from the World Bank definition of poverty. 'Fish catch', 'cultural traditions' and 'access and rights' were added. The first one was added because impacts on fishing catch are closely tied to MPAs. The second was added because local communities often have cultural traditions that play an active role in how they manage marine resources, such as designating an area where no fishing is allowed after a chief dies or traditional decision-making processes about marine resources. The third was added because local communities may have customary access and rights for natural resources (such as for food supply or traditional medicines) and for the maintenance of spiritual and cultural values (such as protecting sacred sites). One focal area, 'social cohesion', was modified from the World Bank's 'strengthening organizations for poor people' because the most important of such organizations in the study sites were expected to be community-led and thus dependent upon the strength of the social fabric. This focal area helps to ascertain how marine protected areas impact the social cohesion of communities – do they tend to unite communities or divide them?

5.2.2 Preparation and site selection

Prior to the detailed design of the study, a comprehensive review of the poverty–environment literature was completed. The literature review makes clear that this study is one of the first to empirically analyse the link between marine conservation initiatives and poverty reduction, including developing and testing specific methodologies to assess the link.

Site selection for this study was limited to the Asia-Pacific region because this is the geographic area of interest for two of the study sponsors. Within Asia-Pacific, sites were selected based on three factors. First, experts who knew a site had to agree that the MPA was likely to have contributed to poverty reduction. Second, the MPA had to be in an area poorer than the national average. Third, the MPAs themselves had to be as distinct as possible to give a wider basis for determining common elements of success in contributing to poverty reduction. The four sites selected were:

Fiji	Yavusa Navakavu Locally Managed Marine Area on Fiji's main island of Viti Levu.
Solomon Islands	The Arnavon Community Marine Conservation Area between the large islands of Choiseul and Santa Isabel.
Indonesia	Bunaken National Marine Park at the northern tip of Sulawesi Island in central Indonesia.
Philippines	Apo Island Marine Reserve near Negros Island in the central region of the Philippines.

This portfolio of sites is roughly representative of small, one-community local MPAs (Fiji), medium-sized, multi-community local MPAs (Solomon Islands), big collaboratively managed national MPAs with lots of people (Indonesia) and small, co-managed national MPAs with few people (Philippines).

In terms of age and population, the sites are a good mix as Table 5.1 shows. In terms of size, the sites are well representative of marine protected areas globally. Of the 4435 formally designated marine protected areas worldwide, 95% are equal to or smaller in area than the largest site of Bunaken and equal to or larger than the smallest site, Apo Island (see www.mpaglobal.org).

Table 5.1 *Basic information on the study sites*

Site	Area (ha)	Local population	MPA age (as of January 2008)	Management regime
Navakavu, Fiji	3 710	600	5 years	One community manages all
Arnavon Islands, Solomon Islands	15 800	2 200	11 years	Co-managed by three communities, the provincial government and an international NGO
Bunaken National Park, Indonesia	89 000	30 000	15 years	Managed by the national government with input from two levels of local government, communities, tourism operators and academia
Apo Island, Philippines	74	700	20 years	Managed by the community until 1994 and thereafter by the national government and the community

5.2.3 Research hypotheses

Once sites were selected, expert advisors for each site helped the team to access site-specific studies relevant to this study. Research hypotheses were formulated for each site based on the literature and talking with knowledgeable experts. The hypotheses, as per the study proposal (July 2006) were:

Navakavu: no-take area has significantly increased shellfish and fish populations and was the primary cause of an average increase in household income. The management of the MPA has also strengthened local traditions and improved the community's ability to address other community problems.

Arnavon Islands: alternative income generating activities for fishers, especially seaweed farming, have increased local incomes and reduced poverty. The creation of the MPA has also improved the empowerment and security of the local communities by creating a management framework that improved community decision-making.

Bunaken: private-sector dive tourism and national park entrance fees have reduced local poverty in the communities of Bunaken National Park.

Apo Island: increased fish catches per unit of effort and increased tourism both resulted from setting aside an area of reef as a no-take zone. The increase in incomes and the improved management of the island resources have enabled the island to establish a sustainable development path.

5.2.4 Data collection

The study team was comprised of an economist, a social scientist and a study manager, supplemented in each site by national counterparts who were part of the research team and helped with data collection, translation, logistics and note taking. In addition, graduate students were also part of the research team in some locations. In total, 68 people helped with the fieldwork. The fieldwork lasted about 30 days per site except in Apo Island where it took only 15 days due to a smaller sample size.

To test the hypotheses, quantitative and qualitative information was 'triangulated' for each study site. Table 5.2 shows the sample size achieved for the various types of assessments.

The first point of the information triangle was a qualitative assessment using focus group discussions and key informant interviews. In each site, the research team partnered with an NGO that was working on the ground in the MPA. Focus group discussions were conducted in small groups of 6 to 12 individuals, with local partner NGOs inviting the groups and organizing the meetings in advance. All meetings were conducted in

Table 5.2 *Sample size*

Site	MPA households interviewed	Non-MPA households interviewed (control)	Total	No. of focus group discussions	No. of key informant interviews
Navakavu, Fiji	200	100	300	4	3
Arnavon Islands, Solomon Islands	175	63	238	6	10
Bunaken National Park, Indonesia	199	101	300	5	14
Apo Island, Philippines	83	37	120	3	6
Total	657	301	958	18	33

the local language with translation for the social scientist, and the information was recorded in English by a bilingual recorder. At each community, key informant interviews, some organized in advance by the local NGO, others as a result of focus group discussions, were held with staff from health clinics, schools, youth groups, representatives of particular economic activities, local government and organizations working in the communities.

The second point of the triangle used structured household interviews to compare MPA-related communities to control communities without an MPA but which were similar to the MPA communities in terms of population size, economic activities, the absence of major development projects in the local area (excluding the MPA), location and market access, and ethnic and religious backgrounds. The control sites were selected by consulting experts with in-depth local knowledge. Local experts in several locations reviewed and helped tailor the draft survey to local conditions. The surveys were then pre-tested in about a dozen households per site and revised as needed to ensure the questions were understood and relevant. Local survey takers in each site were then trained to conduct them. All household surveys were conducted in the local language.

The third point of the information triangle was also part of the household survey but looked at perceived changes over the previous 5 to 10 years (depending on the age of the MPA) and whether people believe these changes were caused by the MPA.

The average household interviewed had 5.0 members. The average age of respondents was 44, and 90% of the respondents had an education level of primary school or higher. The gender split was 91% male and 9% female. This male bias was due to a problem with the survey form design and insufficient emphasis to the local survey staff on the need for a rough gender balance. The gender imbalance, however, does not appear to have skewed the results because there is a strong correlation between the findings of the focus group discussions and key informant interviews that *did* have gender balance and the findings of the household surveys that did not.

Detailed reports on each study site were drafted, circulated to all the study team members for each particular location, commented on by experts who knew the site and then finalized after considering all comments. Individual case study reports are available at www.prem-online. org. The overall results of the four case studies are described in the following section.

5.3 Results and discussion

In the four study sites, the MPAs undoubtedly contributed to poverty reduction. 'People in the community are now better off and this is because of the MPA', one local person explained.

For the residents of Navakavu and Apo Island, their MPA contributed to poverty reduction in a very substantial way, and people perceive there are now few remaining poor families (though both sites have fewer than 700 people). In the Arnavons and Bunaken, with populations of 2200 and 30 000, respectively, the MPA has clearly contributed to poverty reduction, though by no means eliminated it. Across all the study sites, over 95% of local people support the continuation of their MPA. For many, the benefits of the MPA are tangible and apparent.

Looking at how each of the study focal areas contributed to poverty reduction shows why these MPAs benefited local poor people.

5.3.1 Opportunities

Figure 5.1 compares five income-related variables from the 'treated' MPAs and non-MPA 'control' areas: (1) monthly household cash income based on

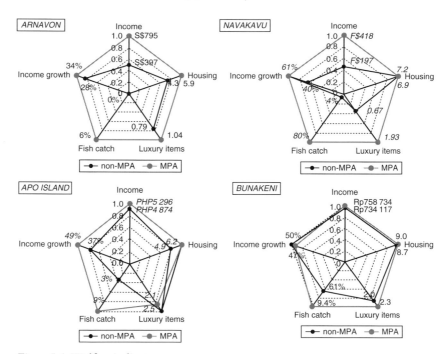

Figure 5.1 Welfare indicators

respondent recall of the last month's expenses; (2) the cost of the house on a scale from two to eight based on the type of walls and roof; (3) the number of 'luxury' items on a scale from zero to ten as defined by local people, such as radios, watches, bicycles, TVs and motorcycles; (4) fish catch, showing the share of respondents who perceived an improvement in fish catches; and (5) income growth, showing the percentage of the respondents who perceived that their economic activities and related income increase compared to 5 or 10 years ago (depending on the age of the MPA). The five welfare indicators were then normalized on the basis of the maximum score: the highest score for each indicator is presented as the maximum score and forms the outer boundary of the figure.

5.3.1.1 Income, housing and luxury goods

Household incomes increased in three of the four sites. In fact, it more than doubled in Navakavu and the Arnavons when compared to the control sites. Not surprisingly, in sites where incomes increased, so too did the number of household 'luxury goods' such as radios, watches, TVs and motorcycles. Housing also improved in two sites but not in the youngest or the oldest MPA. The youngest MPA (Navakavu) was only 5 years old, and it takes time for a household to save the funds required to upgrade their house. The oldest MPA (20-year-old Apo Island) had housing upgrades more than a decade ago, but this was outside the study timeframe reference of 10 years. Regardless, increased incomes plainly contributed to poverty reduction.

5.3.1.2 Fish catches

The MPAs directly benefited fishers in the four sites. All four sites have areas that have been closed to fishing for at least 4 years, and the protected fish have increased in size and abundance to the point that they are spilling over from the no-fishing zones into accessible waters. People in Navakavu fish just outside of the MPA, and 80% of the people there say fish catches are better than before the MPA. The spillover effect was also strong in Apo Island but a bit less so in Bunaken. It was also present in the Arnavons but had less impact because the MPA is an hour by boat from many of the villages, and fuel is expensive in this remote part of the Solomon Islands. MPA fish spillover has clearly contributed to poverty reduction at three of the MPAs. These findings support the increasingly well-documented spill-over effects of MPAs globally.

5.3.1.3 Education

In Apo Island, increased incomes helped fund more schooling for children. There was little impact on formal education attributable to the MPA in the other sites, but environmental awareness, with a strong element of education, increased in all four sites due to the MPA.

5.3.1.4 Alternative livelihoods

While the fisheries benefits were significant in the MPAs, the greatest boost to household incomes came from new livelihoods, especially in tourism. In Bunaken and Apo Island, those who switched to a new occupation in the tourist industry earned approximately twice as much as before. Some of the people who switched were fishers originally: 16% in Bunaken and 52% in Apo Island. In both locations tourism training for local people was done by private-sector tourism operators. The study results suggest that MPAs which have the advantage of infrastructure, access to markets and proximity to larger urban centres have a higher chance of utilizing tourism as a beneficial force than those that lack such conditions. While tourism contributed to a better quality of life for many people in Bunaken and Apo Island, tourist numbers can fluctuate dramatically. Both Bunaken and Apo Island suffered tourism downturns after in-country terrorism attacks and during events such as SARS.

5.3.1.5 Income diversification

The often-used tool of alternative income-generating activities for MPA fishers proved to have lower impacts than expected. Outside of tourism and working for the MPA, most of the alternative income activities in the MPA were not sustainable. Seaweed farming and deep-sea grouper fishing in the Arnavons were both hit by, among other things, dropping commodity prices at the local level. Building clay stoves and making coconut charcoal in Bunaken were hurt by the rising cost of inputs, and mat weaving in Apo Island was hampered by the high cost of inputs and lower quality mats than competitors. At least four of these alternatives produced income for several years before becoming financially unrewarding. The lesson learned is that changes in the price of inputs or outputs can quickly move an alternative income generating activity from success to failure. Such activities appear to often have short 'half lives' because of changes in the marketplace. This suggests that most alternative income-generating activities are better suited for offsetting initial lost income from closing fishing areas rather than as long-term tools to improve incomes or move people away from fishing. It was the larger, capital-intensive investments in tourism that lead to long-term gains in non-fishing income.

5.3.2 Empowerment

5.3.2.1 Governance mechanisms

For all four study sites, new governance mechanisms were established for the management of the MPA, and all four involve communities in management decision-making. These new governance mechanisms made the MPAs more responsive to community needs, and thus contributed to poverty reduction. The MPA management committees also serve as forums for addressing other community issues. This helped to strengthen local governance and reduce conflict. Accountability and transparency from community representatives on these management committees is still an issue, however, as is the general lack of women and youth on the committees.

5.3.2.2 Community participation

MPAs need local communities, just like local communities need MPAs. But the communities need to support the MPA from the beginning (see Figure 5.2). The findings suggest that community support for starting an MPA may be greater if there is a perceived crisis in fisheries. In all four sites, the realization by local communities that marine resources were in steep decline provided the incentive for changing the status quo. Problems with the current marine resources management regime made the communities more willing to try something new such as an MPA even if it had short-term costs.

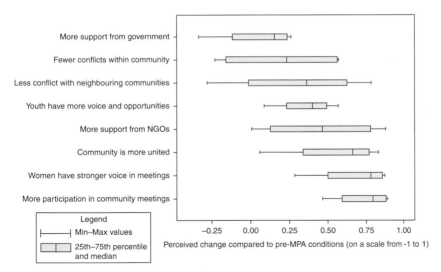

Figure 5.2 Statements about community engagement
Note: The scores show the average score across the MPA communities in the four study sites.

It is also apparent from the fieldwork that local people need to understand the link between a healthy ecosystem and quality of life if they are to support the MPA. Local people made observations such as: 'I never thought that the conservation area has so much to do with how we live as a community.' 'I thought the marine protected area is something to do with helping the environment but did not understand until now how much it also helps people.' Such understanding is crucial to sustainability.

Government policies that provided legal recognition for local community management of marine resources clearly supported community participation in all four sites. Community-led marine resource management is also easier if a neutral actor such as a university or NGO helps the MPA stakeholders to reach consensus about the distribution of costs and benefits. In all four sites, a university or an NGO played a catalytic role in getting the MPA established and helping it move towards sustainability (often with the support of an outside donor). External organizations can also help communities build on local traditions for managing fish stocks. This helped build community support for the MPA and ensures time-tested local fisheries traditions are not overlooked.

5.3.2.3 Benefits to women and youth

In all four sites, the MPA helped empower women economically and in some cases socially. Female-headed households are often among the poorest in a community, and helping to improve the welfare of women can have significant poverty reduction benefits. Women are the reef gleaners in Navakavu and benefited financially by collecting and selling a portion of the large increase in shellfish just outside the marine protected area. They have also increasingly been food catering for researchers and visitors to the MPA and this is a source of revenue for them. In Bunaken and Apo Island, dive tourism created more high-income job opportunities for women (as well as more need for food catering and opportunities for selling artefacts, amongst other things) and residents noted an improvement in women's lives because of the MPA. In the Arnavons, when women became involved in seaweed farming and the making of traditional clothes to earn income, they gained a stronger voice in community meetings. Partially as a way of looking for alternative income generation activities youth dancing groups in the Arnavons have become more widespread; in Navakavu a youth theatre group was formed to promote environmental awareness messages as part of the MPA activities and this group has been sponsored to travel to other parts of Fiji with such messages; a young man has also been trained by researchers at the university to undertake environmental monitoring; in the Arnavons and Navaku, since the MPAs were established, school environmental awareness programmes have been developed and school children often go for excursions to the MPA.

5.3.2.4 Access and rights

Local people perceived that the MPAs made access to marine resources worse in one site, and in two sites, restricted access to resources but gave them long-term rights to the resources. In all four sites, the MPA caused fishers to spend more time travelling to fishing areas than before but often with higher fish catches.

5.3.3 Security

5.3.3.1 Health

In all four of the MPA sites, health was improved because of the MPA. Greater fish catches in Navakavu and Apo Island led to greater protein intake and a perceived improvement in children's health in particular. Greater incomes from the MPA in Apo Island led to more frequent doctor visits and a resident midwife. The MPA speedboat is used as the water ambulance in the Arnavons, and in Bunaken, visitor entry fees funded water-supply tanks, public toilets and washing places in several villages, which improved public health. In three sites, the increased environmental awareness from the MPA operation translated into better understanding and acceptance of solutions to sanitation problems.

5.3.3.2 Social cohesion

It is clear that all the MPAs strengthened the unity and social fabric of the communities. In the Arnavons, stronger social cohesion help to bring together different cultural groups and gave the communities a more unified voice in requesting services from the provincial government, who in turn provided more support for fisheries and basic health care. In Navakavu, greater social cohesion made community members more likely to fulfil their social obligation to help other members in times of need. In three of the sites, conflict with neighbouring communities is perceived to have been reduced because of the MPA. This both contributes to and benefits from the improvements in social cohesion. In short, stronger social cohesion increased security for community members in all four sites.

5.3.3.3 Cultural traditions

In Navakavu the MPA helped revive the traditional practice of closing a portion of the fishing grounds. The MPA also helped strengthen the traditional leadership of the chief. 'Respect for resource rules is fostered and this is a respect for the leadership', explains a local leader. In the Arnavons, people cited a greater sense of belonging and safety because of the greater observance of obedience to the chief, production of traditional clothing and revitalized youth dancing groups. In Bunaken traditional knowledge of species spawning locations and its periodicity contributed to the zoning and temporary closures, which help replenishment and reinforce cultural traditions. There was no conclusive evidence of MPA impacts on cultural traditions in Bunaken and Apo Island.

Finally, as illustrated in Figure 5.3, in all four sites the contributions by the MPA to poverty reduction came from more than the expected better *opportunities* for income in fisheries and tourism. Using MCA, it is clear that the contribution of the three elements of poverty were relatively equal in all but Apo Island. Stronger social cohesion, benefits to health and to women, and better local governance contributed as well to reducing local poverty via greater *empowerment* and *security*. Poverty reduction can clearly come from more than just an increase in income.

Figure 5.3 Relative contribution towards reducing poverty from the three elements of poverty

5.4 Conclusions and policy recommendations

Policy incentives are crucial for maximizing the benefits of an MPA to poverty reduction. Emphasizing the following policies will improve an MPA's contribution to poverty reduction:

5.4.1 Macro level

5.4.1.1 Invest in MPAs

Like a school or a health clinic, an MPA needs financial support, particularly at start up. But also like a school or health clinic, an MPA has significantly more benefits than its modest costs. The investment, for example, in the Navakavu MPA over the 5 years since start up has been less than US$ 5900 equivalent, and this moderate investment has helped to double the income of about 600 people. This is why more than 120 new locally managed marine areas have been started in Fiji since 2004. In all four MPAs it was an external donor agency that provided the transformative funding. A large fund that provided modest amounts to coastal communities for establishing MPAs could have dramatic benefits to local fisheries and in some cases tourism. A number of marine scientists now believe that setting aside 30% of coastal areas in a string of ecologically connected MPAs could well ensure sustainable marine fishing and tourism in perpetuity even in the face of global threats such as overfishing and climate change.

5.4.1.2 Think small

The MPAs with the greatest contributions to poverty reduction were the two smallest. Navakavu and Apo Island are tiny MPAs within sight of the beneficiary villages. Both have low operating costs and high benefits. They also have fewer stakeholders to consult, and the fish spillover from the MPA is easier for local communities to see. This suggests that a network of smaller MPAs each affiliated with a local community may contribute more to poverty reduction than a single larger MPA. In fact, a network of small MPAs – be they 'locally managed marine areas' or 'community conserved areas' – that are ecologically connected may have the greatest potential yet for both reducing coastal poverty and conserving marine biodiversity. Kimbe Bay in Papua New Guinea is one example of how a network of about a dozen MPAs can be designed around community and local government co-management that will help sustain the area's fisheries and ensure individual MPAs are resilient to climate change. For MPAs that already exist, be they large or small, it is worth keeping the key attributes of success in mind: community participation in the management and an emphasis on tangible benefits to local communities from the MPA.

5.4.1.3 Five years or more

Establishing an MPA can take considerable time – several years from conception to start up is not unusual. It may take an equal amount of time for the ecological and socio-economic benefits to materialize. MPAs do not always fit well with the short-term cycle of politics, but they need medium-term commitments. Financial and technical support needs to be for a minimum of 5 years. In Apo Island, it took 6 years for the total financial benefits to exceed the costs since start up.

5.4.1.4 Participation of an external organization
In all four sites, a university or an NGO played a catalytic role in getting the MPA established and helping it move towards sustainability. A neutral actor who is not allied with any interest group helps the MPA stakeholders to reach consensus about the distribution of costs and benefits. The external organizations in all four MPAs also helped channel funding support for the MPA and ensured financial transparency. Policies that encourage the involvement of external organizations are vital to ensuring MPAs start off right and contribute to local poverty reduction in the medium to long term.

5.4.2 Micro level

5.4.2.1 Empowerment of local communities
The four MPAs in the study have empowered local communities in marine resource decision-making. This led to lower costs – especially for enforcement – and greater benefits. MPAs need local communities, just like local communities need MPAs. Government policies that provided legal recognition for community management of local marine resources plainly supported community participation in three of the four study sites. The benefits of community management can be further strengthened by linking MPA communities together via peer-learning networks such as the Locally Managed Marine Area Network. These networks cross-pollinate best practices and provide specialized training and technical resources for local communities.

5.4.2.2 Awareness activities are key
In all four sites people noted that understanding the link between ecosystem health and well-being was important for their continuing support of the MPA. 'We can now better understand the link between the marine protected areas and our lives', noted one local person. This understanding is crucial for sustainability. Incentives for early and frequent environmental awareness activities need to be built into funding for the MPA. Communities that engage in the actual MPA monitoring activities tend to develop environmental awareness more rapidly.

5.4.2.3 Understand and respect customary use and access rights of local communities

This is crucial to gain support *and* ownership from the ones who can in the long term have the most critical role in sustaining that MPA. In the Arnavons this was an issue that the management committee has had to address, for instance, where some communities felt they were the 'owners' of the area where the MPA was established more so than others but restrictions on use needed to be imposed on all communities equally. Understanding and recognizing that people felt that way led to extensive dialogue and processes with those communities to allow them to realize and accept the benefits of an MPA and also to bring other nearby communities to help in the MPA's management.

5.4.2.4 Build on local traditions

In Navakavu, the Arnavons and Bunaken traditional marine resource management systems were incorporated into the MPA's management to varying degrees. This helped strengthen community support for the MPA and builds on time-tested fisheries management mechanisms. In Navakavu the MPA establishment was supported by the traditional custom of temporarily closing a fishing area. In Bunaken people noted that some locally protected fish spawning areas were *incorporated* within more stringent conservation zones in the management plan. In the Arnavons and Navakavu the MPA strengthened customary relationships within the community. Policy incentives are needed that encourage seeking out and building on a community's marine resource management traditions.

References

Agnew, D. J., Pearce, J., Pramod, G. *et al.* (2009). Estimating the worldwide extent of illegal fishing. *PLoS ONE*, **4**: e4570. DOI:10.1371/journal.pone.0004570.

Allen, G. R. (2008). Conservation hotspots of biodiversity and endemism for Indo-Pacific coral reef fishes. *Aquatic Conservation: Marine and Freshwater Ecosystems*, **18**: 541–556.

Allen, G. R. and Erdmann, M. V. (2009). Reef fishes of the Bird's Head Peninsula, West Papua, Indonesia. *Check List*, **5**: 587–628.

Aswani, S. and Furusawa, T. (2007). Do marine protected areas affect human nutrition and health? A comparison between villages in Roviana, Solomon Islands. *Coastal Management*, **35**: 545–565.

Baticados, D. B. and Agbayani, R. F. (2000). Co-management in marine fisheries in Malalison Island, Central Philippines. *International Journal of Sustainable Development and World Ecology*, **7**(4): 343–355.

Ferraro, P. J. and Pattanayak, S. K. (2006). Money for nothing? A call for empirical evaluation of biodiversity conservation investments. *PLoS Biology*, **4**: e105.

Gjertsen, H. (2005). Can habitat protection lead to improvements in human well-being? Evidence from marine protected areas in the Philippines. *World Development*, **33**(2): 199–217.

Katon, B. M., Pomeroy, R. S., Garces, L. R. and Salamanca, A. M. (1999). Fisheries management of San Salvador Island Philippines: a shared responsibility. *Society & Natural Resources*, **12**(8): 777–795.

Le Gallic, B. and Cox, A. (2006). An economic analysis of illegal, unreported and unregulated (IUU) fishing: key drivers and possible solutions. *Marine Policy*, **30**: 689–695.

Maliao, R. J. and Polohan, B. B. (2008). Evaluating the impacts of mangrove rehabilitation in Cogtong Bay Philippines. *Environmental Management*, **41**(3): 414–424.

Oracion, E. G., Miller, M. L. and Christie, P. (2005). Marine protected areas for whom? Fisheries tourism and solidarity in a Philippine community. *Ocean & Coastal Management*, **48**(3–6): 393–410.

Pascale, R. T., Sternin, J. and Sternin, M. (2010). *The Power of Positive Deviance: How Unlikely Innovators Solve the World's Toughest Problems*. Boston, MA: Harvard Business Press.

Pauly, D. (2006). Major trends in small-scale marine fisheries, with emphasis on developing countries, and some implications for the social sciences. *Maritime Studies (MAST)*, **4**: 7–22.

Pomeroy, R. S., Pollnac, R. B., Katon, B. M. and Predo, C. D. (1997). Evaluating factors contributing to the success of community-based coastal resource management: the central Visayas Regional Project-1, Philippines. *Ocean & Coastal Management*, **36**(1–3): 97–120.

Pullin, A. S., Knight, T. M., Stone, D. A. and Charman, K. (2004). Do conservation managers use scientific evidence to support their decision-making? *Biological Conservation*, **119**: 245–252.

Scherl, L. M., Wilson, A., Wild, R. *et al.* (2004). *Can Protected Areas Contribute to Poverty Reduction? Opportunities and Limitations*. Gland, Switzerland: IUCN.

SOFIA (2008). *State of World Fisheries and Aquaculture. World Review of Fisheries and Aquaculture. Part 1*. Rome: FAO Fisheries and Aquaculture Department.

Sternin, J. and Choo, R. (2000). The power of positive deviance. *Harvard Business Review*, January–February 2000: 14–15.

Ravallion, M., Chen, S. and Sangraula, P. (2009). Dollar a day revisited. *World Bank Economic Review*, **23**(2): 163–184.

Tobey, J. and Torell, E. (2006). Coastal poverty and MPA management in mainland Tanzania and Zanzibar. *Ocean & Coastal Management*, **49**(11): 834–854.

Varkey, D. A., Ainsworth, C. H., Pitcher, T. J., Goram, Y. and Sumaila, R. (2009). Illegal, unreported and unregulated fisheries catch in Raja Ampat Regency, Eastern Indonesia. *Marine Policy*, **34**: 228–236.

Veron, J., Devantier, L. M., Turak, E. *et al.* (2009). Delineating the Coral Triangle. *Journal of Coral Reef Studies*, **11**: 91–100.

WDMPA (World Database on Marine Protected Areas) (n.d.) Available at: http://www.wdpa-marine.org/Default.aspx#/countries/about, accessed 23 November 2010.

Webb, E. L., Maliao, R. J. and Siar, S. V. (2004). Using local user perceptions to evaluate outcomes of protected area management in the Sagay Marine Reserve, Philippines. *Environmental Conservation*, **31**(2): 138–148.

World Bank (2001) *Attacking poverty. World Development Report 2000/2001.* World Bank, Washington DC.

Worm, B., Hilborn, R., Baum, J. K. *et al.* (2009). Rebuilding global fisheries. *Science*, **325**: 578–585.

6 · Economics of conservation for the Hon Mun Marine Protected Area in Vietnam

NAM PHAM KHANH AND PIETER J. H. VAN BEUKERING

6.1 Introduction

Marine and coastal resources in Vietnam are under increasing threat from human activities (Burke *et al.* 2002). One way to manage these threats is through Marine Protected Areas (MPAs), which safeguard valuable ecosystems within their confines. Despite the ecological and socio-economic benefits they provide (Whittingham *et al.* 2003), the management of MPAs is often severely constrained by both a lack of funding and a poor relationship with communities living around (or within) them.

Although many efforts aimed at coral reef management and conservation within MPAs in Vietnam have been initiated and implemented, policy-makers repeatedly face questions such as how do local residents feel they have been affected by the establishment of the MPA? Will active management of an MPA provide greater net economic benefits in the long term? How can these benefits be 'captured', both to fund MPA management and provide socio-economic stability for local communities?

Using the Hon Mun MPA as an example, this study explores the relationships between (1) the economic value of coral reefs, (2) coastal livelihoods, (3) coral reef degradation and (4) possible policy interventions in Vietnam. We estimated the economic value of coral reefs in the Hon Mun MPA through a focus on reef fisheries and reef-related tourism, as well as non-use values provided by reef ecosystems. Livelihood aspects of reef use were investigated by engaging with different reef stakeholders. We found out how villagers perceived changes in both quality of life and resources, and uncovered their views on resource impacts on human activities. To derive a

Nature's Wealth: The Economics of Ecosystem Services and Poverty, ed. P. J. H. van Beukering, E. Papyrakis, J. Bouma and R. Brouwer. Published by Cambridge University Press, © Cambridge University Press 2013.

decision-making framework for the MPA management options, we analysed the costs and benefits of a 'with MPA management' scenario and a 'without MPA management' scenario. We also established linkages between economic values of coral reefs, financial sustainability and local socio-economic issues.

Our results showed that tourism benefits of coral reefs are the key source of revenue for park management and local livelihood improvements. We suggested that, in the long term, continued MPA management would provide greater net benefits (particularly in terms of fisheries and tourism) than a 'without MPA management' scenario. Yet, to ensure future management, Hon Mun needs to develop its own sustainable and autonomous financing regime. One way to 'appropriate' the park's potential economic benefits is through a user-fee for ecotourists. Subsequent revenues could be ploughed back into management of the park and its buffer zone, and could also support much-needed alternative livelihood schemes in the region. If implemented successfully, a 'win–win–win' scenario can be realized, where ecological, economic and social needs are fulfilled.

6.2 Data and methodology

6.2.1 Background

The Hon Mun MPA comprises a group of eight islands located in the south of Nha Trang Bay. The distance between the islands and the mainland ranges from several kilometres to approximately 15 km (in the case of the furthermost islands). The Hon Mun MPA can be seen as representative of Vietnam's south central biogeographic zone. Its biodiversity levels are comparable to those of the global centre of coral diversity, within Eastern Indonesia, the Philippines and the Spratly Islands. The site supports a variety of habitats and ecosystems, including fringing coral reefs, mangrove forests and seagrass beds, as well as an adjacent deep-water upwelling. Recent surveys have counted around 350 species of reef-building corals, 220 species of demersal fish, 106 species of molluscs, 18 species of echinoderms and 62 species of algae and seagrass (Vo Si Tuan et al. 2002).

Hon Mun almost entirely supports the tourism industry in the city of Nha Trang. There are six diving schools based in Nha Trang, which use the area around Hon Mun Island as a dive site. Fishing is a major activity in the area, with about 79% of household heads being fishers. The fisheries are still small in scale and use a variety of fishing gears. Most of the fishing boats have limited power (15–45 CV) and trawl in offshore waters for squid by night. Inshore bottom trawling is also common (Hon Mun MPA 2005a, IUCN 2003).

The 5300 inhabitants of the Hon Mun MPA greatly depend upon coral-reef-based resources for their livelihoods. The three most significant productive activities for these inhabitants are small-scale fisheries, lobster cage culture and fishing crew labour. All are considered to be highly dependent on coral reefs. Recent analysis also shows that the standard of living within the MPA is not as low as in other areas (Hon Mun MPA 2005b). However, the livelihoods of the poorest people have still not been guaranteed in terms of food security, vulnerability and unsustainable use of reef resources. As most MPAs around the world, Hon Mun faces high human pressure and demands on its valuable coral reef resources. The biggest threats to the biodiversity of Hon Mun MPA include destructive fishing methods and overfishing.

The establishment of the Hon Mun MPA was approved in January 2001. Its 4-year management project is principally funded by international donors, who have contributed over US$ 2 million to date. This financial support has ensured sound management of its natural resources: it is recognized as a well-run MPA (ICEM 2003). In the long term, its managers face two central questions: (1) whether the continued MPA management model would provide greater net benefits (particularly in terms of fisheries and tourism) than a 'no MPA management' scenario, and (2) how to develop its own sustainable and autonomous financing regime.

6.2.2 Approach

The economic challenge of MPA management is to address both the costs and benefits of reef management, within a context of ecological complexity. Figure 6.1 shows the conceptual net benefits of coral reefs over time. The proposed scenarios yield benefits and incur costs over stakeholders' lifetimes. To ascertain the magnitude of the total net benefits of the scenarios, net present values (NPVs) are usually used. In this study, the net benefits of coral reefs were defined as value added by coral-associated tourism service suppliers, consumer surplus of divers and snorkellers, spillover effects of reefs on the fishing industry and the conservation values that visitors are willing to pay for an improved marine park. Various valuation techniques were used to estimate these costs and benefits.

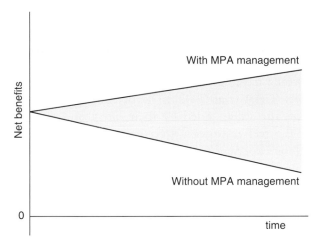

With MPA management

Net benefits

Without MPA management

0

time

Figure 6.1 Conceptual net benefits of coral reefs over time

6.3 Results and discussion

6.3.1 Tourism benefits

The Hon Mun MPA is the most heavily used marine reserve in Vietnam (Pham and Tran 2001), currently attracting around 300 000 visitors per year. However, the number of visitors who directly use the coral reefs by diving and snorkelling is much lower. The number of dive days is 18 000 per years (4500 and 13 500 for domestic and foreign visitors, respectively) served by six dive clubs, which are based in Nha Trang city. The number of snorkel days is nearly threefold this (approximately 52 000 per year), of which 36 400 are domestic and 15 600 are foreign visitors.

Reef-related tourism benefits comprise two components: consumer surplus and producer surplus. From the various travel cost models (TCM),[1] the zonal travel cost model (ZTCM) was selected to estimate the consumer surplus of the recreational activities of coral reef users, who are divers and snorkellers in the Hon Mun MPA. Visitors were selected using a systematic sampling procedure. We interviewed 259 domestic visitors, of which 98 were divers and 161 were snorkellers, and 281 foreign visitors (112 divers and 169 snorkellers). All the respondents were

[1] Classic TCM applications for coral reefs can be founded in Hundloe *et al.* (1987), Leeworthy (1991) and Pendleton (1995).

interviewed in the tour boats on the way back to, or in the port. The estimated consumer surpluses are VND 138 614 and 1 076 460 per domestic and foreign visitors, respectively.

A production approach was used to estimate the producer surplus of those producers who would have been willing to offer their services for divers and/or snorkellers (Cesar *et al.* 2002). The benefits of producers were calculated through the value added by their direct and indirect expenditure related to the reef activities. Direct expenditure related to reef diving and snorkelling experiences includes boat tour tickets to the site, as well as diving/snorkelling equipment hire and services. A total of 25% of this expenditure can be considered as value added. Indirect expenditure includes transportation costs (both international and national), hotel costs and other costs. For the international airfare, we assumed the value added is 3%. We also assumed local travel and hotels yield a value added of 25%. Recreational activities, directly and indirectly, have spillover effects for the local economy through employment generation and development of secondary industries, etc. The multiplier effect of the value added was taken into account and was assumed to be 1.5. Producer surplus or total value added is the sum of value added of expenditure, value added of indirect expenditure and the multiplier effect of these expenditures. The estimated producer surplus is VND 11 381 million for the diving industry and VND 17 056 million for the snorkelling industry.

The welfare gain of the visitors and value added for the economy from the recreation industry constitute recreational benefits of coral reefs in Hon Mun for society. Table 6.1 summarizes these estimated recreational benefits of coral reefs in Hon Mun in 2003.

Table 6.1 *Recreational benefits of coral reefs in Hon Mun in 2003 in million VND (US$ in brackets)*

Consumer surplus		Producer surplus			
Domestic visitors	Foreign visitors	Value added of direct expenditure	Value added of indirect expenditure	Multiplier effect	Total recreational benefits
5669	31 323	9071	9888	9479	65 430
(368 136)	(2 033 969)	(589 011)	(642 046)	(615 528)	(4 248 690)

6.3.2 Fishing and aquaculture benefits

The coral reef ecosystem is characterized by high biodiversity and productivity. These ecosystems support a great diversity of species of demersal fish, octopus, lobsters, molluscs, etc. In the study area, coral reef ecosystems even support lobster and grouper cage culture, which are highly lucrative activities in the coral reef areas of Vietnam (IUCN Vietnam 2003).

The production valuation method has been widely used to estimate the economic value of reef-associated fisheries (Cesar 2000). There are two common estimation approaches. In the first approach, total gross value of reef fisheries was based on reef dependence, fish catch and fish price (Cesar *et al.* 2002). Another approach is to directly estimate the fisheries' value per km^2 of reef by multiplying annual potential reef fisheries yield per km^2 and average market price of reef fish (Cesar 1996). We applied the former[2] and estimated the value added of reef-associated fisheries and aquaculture through calculations of the total gross value as well as costs and labour inputs. The commercial fishery data collected covers both offshore and near shore fisheries. Therefore, when estimating reef fisheries' (near shore fisheries) value, we needed to determine the reef-dependence of these fisheries.

The net fisheries value per km^2 of reef was estimated based on (1) the annual potential fisheries yield per km^2, (2) the average market price of reef fish per kilogram and (3) data on effort and cost to capture fish (Hon Mun MPA 2005a). The annual potential fisheries yield could be estimated as half the average reef fish standing stock (calculated from reef fish visual censuses when detailed stock data is not available; Uychiaoco *et al.* 2004 quoted from Schaefer 1954). The gross fisheries value of the reef was estimated at US\$ 15 538 per km^2 (by multiplying the annual potential fisheries yield per km^2 and the average price of fish). Consultations with experts in the Hon Mun MPA indicated that the spillover effect of the MPA is negligible at the studied time. The possible reason was that the MPA was established in 2001, so there had been little time for the spillover effect to materialize. Therefore in this study, the spillover effect was ignored.

After investigating various types of fishing gear (such as push net, purse seine, lift net, lobster seeding and diving) the recent survey by Hon Mun MPA concluded that 70% of the gross fishing value is based on gains during the main fishing season (Hon Mun MPA 2005a). We took this

---

[2] The choice of estimation method in this case depends heavily on the availability of data.

ratio to calculate fishing costs, valued at US\$ 4661 per km^2 of reef. The estimated value-added fishery value was US\$ 1 740 256 per year, calculated from the gross fishery values, fishing costs and the total area of the MPA.

Lobster cage culture in Hon Mun MPA is becoming a thriving industry. There were a total of 2000 lobster cages with a total production of about 128 tons[3] in the MPA in 2003. We estimated the total value added from coral reef's support function for fisheries and aquaculture. This was based on the value added of each activity and its reef dependencies. Table 6.2 shows the total values of reef-based fisheries and aquaculture in the Hon Mun MPA.

6.3.3 Conservation benefits

In this study, conservation benefit was defined as a visitor's willingness to pay to conserve coral reefs in their current state in Hon Mun MPA. The valuation question was designed to capture non-use values of coral reefs, such as option and bequest value. The contingent valuation with a payment card elicitation procedure was used. We estimated the reefs' value to the population of visitors, both international and domestic. However, our sample was confined to visitors who actually experienced the coral reefs during day visits.

To elicit willingness to pay (WTP) of visitors for conserving biodiversity through a trust fund, bid amounts ranged from VND 5000 to more

Table 6.2 *Values of reef-based fisheries and aquaculture in the Hon Mun MPA 2003*

	Quantity (tons)	Value added – million VND (thousands US\$)	Reef dependence (%)	Value added (US\$)
In-shore fisheries	13 000–70 000 kg/ 1.26–2.50 km^2 [a]	1 740 256	100	1 740 256
Lobster	156	19 313 (1254)	100	1 254 078
Grouper	Negligible[b]	0	–	0
Total	–			1 994 334

[a] Uychiaoco *et al.* 2004
[b] Key informant interviews show that the Hon Mun water quality is not suitable for grouper cages, e.g. groupers become diseased easily.

---

[3] A cage produces an average of 64 kg lobster product.

than VND 150 000 per dive/snorkel and from US$ 1 to more than US$ 10 per dive/snorkel for domestic and foreign respondents, respectively. Rejecters comprised 29.7% and 24.2% domestic and foreign respondents, respectively.

Using the linear utility model, the mean WTP per visit was estimated as VND 48 288 (US$ 3.1) for domestic visitors and VND 60 830 (US$ 3.9) for foreign visitors. The total conservation values of Hon Mun coral reefs were estimated to be approximately VND 1975 million (US$ 28 245) for domestic visitors and VND 1770 million (US$ 14 945) for foreign visitors.

6.3.4 Livelihood aspects of coral reefs uses

Many poor fishing communities live in areas where there are coral reefs. These communities often struggle to obtain the basic necessities of life and are vulnerable to multiple risks (Whittingham *et al.* 2003). However, it is the local people who ultimately decide the sustainable value of coral reefs. Thus, for researchers and policymakers, the vital research questions were: (1) How dependent are local communities on reef resources? (2) Are livelihoods affected by conservation activities? (3) What approaches are needed to solve the conflicts between conservation and exploitation? The answers could be addressed using some analytical tools such as (1) identification of reef dependents; (2) reef benefits to the poor; (3) perceived trends in quality of life and resources and (4) cost–benefit analysis. The data were collected using rapid appraisal techniques, which include mapping, observation, key informant interviews and sample surveys. The sample size for the household survey was 259. Households were selected using a random sampling procedure.

6.3.4.1 Reef dependent stakeholders

'Among those people dependent on coral reefs the numbers living in poverty is significant' (Whittingham *et al.* 2003: 4). Villagers in the Hon Mun MPA could be said to belong to this group of poor stakeholders. Characteristics of reef-dependent poverty in Hon Mun are threefold. First, the coral reefs provide a rich and accessible resource to the poor. Second, due to relying severely on this resource, the poor suffer a high vulnerability. Third, the poor receive poor infrastructure and weak support services. The analyses in this subsection and the next subsections will show how these characteristics operate.

The understanding of resource dependence of the poor is diverse. Some researchers focus on the flow of income to the households while others add the livelihood context into the notion. Gadgil and Guha (1995) defined resource stakeholders as those who depend on the natural environments of their own locality to meet most of their material needs. Whittingham *et al.* (2003) classified reef stakeholders as reef direct users and indirect users. After all, analysis of the reef stakeholders allows better understanding of the interaction between reefs and vulnerable local communities. Occupation structure emerged as an appropriate tool to analyse the reef-stakeholder context. Analysis of occupation structure allows an understanding of the relative importance of different components of the coastal resources and the role of coral reefs in creating jobs and livelihoods for the coastal communities (Pollnac and Crawford 2000).

Table 6.3 shows the distribution of ranking of productive activities undertaken by the MPA villagers. The reef-related aquaculture was ranked second by 24% of respondents while near-shore fishing was ranked first by 47%. The diversification of productive activities in the table implies a narrow range of activities, mainly in fishing, aquaculture and as crew members. These results also implied that the local community's livelihoods depend heavily on coral reefs.

This extensive resource dependence of reef-based fishing and aquaculture activities shows a high vulnerability situation. Given the urbanization in Nha Trang city, continued development of tourism estates and extending impacts of global warming, the pressures on coral reefs' wealth are inevitably increasing. As a result, the reef's capacity to provide benefits

Table 6.3 *Percent distribution of ranking of productive activities in Hon Mun (%)*

Activity	Rank				Total
	1st	2nd	3rd	4th	
Fishing	47	10	–	–	57
Aquaculture	24	22	1	–	47
Crew member	20	10	1	–	31
Teacher	1	–	–	–	1
Fish trading	5	11	8	–	24
Mechanic	1	4	3	1	9
Husbandry	–	–	1	–	1
Tailor	–	2	1	–	3
Official	–	–	1	–	1
Processing/handicraft	–	5	5	–	10
Total	98	64	21	1	

to the poor is reduced, increasing livelihood vulnerability. The situation has worsened in Hon Mun where more conservation efforts, which tended to exclude the poor, have been introduced.

6.3.4.2 *What benefits do coral reefs provide to the poor?*

The benefit flows of coral reefs to their stakeholders' livelihood assets is an important factor to derive how people depend on reefs and whether the changes in the reef resources have positive or negative effects on live-lihood's assets. Among many analytical approaches, the British Department for International Development's Sustainable Livelihood Framework, which identifies five core asset categories (natural, physical financial, human and social) on which livelihoods are built, appears suitable to analyse the benefit flows (Whittingham *et al.* 2003). This framework is shown in Table 6.4.

The coral reef's near-shore location allows easy access and also requires simple technology and financial investment. Reef-based fishing activities are the main source of income for the Hon Mun MPA villagers. Reef-based fisheries not only provide income, employment and foreign exchange, but also are a major source of animal protein for local people.

Lobster and grouper culture is providing a new form of income for coastal communities (Hon Mun MPA 2005b). Results from key inform-ant interviews show that an average household with two lobster cages can earn a net income ranging from 23 million up to 60 million VND in 20 months. This income source plays an important role in improving local communities' livelihoods and alleviating poverty. It was argued that small-scale fishermen have been forced out of the market simply because they have not been able to cover investments for off-shore fishing boats. Aquaculture has become an important factor in supporting coastal live-lihoods. Aquaculture also helps to reduce pressure on coastal fish stocks. However, the relationship between aquaculture and the protection of coral reefs needs to be seriously considered. It is evident that coral mining occurs in adjacent areas for trapping lobster seed; trash fish such as shellfish and small fishes are also overexploited to feed the lobsters.

Income from reef-related tourism activities was quite low. The level of involvement in this industry by the communities was negligible. Local people participate in the tourism industry by boat services and supplying seafood. However, these services are limited.

Women in Hon Mun communities are involved to some extent in reef-related productive activities. Most of them are concerned with doing household work, giving birth to children, and play a key role in child

Table 6.4 *Summary of reef benefits to household resources*

Resources	Benefits from reefs	Quantitative description
	Reef's diversity supports small-scale and subsistence fisheries	Boats of length less than 10 m and horsepower less than 20 take 65% of total fleets; 72% of interviewed households fish near the reefs and adjacent areas
Natural	Seeding for lobster and grouper cultures	Total value of seeding per crop is approximately VND 46.4 billion. However, wild-caught seed stock from Hon Mun is unknown
	Trash fish and shellfish for aquaculture	The area provides around 2500 tonnes of trash fish per year (approximate value of VND 11.4 billion per year)
Physical	Gastropod shells used for house construction	Negligible
	Reef provides key reference for position and fishing grounds	
	Cash sale from fish	Average VND 527 000 per fishing gear per day
Financial	Cash sale from aquaculture	54% of cash income of those engaged in aquaculture (roughly VND 1 million/month/household)
	Income from tourism activities	Potential source, a villager operating a glass-bottom basket boat can earn VND 80 000 a day
	Protein from fish	85% of interviewed households eat fish every day while almost 60% of them eat meat once a week
Human	Fishing valued as skill and knowledge	
	Near-shore fishing is safer than off-shore fishing	
	Cultural exchange with tourists	Negligible
Social	Female participation	Negligible: 2.7%, 5.0%, 11.6% and 3.9% of women participate in reef-related fishing, aquaculture, fish trading and handicraft, respectively.

nutrition in early infancy. However, women's status in fishing communities is generally low since the main economic activity, fishing, is often considered to be a male activity. Unlike other coral reefs areas (Whittingham *et al.* 2003), Hon Mun coral reefs do not provide many benefits in terms of social and physical resources. This is due to the topographical characteristic of reefs in Hon Mun. Hon Mun MPA has few shallow water areas,[4] which can provide jobs, such as harvesting inter-tidal resources, and therefore status to women.

One of the interesting questions regarding reef use and conservation is whether the poor perceive the benefit flows of coral reefs and why they are short-sighted in using such a precious resource. Part of the answer comes from the analysis of individual perceptions of resource impacts on human activities (Pollnac and Crawford 2000). Information concerning villagers' perceptions of coastal resources and potential human impacts on these resources was gathered from a sample of 259 respondents. These respondents were requested to indicate the degree of their agreement or disagreement with 10 statements (see Box 6.1). Each of the 10 statements involved some aspect of the relationship between coastal resources and human activities. Using factor analysis and multiple regression, analysis of the responses to these statements showed an increasingly strong and accurate belief concerning the content of the statement, implying that fishermen actually perceive the role of coral reefs and under some certain circumstances they can sustainably use their resources.

6.3.4.3 Perceived trends in quality of life and resources
Despite the fact that coral reefs do support productive activities of the local community, 70% of Hon Mun MPA households were ranked as poor.[5] Moreover, most of the villages in the MPA lack access to electricity, fresh water, the healthcare system and secondary school. The unemployment rate of women was extremely high at 90%.

Considering these issues in a dynamic context, researchers and policy-makers may want to know how the poor perceive trends in quality of life, which is an important indicator for deriving behaviour aspects of resource uses (Pollnac and Crawford 2000). Respondents were asked how they compare their well-being today with that of 5 years ago and 5 years in the future (better off, worse off and the same) and to provide reasons for the changes. Around 30% of respondents indicated a better-off situation.

---

[4] The average depth of the Hon Mun MPA waters is 10 to 20 m (Nguyen Chu Hoi *et al.* 1998).

[5] A self-ranking criteria system, of which the poor are considered as those who work as hired labour, have many children, a shortage of food, houses with a thatched roof and rattan walls, no TV and generator, no fishing boat and no consistent job (Nguyen and Adrien 2002).

Box 6.1   *Perceptions of resource impacts from human activities – attitude scale construction*

1. We have to take care of the land and the sea or it will not provide for us in the future.
2. Fishing would be better if we cleared the coral where the fish hide from us.
3. If our community works together we will be able to protect our resources.
4. Farming in the hills behind the village can have an effect on the fish.
5. If we throw our garbage on the beach, the ocean takes it away and it causes no harm.
6. We do not have to worry about the air and the sea, God will take care of it for us.
7. Unless the coral is protected we will not have any small fish to catch.
8. There are so many fish in the ocean that no matter how many we catch, there will always be enough for our needs.
9. Human activities do not influence the number of fish in the ocean.
10. There is a limit to the area of the sea that can be used by the village.

Source: Pollnac and Crawford (2000)

There are two main reasons for this: (1) lobster cage culture (31%) and (2) more productive fisheries (21%). The MPA's conservation activities were considered to create good sources of hatchery and feeding for lobster culture. The more productive fishery may be explained by investments in modern fishing gear and/or the spillover effect of the MPA.

Table 6.5 provides a breakdown of respondents' perceived quality of life in the Hon Mun MPA. The reef conservation activities of the MPA have improved its ecological condition, attracting more tourists to the site (i.e. creating more economic benefits to society). Nevertheless, some local people have become worse-off, which might lead to increasing threats to coral reefs and management efforts. The main reasons included limited fishing grounds due to the presence of the MPA. However, our in-depth interviews showed that the deterioration of fish production should not be blamed entirely on conservation activities of the MPA. The daily catch has decreased in part due to small-scale fishing boats having to increasingly compete with more modern, well-equipped vessels.

Table 6.5 *Percent distribution of perceived quality of life*

Well-being	Compared to 5 years ago	Compared to 5 years in the future
Worse-off	36.7	4.6
Same	32.8	68.3
Better-off	30.1	26.6
Don't know	0.0	0.4

As a result, destructive fishing activities are still taking place within the MPA. This may be compounded by inadequate efforts for alternative income-generation programmes and inadequate livelihood improvement efforts. The inadequate efforts are usually blamed on inappropriate policies and inadequate resources for such efforts.

Livelihoods of fishing communities within the MPA were unstable and heavily dependent on fishing. The establishment of an MPA probably had negative effects on their livelihood in the short run and may have undermined that MPA in the long run. The role of marine systems in supporting fishing has been deteriorating due to overexploitation, degraded coral reefs and poor management. The use of modernized fishing gear may have improved catches for certain fishermen, yet this is not a sustainable, long-term option. Conservation may provide some 'spillover effects' in the form of an enhanced source of fry/feed for aquaculture and increased abundance of target fish stocks; yet these 'spillover effects' are rarely instantaneous, and may take years to fully materialize.

6.4 Decision-making framework

In Hon Mun MPA, and possibly in other potential MPAs in Vietnam, the fishing villages are most affected by the MPA establishment and their self-perceived well-being has decreased since the MPA went into operation. Another fact is that while the community is worse-off, the coral-related tourism industry and its tourists are better-off. While the Hon Mun MPA is at the end of its funding and other MPAs in Vietnam are in the planning process, several questions have emerged. Given the above problems, is it economically feasible to continue the MPA management model? If the answer is yes, how can livelihood issues be solved (i.e. conflicts between conservation and exploitation)? We based our answers on two main approaches: (1) a cost–benefit analysis for the MPA management scenario versus without MPA management scenario; and (2) a proposed benefit transfer from tourists to villagers.

6.4.1 Cost–benefit analysis for MPA management scenarios

6.4.1.1 Without MPA management scenario

Given the fact that the current financial support for Hon Mun MPA ended in 2005, it made sense to assume that management efforts at the present level will cease. Before the establishment of Hon Mun MPA, the local government used to invest in the Hon Mun area. The investment amounted to about US$ 28 000 per year and funded activities related to fisheries management, including aquaculture development (GEF 1999). Although this addressed immediate national economic and social priorities, it failed to tackle the protection of the important coral reef, mangrove and lagoon ecosystems, or the incremental benefits to global biodiversity. The current enforcement would trim down. The natural functions of coral reefs, therefore, would decline, leading to a deterioration of the fishery industry in the long term. Ecotourism would not be developed either.

6.4.1.2 With MPA management scenario

Financial investment was assumed to continue, at least at the current level. Threats to coral reefs, such as destructive fisheries, coral mining and unsustainable tourism would be eradicated. Ecotourism would be developed to its maximum potential. Management of Hon Mun could clearly generate significant domestic benefits, especially from tourism. The hypothetical cost of operating Hon Mun as an MPA with the purpose of improving local communities' livelihoods, sustainable tourism development and conservation of marine biodiversity was estimated at US$ 230 500 start-up costs plus US$ 300 000 annual operating costs (GEF 1999). The MPA management scenario was based on the infrastructure established under the previous management system. As such, set-up costs were not taken into account. Based on van Beukering *et al.* (2003: 48) who argued that 'this period leaves enough time for the main environmental impacts to come into effect, while it is sufficiently short to estimate future development', a time frame of 25 years, from 2005 to 2030, was used for the cost–benefit analysis.

Tourism benefits It is clear that the 'with MPA management' option causes the total annual benefits from coral-associated tourism to increase quickly right after the introduction of a new financial source in 2005.

Since this financial source comes from beneficiaries, i.e. divers and snork-ellers, it probably excludes some tourists (i.e. those who protest against user fees). However, these protesters are negligible in number and do not significantly affect the rise of total values, which are mainly due to more beautiful coral reefs and the increasing numbers of tourists.

In the 'without MPA management' scenario, due to inadequate con-servation, net annual benefits of divers and the related service providers decline after 2010. However, due to the net benefits of snorkeller com-ponents outweighing the benefits of the divers, the total annual benefits still increase slightly over time. This rise in values could be explained by the continued growth in numbers of tourists visiting the site (as it has not yet reached its maximum capacity).

Fishery benefits   The main difference between fishery benefits associated with the 'with MPA management' and 'without MPA management' scenarios was the spillover effects. In the 'without MPA management' scenario, fishery values were high and sustained over time because lobster culture was supported (through the provision of feed, seed and water filtering). The fisheries value of the 'with MPA management' scenario at the steady state in year 2015 was estimated at US$ 3.43 million.

Conservation benefits   As mentioned, the estimation of conservation ben-efits depends on two main variables: average visitor WTP and number of visitors to the site. In the 'without MPA management' scenario, annual biodiversity benefits (i.e. conservation values that visitors are willing to pay for an improvement of the marine protected area) were zero because there will be no improvement. MPA management would attract more tourists to the resource and thus yielded increasing biodiversity benefits over time. These benefits were estimated to reach a maximum level of US$ 479 767 in 2015.

6.4.2 Total economic values of scenarios

Figure 6.2 shows the net annual benefits for both 'with' and 'without MPA management' scenarios over the period 2005–30. Net benefits of the 'without MPA management' scenario, surprisingly, slightly increase until 2015 and are then sustained. In the total economic value, the tourism and fisheries benefits of coral reefs for one year at the present time take the largest shares of 51% and 46%, respectively, and therefore would deter-mine the trend of the benefit flow over time. While the fisheries benefits

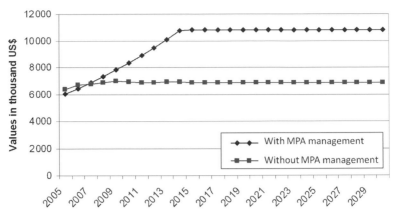

Figure 6.2 Net annual benefits over the period 2005–30

remain unchanged over time, the tourism benefits increase due to the overall growth of the coastal tourism market (irrespective of unmanaged coral reefs). As mentioned, this increase mainly relates to snorkellers who enjoy coral reefs but are not sensitive to changes in coral biodiversity. The number of divers decreases over time but this does not significantly affect the total values.

In the first 2 years of the analysed period, the annual net benefits of the 'without MPA management' scenario outweigh the benefits of the 'with MPA management' scenario. This is due to the 'with MPA management' scenario's higher costs. Until 2015, the net benefits of the 'with MPA management' scenario increase gradually. This is the result of two main effects: (1) better functioning coral reefs support tourism and fisheries, and (2) the overall growth of the coastal tourism industry of Vietnam and Nha Trang (in particular).

Based on annual benefits of scenarios and the choice of social discount rate, the net present values of management scenarios can be calculated. The net present value for the Hon Mun area over the period 2005–30 is US$ 70.3 million under 'with MPA management' scenario and US$ 53.8 million under 'without MPA management' scenario (Table 6.6).

As highlighted above, net present values of both options are positive. 'Without MPA management' means the marine protected area will not exist at the site, but the site will still be under management of local government. The main contribution of the NPV of the MPA management scenario is tourism benefits. While the cost side is easy to see as monetary outflows, the benefits of MPA management are set in a wider

Table 6.6 *Net present benefits and costs of management options for Hon Mun MPA (in US$ million)*

Present values	Tourism benefit	Fishery benefit	Conservation benefit	Total benefits	Costs	NPV
With MPA management	44.30	25.50	2.88	72.68	2.37	70.31
Without MPA management	30.47	23.64	0.00	54.12	0.22	53.89

Note: The discount rate is 12% over the period 2005–30. This rate was averaged from discount rates used for public projects in Vietnam.

context. These include, for example, profits of international travel agents or fisheries spillover effects, and sometimes intangible benefits, such as consumer surplus of divers/snorkellers or use of future generations, etc.

6.4.3 Benefit transfer for conservation and livelihood improvement

The cost–benefit analysis results showed that coral-based tourism, which relies on both local and international tourists, provides the largest share in total reef's benefits. The problem is that these beneficiaries of reef's recreational value, who are tourists and service providers, are non-resident and do not pay for the coral reef itself. They take the existence of coral reefs for granted, while the villagers, who are living with and saving coral reefs, are facing declining benefit flows from coral reefs and at the same time living in poverty. Therefore, there emerges a need to transfer reef's recreational benefits from tourists to villagers. A common tool to facilitate this transfer in Hon Mun context is an 'earmarked conservation fee'. The large tourism benefit implies adequate justification for this conservation fee.

6.5 Conclusions and policy recommendations

Threats to Vietnam's MPAs are largely a consequence of policymakers giving inadequate attention to the wider socio-economic context in which parks are set. MPA managers need to shift their focus to the interrelationships between marine resource dependency, local livelihoods and coastal poverty. Consideration also needs to be given to the economic values of marine resources, and how these can be 'captured' for the benefit of parks and their residents.

At the time of study, Hon Mun enjoyed a US$ 2 million (largely international) funding programme, which is quite exceptional for a Vietnamese MPA. Because this external financing will come to an end, alternative funding must be secured to guarantee the park's long-term future. In this study, the economic outcomes of two different policy scenarios were simulated in order to evaluate the long-term gains of MPA management. Impacts on tourism, fisheries and biodiversity were considered individually, and then combined to generate an overall economic value for the scenarios.

The net benefits of the 'with MPA management' scenario increase until 2015, after which they remain stable and considerably higher compared to the 'without MPA management' scenario. This is a result of three main factors (1) healthier marine ecosystems sustain more productive capture fisheries/aquaculture, (2) preserved marine life (particularly reefs) attract greater numbers of tourists and (3) Vietnam's coastal tourism increases regardless of marine conservation initiatives (at least during the period under consideration).

This study indicated that tourism would generate the greatest economic benefits for Hon Mun over the next two decades. Policymakers need to 'capture' a proportion of these benefits in order to fund operating costs and support communities who, as yet, do not sufficiently gain from marine conservation. One way of doing this is through a 'conservation fee' for MPA visitors.

A proportion of these funds could be allocated to improving access to public services and infrastructure as well as improving income generation programmes for affected fishermen in order to reduce their vulnerability. Residents could also be encouraged to become involved in the tourism industry directly, for example, if they were given appropriate training. Top-down management is by no means made redundant by community involvement or by the development of tourism. Uncontrolled tourism development can itself cause ecological degradation. Similarly, a community-based management approach would not deter illegal fishing in its entirety. As such, a regulatory approach to marine conservation still has a critical role to play.

A conservation fee is one prevalent way of appropriating the benefits of MPAs, as visitors are typically willing to contribute to park management. However, alternative revenue-generating mechanisms might work more successfully in Vietnam's other MPAs, particularly those that are not popular tourist destinations. Funding opportunities include: government appropriations, levies, surcharges, leases and concessions, bio-prospecting, trust funds, donations, corporate sponsorship, debt-for-nature swaps and international donors. All of these options (individually or in combination) could be explored by MPA management authorities along the coast. Without the socio-economic stability that this funding could provide, the long-term conservation of Vietnam's marine ecosystems is uncertain.

References

Burke, L., Selig, E. and Spalding, M. (2002). *Reefs at Risk in Southeast Asia*. Washington DC: World Resources Institute.

Cesar, H. (1996). Economic analysis of Indonesian coral reefs. Working Paper Series. World Bank, Washington, DC.

Cesar, H. (ed.) (2000). *Collected Essays on the Economics of Coral Reefs*. Stockholm, Sweden: CORDIO.

Cesar, H., van Beukering, P., Pintz, S. and Dierking, J. (2002). Economic valuation of the coral reefs of Hawaii. Final report. Cesar Environmental Economics Consulting, Hawaiian Coral Reef Initiative Research Program (HCRI-RP), University of Hawaii, Manoa, Hawaii.

Gadgil, M. and Guha, R. (1995). *Ecology and Equity. The Use and Abuse of Nature in Contemporary India*. London and New York: Routledge.

Global Environmental Facility (GEF)'s Project Database(1999). Hon Mun Protected Area Pilot Project. Available at: http://www.gefonline.org.

Hon Mun MPA (2005a). Bao Cao Ket Qua Giam Sat Danh Bat Ca Trong Khu Bao Ton Bien Vinh Nha Trang – vu Bac. Bao cao da dang sinh hoc so 14. Nha trang [Report on fishing monitoring in the Nha Trang Bay MPA: the north season. Biodiversity report number 14].

Hon Mun MPA (2005b). *Socio-economic Impact Assessment of the Hon Mun MPA Project on Local Community Within the MPA*. Nha Trang: Hon Mun MPA.

Hundloe, T., Vanclay, F. and Carter, M. (1987). Economic and socio-economic impacts of crown of thorns starfish on the Great Barrier Reef. Report to the Great Barrier Reef Marine Park Authority, Townsville.

ICEM, (2003). Vietnam national report on protected areas and development. Review of Protected Areas and Development in the Lower Mekong River Region, Indooroopilly, Queensland, Australia.

IUCN (2003). Improving local livelihoods through sustainable aquaculture in Hon Mun marine protected area, Nha Trang Bay, Vietnam: a stream case study. In G. Haylor, M. R. P. Briggs, L. Pet-Soede *et al.* (eds.), *Improving Coastal Livelihoods Through Sustainable Aquaculture Practices: A Report to the Collaborative APEC Grouper Research and Development Network (FWG/01/2001)*. Bangkok: STREAM Initiative, NACA.

Leeworthy, V. R. (1991). *Recreational Use Value for John Pennekamp Coral Reef State Park and Key Largo National Marine Sanctuary*. Silver Spring, MD: Strategic Environmental Assessments Division, National Oceanic and Atmospheric Administration.

Nguyen Chu Hoi, Nguyen Huy Yet and Dang Ngoc Thanh, eds. (1998). *Scientific Basis for Marine Protected Areas Planning*. Hai Phong: Hai Phong Institute of Oceanography. In Vietnamese.

Nguyen, T. H. Y and Adrien, B. (2002). Socio-economic assessment of the potential of the establishment in the Hon Mun MPA. Community Development Report 2. Hon Mun MPA pilot project. Nha Trang.

Pendleton, L. H. (1995). Valuing coral reef protection. *Ocean and Coastal Management*, **26**(2): 119–131.

Pham, K. N. and Tran, V. H. S. (2001). Analysis of the recreational value of the coral-surrounded Hon Mun Islands in Vietnam. Report for the Economy and Environment Program for Southeast Asia.

Pollnac, R. and Crawford, B. R. (2000). Assessing behavioral aspects of coastal resource use. Proyek Pesisir Publication Special Report. Coastal Resources Center Coastal Management Report #2226. Coastal Resources Center, University of Rhode Islands, Narragansett, Rhode Islands.

Uychiaoco, A. J., Gomez, E. D., Cesar, H. S. J. *et al.* (2004). Economic value of three South-east Asian coral reef sites with different biological diversities. TOTAL project.

van Beukering, P., Cesar, H. and Janssen, M. (2003). Economic valuation of the Lauser National Park on Sumatra, Indonesia. *Ecological Economics*, **44**: 43–62.

Vo Si Tuan (2002). National report of Vietnam for GCRMN EA SEA workshop. *Proceedings of the Global Coral Reef Monitoring Network*. Regional workshop for the East Asian Seas. Ishigaki, Japan, 27–30 March 2002. Ministry of the Environment, Government of Japan: 124–130.

Whittingham, E., Campbell, J. and Townsley, P. (2003). *Poverty and Reefs*. Paris: DFID–IMM–IOC/UNESCO.

7 · A multi-criteria approach to equitable fishing rights allocation in South Africa's Western Cape

RON JANSSEN, ALISON R. JOUBERT AND
THEODOR J. STEWART

7.1 Introduction

Fisheries resources are vulnerable to overexploitation, in large part because of their open-access nature. For long-term ecological and socio-economic sustainability, fisheries therefore need to be regulated by limiting TAC and/ or Total Allowable Effort (TAE). It can be argued that to maximize the efficiency of the fisheries sector tradable fishing rights is the way to go. This is the solution implemented successfully in countries such as Iceland, New Zealand, etc. (Arnason 2005, Scott 2000). In many developing countries, however, protection of traditional fishing communities with their subsistence fisheries is added. Objectives of fishing rights allocation can then include poverty reduction and preservation of traditional culture.

This study deals with the fishing rights allocation in South Africa. South Africa's fisheries yields peaked in the 1960s and 1970s, but since then many stocks have declined due to overexploitation. Although the fishing industry has historically been dominated by a few white-owned companies, since the end of apartheid new policies have been introduced to (1) rectify this inequitable distribution of fishing opportunities and (2) improve the sustainability of fisheries (Cockcroft *et al.* 2002, RSA 1998).

At the end of the apartheid era, the profound need to change South Africa's economy and society became the focus of government policy. A key part of this so-called 'transformation' involved redistributing resources more equitably amongst the country's different demographic groups. Although access to fishing rights was not officially affected by apartheid, the political system limited non-whites' ability to acquire sufficient capital

Nature's Wealth: The Economics of Ecosystem Services and Poverty, ed. P. J. H. van Beukering, E. Papyrakis, J. Bouma and R. Brouwer. Published by Cambridge University Press, © Cambridge University Press 2013.

or expertise to enter the fishing sector or make full use of rights. In 1994 less than 1% of the TAC for commercial fisheries was allocated to so-called 'historically disadvantaged persons' (HDPs) of black and coloured origin. With the coming of democracy, HDPs sought greater access to fisheries resources. Yet their expectations have not been met, causing mistrust and dissatisfaction with the current management system. Inadequate fishing rights and declines in key stocks have contributed to growing poverty and unemployment, and a loss of social cohesion, in the fishing communities of South Africa's Western Cape.

In South Africa an important objective of the fishing rights allocation policy was to adjust inequitable allocations from the past. The 2001 allocation system tried to address these equity issues in combination with efficiency and preservation issues. This system was considered unfair and too autocratic by the local fisherman.

This study analysed the 2001 allocation system and developed an alternative approach based on cooperation with all stakeholders. This chapter explores the rights allocation system used to operationalize the quota system and describes an MCDA that would better meet the multiple objectives of the allocation system. A full account of the study can be found in Joubert et al. (2005, 2008).

This chapter starts with the policy context and an analysis of the scoring method of the 2001 allocation system (Section 7.2). The project results are included in Section 7.3. This section describes the stakeholder consultation and the design of an alternative allocation system. Conclusions and policy recommendations are offered in Section 7.4.

7.2 Policy context

The starting point of this study was the 2001 allocation system. This section provides a description and analysis of this system as background to the alternative system described in the next section.

At present, there are serious inconsistencies and a lack of transparency in the allocation of fishing rights. This has caused high levels of discontent (and provoked illegal fishing) in South Africa's Western Cape, where fishers feel their access to rights remains constrained. Disadvantaged communities in this region specifically seek greater access to hake, traditional line fish, West Coast rock lobster and abalone stocks. These small-scale fisheries are particularly important for poorer fishers, as they do not require high capital investment.

In South Africa, the allocation of fishing rights is carried out by the Marine and Coastal Management Directorate (MCM) of the Department of Environmental Affairs and Tourism. The total TAC/TAE for each fishery is allocated to various sectors (full commercial, limited commercial, recreational, subsistence, etc.), and then to specific companies and individuals within these sectors. The most recent allocation system (2001) focuses on 'the need to balance the sustainability of the industry while enhancing the capacity of historically disadvantaged communities to establish commercially viable businesses' (DEAT 2002, Kleinschmidt *et al.* 2003).

In the study the focus is on the small-scale commercial fishermen operating from the traditional fishing communities (limited commercial). Application fees for this group are relatively small as are the quota allocated. The allocation of West Coast lobster is used as an example. Other fisheries have similar approaches.

As a first step, people who did not have access to a fishing vessel or who had violated the Marine Living Resources Act in the past were excluded from the process. Next the applicants were evaluated by scoring them according to three criteria.

1. Degree of investment and *involvement in the industry*, measured by ownership of (or access to) a vessel, and in some cases, previous fishing rights. For some fisheries, evidence of a business plan for fishing operations (indicating both financial viability and 'business acumen' of the fisher/company) is required.
2. *Historical involvement*, measured by the type of allocation rights or participation in the past.
3. Degree of '*transformation*', measured by the HDP status of the applicant and/or by the percentage of HDP ownership/management of the enterprise.

The total score for each applicant was determined by summing the scores for each criterion. Table 7.1 shows the scores for the 2001 West Coast rock lobster allocation system.

In mathematical notation a score is given to applicant j for his/her performance according to criterion i. This score can be referred to as s_{ij}. The scores for each criterion are added to get a total score S_j for applicant j. The system is, therefore, basically a multi-criteria approach, using a sum of scores:

$$S_j = \sum_{i=1}^{n} s_{ij}, \tag{7.1}$$

Table 7.1 *Criterion scores for 2001 West Coast rock lobster allocation system, east of Hangklip region (DEAT, unpublished data)*

Vessel access	Score	Previous involvement	Score	HDP status	Score	Wt★
75–100% vessel owner	5	Experimental permit	3	HDP female	5	
40–74% vessel owner	4	Experimental crew	2	HDP male	4	5
25–39% vessel owner	3	Commercial crew	1.5	Non-HDP female	2	
1–24% vessel owner	2	Subsistence permit	1.5	Non-HDP male	0	
50–100% purchase agreement	1	Processing / Marketing	0.5			
1–49% purchase agreement	0.5	Recreational permit	1			
Bareboat charter agreement	1	No permit / No crew	0			
Charter / catching agreement	0.5					
No access	VETO					

This is similar to the well-known multi-criteria method usually referred to as weighted summation (Janssen 1992). However, weights do not explicitly appear in this approach but are hidden in the maximum values of the three criteria. Since the maximum of the criterion historical involvement is lower, the maximum contribution of the criterion to the total score is lower, which is similar to assigning a lower weight using weighted summation. To justify the weighted summation of an applicant's scores for different criteria, the scores need to be on an interval scale, i.e. a change in score from 0 to 1 should have the same worth as a change from 9 to 10 (Wolman 2006). It was not clear that this was the case in the association of scores with achievement levels.

Figure 7.1 shows the total scores for a subset of the West Coast rock lobster limited commercial applicants and indicates whether the right was granted (triangle). The figure shows a selection of 170 applicants that scored higher than 6 from a total of 904 applicants. Because no HDP females owning a boat and holding an experimental permit applied, the maximum score of 13 is not reached by any of the applicants. Applicants who were considered to be applying for a quota only to sell it to somebody else (a paper quota risk) were vetoed with a score of –5.

Fishermen complained that the application forms were difficult to understand and fill in. This apparently led to a thriving business for lawyers or consultants who assisted people in filling in their forms. Once allocations had been carried out, fishers and communities complained that in many cases people with other jobs or with no experience in fishing had been given the rights, while they had not. Fishers were concerned that traditional ways of life and communities were being eroded and this was partly attributed to the allocation system because of the relative weights applied to criteria.

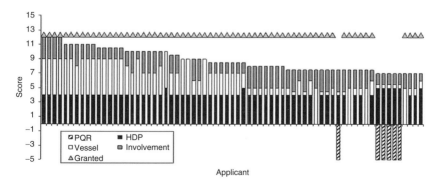

Figure 7.1 Total scores of a subset of the West Coast rock lobster limited commercial applicants

The emphasis in the criteria and scores on vessel ownership and investment in a vessel means that the rights might not have gone to traditional fishers and that the allocation process was not necessarily equitable or redistributive. It was felt that in small-scale fisheries, there seems no reason why a bona fide fisher (who has the appropriate fishing skills) should be required to be a businessperson. In addition it was evident that there was not always a link between the applicant's score and whether or not he/she was given a right.

7.3 Data and methodology

7.3.1 Stakeholder workshops

The project concentrated on the Western Cape fisheries (which accounts for about 90% of the total South African fisheries value (Wesgro 2001)) and fishing communities, the geographical position of which is illustrated in Figure 7.2. Fishing in the Western Cape consisted of a relatively small

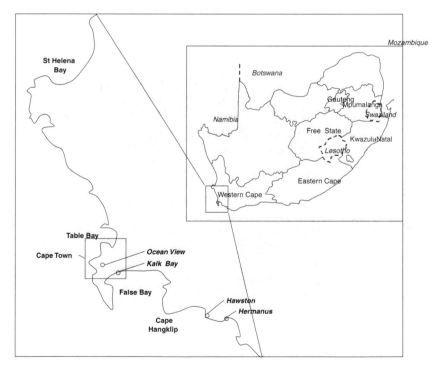

Figure 7.2 Western Cape showing the location of the fishing communities

number of large, predominantly white-owned fisheries, which employed many people, as well as smaller individual fishing companies or individuals involved formally or informally in less capital-intensive fisheries such as line-fish, West Coast rock lobster and abalone. The latter group included white and coloured people. Coloured fishing communities are among the poorest in the Western Cape.

Two study sites were selected for the purpose of this project, namely a relatively rural fishing community (which we shall designate as the Hawston/Hermanus community) located along a stretch of coast about 100–150 km east of Cape Town, and a more urbanized community within the boundaries of the City of Cape Town. For practical purposes, it was convenient to split the urban community into two subgroups denoted as Kalk Bay (referring to a small fishing harbour) and Ocean View (a fishing village created under earlier apartheid legislation). The relevant government department (MCM) is located in the centre of Cape Town. The Western Cape and the three fishing communities studied are presented in Figure 7.2.

In developing a new system, several workshops were held on location with the fishers from the three communities (Hawston, Kalk Bay and Ocean View). The MCM process followed a similar but more condensed format. The workshops involved three steps: (1) problem structuring, (2) definition and (3) ranking. Problem structuring started with a post-it session to collect views about the current allocation system. The results were used to create causal maps and value trees. In step 2, definition, performance levels and weights were set and finally in step 3 a 'trial runs' of the new system provided a ranking of the applicants. The next sections describe the results from the workshops and associated investigations. The east of Cape Hangklip West Coast rock lobster allocation is used as an example.

7.3.2 Problem structuring

In each case, workshops were conducted with representatives from the community or group. The meetings started with opportunities for free expression of concerns by the participants, after which a more formalized brainstorming session using 'post-its' took place (as described in Belton and Stewart 2002). These post-it sessions were used to capture perceptions of the problems, goals and potential courses of action, and were subsequently summarized in the form of causal maps, which were used during subsequent discussions to confirm that views of these communities had properly been captured and represented (see also Stewart et al. 2010)

The causal map constructed from workshops with the Hawston community is shown in Figure 7.3. The map was drawn using the Decision Explorer software (www.banxia.com), which although designed primarily for cognitive mapping is well suited to this problem. It is, for example, easy to identify the 'tails' (concepts with no incoming arrows, and which therefore may be associated with external driving forces and/or policy actions) and 'heads' (concepts with no outgoing arcs, and which therefore may be associated with consequences perceived by the group to be of fundamental importance). The central theme is probably that of concept 103, namely that the 'wrong' people were receiving allocations of fishing rights, rather than the traditional fishing communities. Much of the remainder of the map provides an indication of perceived reasons for this theme, but the map also identifies a potential new course of action (concept 117).

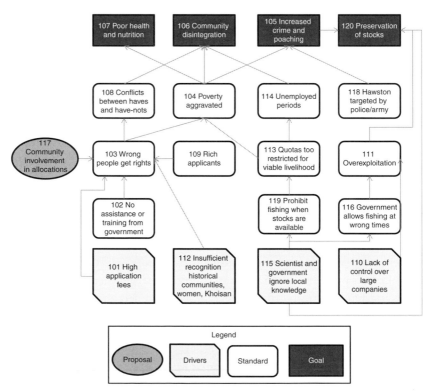

Figure 7.3 Causal map from Hawston workshops

From these considerations it was concluded that the primary goals as perceived by the community could be summarized as:

- Elimination of poaching-related gangsterism and crime (concept 105), which is viewed to arise because the 'true' fishermen were not receiving rights (concept 103).
- Prevention of community disintegration (concept 106). The other links in the map suggest that the goal may be achieved by addressing the following issues, which may be viewed as means-ends objectives in the sense of Keeney (1992): reduction in poverty; reduction in unemployment; fair division of rights across the community (cf. concept 108).
- Improvement of health and living standards (concept 107), which is also seen to be driven by poverty levels.
- Preservation of stocks (concept 120).

Since the goal of stock preservation is addressed by the setting of the TAC for the fishery, from the point of view of this community, the criteria relevant to allocation of rights are represented by concepts 105, 106 and 107. These concepts can be grouped together into a fundamental objective of preserving the social life and structure of the community.

To be able to use the information from the cognitive map in the new allocation system it was transformed into a value tree. A value tree links objectives with attributes or criteria that can be used for evaluation of alternatives, in this case the applicants for fishing rights.

Value trees were also constructed within the different communities and with MCM. The objectives of the communities related very much to historical involvement in the fishery, dependence on the fishery for subsistence and living, and links of the applicant to traditional fishing communities, all of which are incorporated into the integrated value tree of Figure 7.4. The integrated value tree was also presented at final workshops with the fishing communities, and no substantial objections were raised. Opinions differ in the degree to which each applicant satisfies these criteria. For example, MCM focused on boat ownership where the local communities focused on access to a boat. MCM would focus on business qualities of fishers while the communities would focus on fishing skills.

7.3.3 Definition

As with the 2001 system, levels of performance need to be defined for each criterion. In the 2001 system, the range of scores (veto to 5 for 'access to vessel', 0–3 for 'historical involvement' and 0–5 for 'HDP status' see

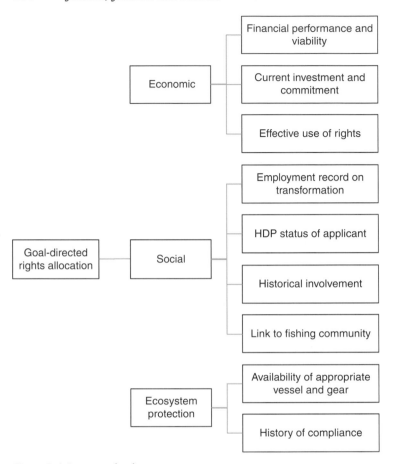

Figure 7.4 Integrated value tree

Figure 7.1) indicated the relative importance of a criterion. The score (s_{ij}) of Eq. (7.1) therefore combines in one number the relative value of the applicant according to that criterion (i.e. a score of 3 is better than 2) with the importance of that criterion relative to others. In the proposed new system, it is recommended that the definition of scores is separated from the assessment of weights. Thus, for all criteria, the performance levels would be associated with scores on a 0 to 100 scale. Subsequently, the weights for each criterion would be defined. Thus, in contrast to the formula applied in the 2001 system, Eq. (7.1), the following would apply:

$$V_j = \sum_{j=1}^{n} W_i u_{ij} \qquad (7.2)$$

where V_j is the overall score of applicant j, u_{ij} is the score of applicant j according to criterion i, and w_i is the weight of criterion i.

Participants were provided with coloured stickers which they could paste against the criteria and objectives identified in the value tree. They were asked to allocate stickers four times: three times to assess weights within the groups and once to indicate between the groups. They were given 12 stickers each time, though they did not always use them all. Weights were rescaled to sum to one.

The weights allocated to the main groups of criteria, as assessed independently by the two groups (see Joubert *et al.* 2005), were substantially similar (see Figure 7.5). Although no absolute guarantee of validity, this correspondence at least satisfies the necessary condition for between-group validation of the consistency of the procedure. It is interesting to comment that this process demonstrates how simple and unsophisticated procedures can nevertheless be formally correct.

During the workshops, scores and weights were elicited for the criteria used in the 2001 East of Hangklip allocation from MCM and fishers. This demonstrated both (a) that MCM's priorities had changed in the couple of years since the development of the allocation system (or else had not been well captured at that time) and (b) that for some criteria MCM and fishers showed more agreement than had previously been apparent (10). For

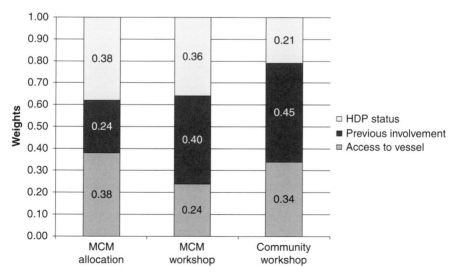

Figure 7.5 Weights derived for the three criteria used in the 2001 east of Hangklip allocation from the community workshops, the MCM workshop and those actually used in the 2001 allocation.

example, previous involvement was considered to be the most important issue to both fishers and MCM.

7.3.4 Ranking

Table 7.2 gives an example of 20 West Coast rock lobster applicants. Details of the applicants have been changed for illustrative purposes and to

Table 7.2 *Performance of a selection of the Hermanus/Hawston applicants for the east of Cape Hangklip West Coast rock lobster allocation using the scores and weights from the MCM workshops*

1 – Applicant No	2 – HDP status	3 – Vessel ownership or access	4 – Historical involvement	5 – Total Score (all three criteria)	6 – Total Score (original MCM scores)
		Scores from this study – weighted summation approach			Scores from actual allocation
1	90	100	100	96.4	12
2	90	80	100	91.6	10[*]
3	90	100	80	88.4	11
4	90	100	80	88.4	11
5	90	100	80	88.4	11
6	90	80	80	83.6	10
7	90	80	80	83.6	10
8	90	100	60	80.4	11[**]
9	90	100	60	80.4	11[**]
10	90	100	60	80.4	10.5
11	90	100	60	80.4	10.5
12	90	100	60	80.4	10.5
13	90	95	60	79.2	10
14	0	100	100	64	8
15	90	100	10	60.4	10[**]
16	0	95	80	54.8	7
17	0	100	60	48	7
18	0	100	60	48	6.5
19	0	100	60	48	7
20	0	100	60	48	6.5
Weights	0.36	0.24	0.4	Range 0–100	Range 0–13

[*] Ranking differs because of weighting (lower weight for vessel ownership relative to previous involvement).
[**] Ranking differs because of relative difference in score for historical involvement.

preserve confidentiality. To determine the ranking of the applicants the criterion scores are multiplied by their weights and added to obtain a total score (Table 7.2, column 5). The scores from the 2001 allocation system are shown in column 6 for comparison. This shows, for example, the differences in rank obtained due to differences in weights between criteria derived at the MCM workshop and those used in the actual allocation (e.g. applicant 2) or due to differences in scores within criteria (e.g. Historical involvement, applicants 8, 9 and 15).

An example extract of 60 applicants is shown in Figure 7.6 using the scores developed in the MCM workshop. These applicants consist of the 20 from Hawston/Hermanus and 20 each from Kleinmond and Gansbaai, which are towns in the same allocation area. The scores are displayed as the weighted contribution of each criterion to the total score (i.e. the total score is given in column 5 of Table 7.2). The total score determines whether the applicant is ranked high enough to be granted a (trial) right. A trial allocation was made based on the applicant's rank order and the TAC constraint (all were given the same quantum in this example).

The levels for HDP status were defined as HDP female, HDP male, non-HDP female and non-HDP male (in descending order of score) with non-HDP females still receiving a positive score. In this trial allocation, 97% of those granted rights were HDP applicants. The sum of all applicants' total scores is used as an indicator of the overall performance of the trial allocation (or of those granted rights).

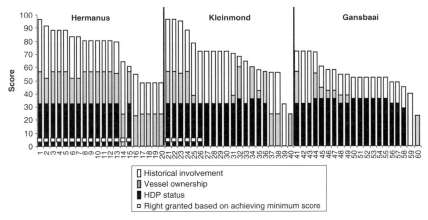

Figure 7.6 Trial allocation resulting from using the total score and granting the rights to the higher-ranking fishers. Weighted contribution of each criterion is shown. Scores are ranked from the highest to the lowest within zones according to the total score.

The process of rights allocation includes two distinct problems in the sense of Roy (1996), namely both a ranking and a sorting (classification) problem. In other words, the new allocation system needs to recognize and to support a two-stage process, first a simple ranking of applicants followed by a classification of applicants into those receiving or not receiving rights.

7.4 Conclusions and policy recommendations

As effective fisheries management requires the support of fishers, it is critical that stakeholder feedback be taken into account when developing criteria and objectives. The use of MCDA helps to achieve this, and its value has been recognized in many fisheries management contexts internationally. MCDA provides structured decision-making support to policymakers when informed trade-offs need to be made. An important feature of MCDA (and one which is often missing in fisheries management) is that of establishing clear objectives and linking criteria to these objectives.

The current fishing rights allocation system used by the MCM can be described as a form of multi-criteria analysis: the system is based on policy objectives such as 'transformation', and criteria linked to these objectives are scored. The MCM then evaluates applications on the basis of the sum of scores. However, there are flaws in this method. In particular, the link between policy objectives and criteria needs to be made more explicit and organized in a more consistent way. A new fishing rights allocation system would therefore maintain the same overall structure as the current approach, but would be improved, for example, by applying scores to performance levels of criteria and weights to different criteria in a systematic manner. It is not realistic for all stakeholders to be entirely satisfied with any given fishing rights allocation system, yet at least a transparent, fair and competent system would inspire more respect and trust.

The MCM needs to reflect on its multiple goals for South Africa's fisheries. Within the boundaries of ecological sustainability, some fisheries need to be managed with poverty alleviation in mind and others for economic efficiency. This means that, where poverty reduction is a principal objective, the requirement for fishers to have 'business skills' may be inappropriate.

A number of workshops were held with both grassroots fishing communities and with the government department responsible for allocations. Considerable success was obtained in establishing clear hierarchies of objectives even with semi-literate groups. In spite of a high level of distrust between the communities and the officials, it appeared that there was a high degree of correspondence between these objectives' hierarchies. Key differences included the concern of fishers to maintain their traditional community structures, and of the state to seek sustainable economic development and the inclusion of other disadvantaged groups (not traditional fishermen) in the industry.

Certain fisheries management and allocation issues still remain to be addressed in South Africa's Western Cape. These include determining the 'minimum viable quota' for small-scale fisheries and supporting alternative livelihood strategies where necessary. This is particularly important in seasonal fisheries, where fishers need other income-generating opportunities for part of the year. Sharing fishing rights within a community through reciprocal crewing arrangements is one means of doing this, though communities need further assistance and advice from the MCM on how to put such strategies into practice.

Much can be learnt from this project. The central thesis is perhaps the use of formal MCDA methodology as both a problem-structuring tool and as a framework for developing decision support systems. It has been demonstrated that it is possible to apply such an approach in working with stakeholders having widely divergent educational and cultural backgrounds. The fact that decision trees and associated weighting of criteria emerging from workshops with different fishing communities demonstrated a high level of consistency is at least prima facie evidence for the validity of the process.

In a context such as this, in which different groups have a history of considerable distrust and antagonism, it is also unlikely that a unitary workshop involving all stakeholders together will be productive. Still, the use of tools of problem structuring and MCDA (such as cognitive maps and value trees) as a means of capturing views expressed in workshops has been shown to be a valuable means of communicating these views and values between stakeholder groups. The same tools lead directly to the design of simple transparent decision support tools, incorporating clear demonstration of how the goals of different groups have been taken into account.

7.5 Epilogue

Despite the general consensus among fishermen and MCM that the system proposed in this study was acceptable for all parties MCM have implemented a considerably more complex system. MCM's new scoring system is not based on *a priori* defined benchmarks for each criterion (as proposed in this study) but on performance *relative* to other applicants. The methods of scoring applicants in terms of these criteria, however, follows a complicated process (based on percentiles of the distribution of inputs from all applicants) that in our view violates the value measurement principles underlying an additive scoring system, and can be highly sensitive to data input errors. The resulting allocations have resulted in continued unhappiness among fishermen and many of the allocations are now facing legal challenges in the courts.

References

Arnason, R. (2005). Property rights in fisheries: Iceland's experience with ITQs. *Reviews in Fish Biology and Fisheries*, **15**(3): 243–264.

Belton, V. and Stewart, T. J. (2002). *Multiple Criteria Decision Analysis: An Integrated Approach*. Boston, MA: Kluwer Academic Publishers.

Cockroft, A. C., Sauer, W. H. H., Branch, G. M. *et al.* (2002). Assessment of resource availability and suitability for subsistence fishers in South Africa, with a review of resource management procedures. *South African Journal of Marine Science*, **24**: 489–501.

Department of Environmental Affairs and Tourism (DEAT) (2002). *Where Have All the Fish Gone? Measuring Transformation in the South African Fishing Industry.* Pretoria, South Africa: Department of Environmental Affairs and Tourism.

Janssen, R. (1992). *Multiobjective Decision Support for Environmental Management*. Dordrecht, the Netherlands: Kluwer Academic Publishers.

Joubert, A. R., Stewart, T. J., Scott, L. *et al.* (2005). *Fishing Rights and Small-scale Fishers: An Evaluation of the Rights Allocation Process and the Utilization of Fishing Rights in South Africa*. Cape Town, South Africa: Department of Statistical Sciences, University of Cape Town. Available at www.premonline.nl.

Joubert, A. R., Janssen, R. and Stewart, T. J. (2008). Allocating fishing rights in South Africa: a participatory approach. *Fisheries Management and Ecology*, **15**: 27–37.

Keeney, R. L. (1992). *Value-focused Thinking: A Path to Creative Decision Making*. Cambridge, MA: Harvard University Press.

Kleinschmidt, H., Sauer, W. H. H. and Britz, P. (2003). Commercial fishing rights allocation in post-apartheid South Africa: reconciling equity and stability. *African Journal of Marine Science*, **25**: 25–35.

Roy, B. (1996). *Multicriteria Methodology for Decision Aiding*. Dordrecht, the Netherlands: Kluwer Academic Publishers.

RSA (1998). Marine Living Resources Act, Act Number 18, 1998, Republic of South Africa.

Scott, A. (2000). Moving through the narrows: from open access to ITQs and self-government. In R. Shotton (ed.), Use of property rights in fisheries management. Fisheries Technical Paper 404/1, Rome: FAO, pp. 105–117.

Stewart, T. J., Joubert, A. R. and Janssen, R. (2010). MCDM Framework for fishing rights allocation in South Africa. *Group Decision and Negotiation*, **19**(3): 247–265.

Wesgro (2001). TheWestern Cape fishing industry. Cape sector factsheet: Business sector updates from South Africa's Western Cape. Available at: www.wesgro.org.za.

Wolman, A. G. (2006). Measurement and meaningfulness in conservation science. *Conservation Biology*, **20**: 1626–1634.

Part III
Forest–related ecosystem services

PIETER J. H. VAN BEUKERING

Compared to many other ecosystems, the link between forest and poverty is well researched (see Angelsen and Wunder 2003, Arnold and Bird 1999, Kerr and Pfaff 2004, Kumar *et al.* 2000, Scherr *et al.* 2003). One perspective on the relationship between poverty and forest is the level of dependence of people on forest ecosystems. Around 410 million people (including 60 million indigenous people) live in, or at the fringes of, tropical forests (Wiersum and Ros–Tonnen 2005), and more than 1.6 billion people living in extreme poverty are estimated to depend on forests for some part of their livelihoods (World Bank 2004, WRI 2005). Alleviating poverty in these areas is therefore unachievable without taking into account the forestry–poverty nexus.

Another perspective on the link between poverty and forest is the possible existence of a negative downward spiral of poverty and environmental degradation. This negative spiral has long been considered valid for the forest sector. Micro-level studies illustrated that poverty may leave people no other option but to clear forest cover in order to gain access to land for cultivation or to use natural resources in an unsustainable manner (Deininger and Minten 1999, Kerr and Pfaff 2004). According to the MEA (2005), more forest was converted to cropland in the 30 years after 1950 than in the 150 years between 1700 and 1850. This conversion rate accelerated further after the 1980s and mainly takes place in developing countries. However, macro-economic studies failed to prove the validity of this generalization for the relationship between poverty and deforestation

Nature's Wealth: The Economics of Ecosystem Services and Poverty, ed. P. J. H. van Beukering, E. Papyrakis, J. Bouma and R. Brouwer. Published by Cambridge University Press, © Cambridge University Press 2013.

(Brown and Pearce 1994a, Sunderlin *et al.* 2006). Claiming poverty to be the main underlying cause of deforestation is inaccurate, particularly since deforestation is sometimes undertaken by wealthy commercial interests (Chomiz 2006, van Beukering *et al.* 2003). Moreover, various studies also show that poor people do invest considerable time and resources in forest management (Arnold and Bird 1999).

Although both perspectives on the mutual link between forest ecosystems and poverty shed a light on this complex relationship, more nuanced insights have evolved over time. Wiersum and Ros-Tonnen (2005) identify important new insights recognizing that forests may contribute to poverty alleviation. First, poverty alleviation nowadays is not only considered to be a matter of increased employment and income generation (e.g. poverty reduction), but is also considered in terms of reducing vulnerability (e.g. poverty mitigation). In the case of poverty mitigation, forest resources are used for subsistence needs and as a 'safety net' to meet occasional shortfalls in production or income. In the case of poverty reduction, forest is used to generate cash income and thus may lead to lasting welfare improvements. Second, since the 1980s, timber is no longer considered the one and only service provided by forest as a livelihood asset. Forests also provide a large variety of non-timber forest products that due to seasonal variations combine well with other livelihood sources. Also, several forest products and services that were formerly freely available are acquiring a financial value. These include (eco)tourism, watershed services and carbon sequestration. These new markets potentially create new opportunities for forest dwellers to diversify their livelihood sources.

The relationship between forest resources and poverty is very location-specific, depending on factors such as the quality of the forest, distance to markets, available infrastructure and transport facilities, access to capital, alternative livelihood options and the degree of organization into producer groups. This context variability is illustrated by Chomitz (2006) in the famous Loggershead report describing the various deforestation trajectories recorded around the world (see Figure III.1):

Intensification with deforestation is typically seen in agriculturally favourable areas such as the soybean areas of the Brazilian savannas where changes in markets or roads increase the value of both standing timber and agricultural land in areas with favourable soils and climate leading to rapid deforestation. The agricultural development and timber harvesting stimulate the growth of market towns with forest-related economy increasing the local population and demand for land. Forest cover stabilizes at a low level, with remaining forest occupying slopes or poor quality land.

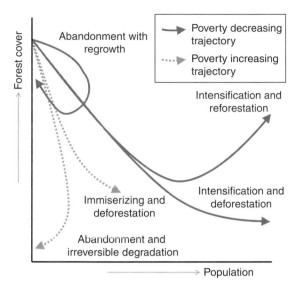

Figure III.1 Poverty–deforestation trajectories.
Source: adapted with permission from unpublished work by Kenneth Chomitz.

The *intensification with reforestation* trajectory reveals similar dynamics as the previous trajectory, but here forest depletion leads to wood scarcity, and better tenure makes it possible for households and communities to manage forests and in some cases fields and pastures are converted back to woodlots. The result is a mosaic of croplands and managed forests. For example, India's agricultural intensification has had a major positive impact, relieving pressure on marginal lands on which most of the forests remain (Kumar *et al.* 2000).

Abandonment with regrowth is the most familiar scenario triggered by population expansion onto marginal lands where rents are low, barely providing sufficient livelihoods for landholders. With development in other sectors in the economy, local populations migrate to better opportunities leaving marginal areas for natural forest regeneration. Costa Rica's strong forest regrowth during the 1990s may be an example of the latter.

The *abandonment with irreversible degradation* trajectory is similar to the previous one, except that the land uses of immigrants prove unsustainable. Soil fertility collapses due to unsustainable management leading to low rents without natural regrowth. Examples include millions of hectares of imperata grasslands in South East Asia.

The *immiserizing and deforestation* trajectory is mainly triggered by population expansion in combination with stagnant technologies and immobile labour. Poor agronomic conditions and practices further reduce incomes and increase pressure for nutrients from fresh deforestation.

In recent years, the recognition that forests may contribute to poverty alleviation has grown (Nasi *et al.* 2002, Wiersum and Ros-Tonen 2005). It has become increasingly clear that forests provide not only productive assets but instead are a source of multiple livelihood assets (Smail and Lewis 2009). Also, it has become increasingly clear that forest-based livelihood activities usually form a part of multiple-component livelihood strategies, which may include farming, animal keeping, wage labour and migration. Also, awareness has increased about the new opportunities for trading forest products and services such as ecotourism, the provision of regular water supplies for domestic needs and carbon sequestration. A recent TEEB study estimated the average annual value of forest ecosystem services to be almost US$ 6000 per hectare, with provisioning, regulating and cultural services measuring, US$ 1285, US$ 3890, and US$ 381, respectively (De Groot *et al.* 2010).

Despite the high value attributed to forest ecosystems, one should not be overly optimistic about the role of forest in reducing poverty. Angelsen and Wunder (2003) highlighted the limitations of forests in alleviating poverty in developing countries simply because markets in these often isolated areas are generally poorly developed and characterized by weak producer organizations and high transportation costs. For example, in the case of non-timber forest products (NTFPs), ecosystem services often play an important part in poverty prevention and helping people survive, yet, forest services are rarely a tool for genuine and structural poverty reduction. In some cases, the role of forests as key safety nets may only be transitory, as poor people diversify and build other assets to insure themselves against a variety of risks.

Various policy interventions have been proposed in the literature that help to alleviate poverty through the management of forest ecosystems. These can be subdivided into several categories. First, policies can be directed at making markets work for the poor. Scherr *et al.* (2003) and FAO (2006) stress that for people to invest in forestry they need security, capacity and ownership in forest-related activities. By its very nature, forestry requires medium- to long-term investment for its returns to be sustainable. Research shows that people are reluctant to invest in sound forest management unless they have secure rights and control over the resources. Moreover, even if tenure and access are grounded in law and policy, there is a need for good governance. This implies that decision-making should be transparent and all relevant information be disseminated to the stakeholders involved. Also, when the conditions to invest in forest are favourable, people still need the skills to manage and use these resources sustainably. Individuals and institutions within the community require the capacity to support such management and to guarantee that the marginalized community members share the benefits from forestry in a fair manner.

Second, a series of instruments have been described in the literature which specifically facilitate income-generating activities in the forestry domain (FAO 2006, Larson and Ribot 2007, Peskett *et al.* 2006). At the micro-economic level, circumstances can be improved at which stakeholders operate, such as micro-finance schemes linked to other interventions such as training in technical and simple business skills that add value to forest services provided. At the macro-economic level, national policies and programmes can be linked more directly to the local needs of people that depend on forest ecosystem services. In spite of the recognized importance of forests to tropical countries, the value of forest to local livelihoods is rarely captured in their national development plans (Brown and Pearce 1994b), such as the Poverty Reduction Strategy Papers (PRSP). In this regard, politicians and other decision-makers need to develop comprehensive and coordinated policies, legislation, strategies and programmes, while advocating for simple and enforceable laws that can build on existing rights (Larson and Ribot 2007). Finally, a more hybrid instrument that can function both at the micro and at the macro-economic scales is the Payment for Environmental Services (PES) and the Reducing Emissions from Deforestation and Forest Degradation (REDD) concept which is aimed at creating markets for the benefits that people obtain from ecosystems (Nasi *et al.* 2002, Peskett *et al.* 2006).

Organization of Part III

This part of the book documents four case studies addressing various aspects of the link between forest ecosystem services and poverty allevia-tion. In Chapter 8, the charcoal chain in Tanzania is systematically analysed with the aim of identifying the most effective entry point for interventions. By analysing the production, trade and consumption of charcoal, it addresses causes of deforestation as well as potential solutions to avoid further unsustainable use. The study contributes to the develop-ment of a comprehensive policy with regard to the role charcoal plays in Tanzania's energy strategy without increasing poverty among rural and urban communities. More specifically, it proves that the forest service 'charcoal' is a means of last resort to escape extreme poverty rather than a profit-generating activity for the people that operate in the charcoal chain. The case study also convincingly shows that current policies directed at the charcoal chain are inefficient. Most of the existing regulations and policies are patchwork in nature and are generally poorly enforced. The study concludes by identifying a number of key improvements that can be made across the charcoal chain in Tanzania. On the supply side, the need for forest resources can partially be met by further expanding production forest and plantations. On the demand side, targeted policies and technological innovations are needed to enhance the attractiveness of substitutes of charcoal. At the chain level, PES provides a promising route for the creation of incentives throughout the charcoal chain to move away from charcoal production to more sustainable forms of livelihood activities.

Chapter 9 presents a feasibility study for a PES scheme in the protected area of Peñablanca in the Philippines, which should achieve the combined goals of resource management and poverty alleviation by providing an incentive for upland dwellers to engage in much needed forest conservation activities. The case study evaluates the potential for such a mechanism in Peñablanca watershed area, paying specific attention to the scientific, social, economic and institutional requirements of a successful PES system. The science component of this study has made a clear connection between deforestation and a change in the streamflow regime of the Pinacanauan River. The economic analysis proves that those depending on and benefiting from a stable stream-flow regime are willing to pay for conservation of the watershed to prevent further degradation and loss of watershed services. The institutional assessment shows that, at least at the national level, important requirements for PES are in place. At a local level, however, things are more complicated. The vast majority of upland dwellers do not have secure land rights at this time. Moreover, many settlers in the forest have been there for less than 5 years, and new settlers keep arriving. Taking all these matters into consideration, the study concludes that a PES scheme is viable in the Pinacanauan watershed, and it should be given priority by the relevant local institutions governing the protected area (Department of Environment and Natural Resources (DENR) and PAMB) as a strategy to halt the ongoing degradation and to improve the fate and prospects of the upland dwellers.

Chapter 10 concentrates on the relationship between the copper curse and forest degradation in Zambia. Rich in mineral deposits, Zambia's Copperbelt province was once the site of relatively prosperous copper mining. Among the world's top ten copper producers, Zambia should have been an example of economic success among developing sub-Saharan African nations. But over-dependence on copper has turned prosperity into economic and environmental hardship. Now the Copperbelt has the largest share of poor in Zambia, in turn causing a rapid increase in deforestation. With no immediate options for alternative livelihoods, many unemployed miners have had to find work by exploiting the area's natural forest resources. Clearing wood for charcoal production to meet high urban demand for cheap energy has resulted in excessive rates of deforestation. This study shows why and how the provision of alternative livelihoods is crucial for halting deforestation in the Copperbelt region and escaping the negative poverty–environmental spiral. Projects that help to achieve improvements in living standards over time are funded by public revenues, i.e. tax from the mining industry. Without access to a significant amount of public revenues to fund targeted poverty alleviation projects, Zambia's sustainable development is likely to be further hindered.

The last case study of this forestry part is presented in Chapter 11 in which institutions and forest management in the Swat region of Pakistan are analysed. This study explores the mechanisms behind deforestation in the north-western part of Pakistan, uncovers mechanisms to reverse the process and provides a framework for determining the bottlenecks in the management of common resources from the perspective of institutions. The study shows that in circumstances where institutional change is necessary, we are faced with a trade-off between the transaction costs related to the enforcement of 'improved' institutional arrangements and the transaction costs improving enforceable institutional arrangements. Different stakeholders in the Swat region have their own ideal management system to put forward: (1) private property rights for the local large landowners; (2) state ownership (effectively enforced by the Forest Department) as the goal of the government; (3) community-based management by local villages (guaranteeing socially and environmentally sustainable outcomes) as common pool resources is identified by the local communities characterized by higher level of trust and most NGOs and donors. The current situation is an intermediate form with an incoherent set of external interventions and strategic reactions by different agents in the local communities. The emergent system of management is the one producing the present dismal outcome. The study concludes that joint forest management seems to have significant potential; the authors argue that the preconditions in terms of well-defined institutional arrangements regarding property rights, control and financial remuneration for management activities should be addressed prior to embracing any scheme that is considered some sort of silver bullet.

References

Angelsen, A. and Wunder, S. (2003). Exploring the forest–poverty link: key concepts, issues and research implications. CIFOR Occasional Paper No. 40. Center for International Forestry Research, Bogor, Indonesia.

Arnold, J. E. M. and Bird, P. (1999). Forests and the poverty-environment nexus. Paper prepared for the UNDP/EC Expert Workshop on Poverty and the Environment, Brussels, Belgium, January 20–21 1999.

Brown, K. and Pearce, D. W. (1994a). *The Causes of Tropical Deforestation: the Economic and Statistical Analysis of Factors Giving Rise to the Loss of the Tropical Forests.* London: UCL Press.

Brown, K. and Pearce, D. W. (1994b). The economic value of non-market benefits of tropical forests: carbon storage. In J. Weiss (ed.), *The Economics of Project Appraisal and the Environment.* Cheltenham, UK: Edward Elgar, pp. 102–123.

Chomitz, K. M. (2006). At loggerheads? Agricultural expansion, poverty reduction, and environment in the tropical forests. World Bank policy research report, Washington DC.

de Groot, R., Fisher, B., Christie, M. *et al.* (2010). Integrating the ecological and economic dimensions in biodiversity and ecosystem service valuation. In P. Kumar, (ed.), *The Economics of Ecosystems and Biodiversity: The Ecological and Economic Foundations.* London: Routledge, pp. 20–40.

Deininger, K. and Minten, B. (1999). Poverty, politics, and deforestation: the case of Mexico. *Economic Development and Cultural Change*, **47**(2): 313–44.

FAO (2006). Better forestry, less poverty. A practitioner's guide. FAO Forestry Paper 149. Food and Agriculture Organization of the United Nations. Rome.

Kerr, S. and Pfaff, A. S. P. (2004). Effects of poverty on deforestation: distinguishing behavior from location. ESA Working Paper No. 04–19. Food and Agriculture Organization of the United Nations, Rome.

Kumar, N., Saxena, N., Alagh, Y. and Mitra, K. (2000). *India: Alleviating Poverty Through Forest Development.* Evaluation Country Case Study Series. Washington DC: The World Bank.

Larson, A. M. and Ribot, J. C. (2007). The poverty of forestry policy: double standards on an uneven playing field. *Sustainability Science*, **2**(2): 189–204.

Millennium Ecosystem Assessment (MEA) (2005). *Ecosystems and Human Well-being: Biodiversity Synthesis.* Washington DC: World Resources Institute.

Nasi R., Dennis R., Meijaard E., Applegate G. and Moore P. (2002). Forest fire and biological diversity. *Unasylva*, **209**(59): 36–40.

Peskett, L., Brown, D. and Luttrell, C. (2006). Can payments for avoided deforestation to tackle climate change also benefit the poor? Avoided deforestation could benefit the environment and forest dependent populations. Forestry Briefing, 12. November 2006. ODI, London.

Scherr, S. J., White, A. and Kaimowitz, D. (2003). *A New Agenda for Forest Conservation and Poverty Reduction: Making Markets Work for Low Income Producers*. Washington DC: Forest Trends.

Smail, R. A. and Lewis, D. J. (2009). Forest-land conversion, ecosystem services, and economic issues for policy: a review. PNW-GTR-797. Portland, OR: U.S. Department of Agriculture, Forest Service, Pacific Northwest Research Station.

Sunderlin, W. D., Dewi, S. and Puntodewo, A. (2006). Forests, poverty, and poverty alleviation policies. Background paper, World Bank Policy Research Report.

van Beukering, P. J. H., Cesar, H. S. J. and Janssen, M. A. (2003). Valuation of ecological services of the Leuser National Park in Sumatra, Indonesia. *Ecological Economics*, **44**(1): 43–62.

Wiersum, K. F. and Ros-Tonen, M. A. F. (2005). The role of forests in poverty alleviation: dealing with multiple millennium development goals. *North-South Policy Brief*, **6**: 1–7.

World Bank (2004). *Sustaining Forests: A Development Strategy*. Washington DC: The World Bank.

World Resources Institute (WRI) (2005). *The Wealth of the Poor: Managing Ecosystems to Fight Poverty*. Washington DC: United Nations Development Programme, United Nations Environment Programme, World Bank, World Resources Institute.

8 · Greening the charcoal chain in Tanzania

PIETER J. H. VAN BEUKERING,
SEBASTIAAN M. HESS, ERIC E. MASSEY,
SABINA L. DI PRIMA, VICTOR G.
MAKUNDI, KIM VAN DER LEEUW AND
GODIUS KAHYARARA

8.1 Introduction

With a population of 34 million and an extremely high reliance on charcoal, Tanzania is a classic example of the social and environmental risks faced by many developing countries. About 85% of the total urban population uses charcoal for household cooking and energy provision for small and medium enterprises (Sawe 2004). In 1992 the total amount of charcoal consumed nationwide was estimated to be about 1.2 million tons (Sawe 2004). In 2002, the charcoal business generated revenues of more than 200 billion TShs (US$ 200 million), with more than 70 000 people from rural and urban areas employed in the industry (TaTEDO 2002b). Dar es Salaam, Tanzania's largest city, accounts for more than 50% of all charcoal consumed in the country.

The charcoal sector is far from sustainable. The forest resources that the industry is relying on are disappearing rapidly and the productivity of the sector has not seen any improvement either. The charcoal sector in Tanzania is operating economically, socially and environmentally in a suboptimal manner. However, solutions that safeguard the charcoal sector's future are not straightforward.

The complexity of the charcoal industry chain is enormous. The recent attempt of the Tanzanian government to stop deforestation by introducing a sudden ban on charcoal production is an example of a one-dimensional short-sighted intervention that failed because it ignored the complexity of the chain. For policies to be effective, a comprehensive

Nature's Wealth: The Economics of Ecosystem Services and Poverty, ed. P. J. H. van Beukering,
E. Papyrakis, J. Bouma and R. Brouwer. Published by Cambridge University Press,
© Cambridge University Press 2013.

approach is needed that recognizes the interdependencies between the different segments in the charcoal chain (i.e. production, trade, consumption) as well as the multitude of dimensions to be taken into account (e.g. technologies, economics, social, etc.).

This chapter provides a comprehensive integrated overview of all three segments of the charcoal sector in Tanzania with the aim of identifying effective interventions to improve the sustainability of the charcoal chain. The outline of the chapter is based on the chain approach, following the physical flow of charcoal moving from producers (Section 8.2) through traders (Section 8.3) to consumers (Section 8.4), while simultaneously monitoring the monetary flows that move in the opposite direction (Figure 8.1). Policy recommendations are given in Section 8.5. To increase our understanding of the developments in the charcoal chain in Tanzania, we systematically address a number of dimensions in each segment of the chain. These dimensions either represent *drivers* that influence the course of the charcoal flows, or denote indicators that reveal the *impact* of certain developments within the chain.

While this chapter deals with the issue of charcoal in Tanzania in general, much of the data presented come from our work in the coastal forests surrounding Dar es Salaam, specifically the Pugu and Kazimzumbwi forests. Data collected consisted of both primary and secondary data. A broad range of methods was employed for primary data collection: semi-structured

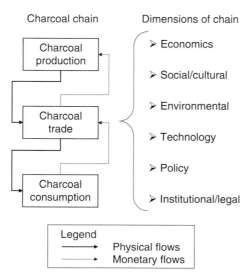

Figure 8.1 Analytical framework for the charcoal chain

interviews, questionnaire surveys among producers (400 households) and consumers (250 individuals), choice experiments, contingent valuation and use of the geographic information system (GIS).

8.2 Charcoal production

Almost all charcoal in the country is produced in rural areas, with the largest shares of raw materials extracted from: open miombo woodlands (owned by local governments); reserved forests; bushland forests (publicly owned); mangrove forests; and farmlands (CHAPOSA 2002). The radius of the area from which these raw materials are collected is steadily increasing, with charcoal makers needing to travel progressively further to obtain the resources needed.

Two types of wood harvesting dominate: clear felling of forestland for agriculture purposes and subsequent carbonization of the felled trees; and selective cutting done by the charcoal makers who aim at generating income from charcoal business only. Selective harvesting is in principle the least destructive form as it allows young trees to grow. However, due to inadequate management skills young trees are also chopped down and used to cover the charcoal kilns, resulting in higher than necessary forest destruction.

The entire production cycle takes almost 2 months, but the burning itself, using the traditional earth mound kiln, takes around 18 days. The efficiency of the process depends on the construction of the kiln, moisture content of wood and perhaps for the largest part on the skills of the producers. Between three and ten people are usually involved in the production. The steps involved in charcoal making include:

Material preparation:	felling of trees, cross-cutting into short logs, wood drying;
Kiln manufacture:	kiln base/structure, piling of the logs into a clamp, covering the clamp with grass/sand or grass/stones;
Carbonization:	initiate the fire, carbonization control, cooling period;
Packing and selling:	packing into bags, transport to roadside.

The most common is the traditional (basic) earth mound kiln. The efficiency of this type of kiln is typically between 10 and 20% (CHAPOSA 2002). There are a variety of kilns with much higher efficiencies of up to 35% (TaTEDO 2002b). Switching to more efficient kilns would greatly reduce the pressure on forests, because less wood would be required.

8.2.1 Social and cultural domain

Population grew rapidly in the study area with an average growth rate of 3% per year. As the population increases, more charcoal is needed for cooking, more fodder is needed for animals, more of the forest is converted to pasture and more forest is cleared for crops. Excessive deforestation in response to the rapidly expanding population can increase stress on the poorer sections of the society and on women, as they have been primarily involved in gathering fuelwood, fodder and water in the traditional village economies.

Charcoal making is the most remunerative among the forest-related activities. It is labour intensive, with very low input of capital, making it more or less a zero-cost activity. The most important factor encouraging people to start producing charcoal is the lack of alternative income-generating activities and the fact that charcoal is a cash product, with a large market ready to absorb the entire production. Producers are generally organized in informal groups, clustered into geographical areas particularly suitable for production. The choice of sites depends primarily on the quantity and quality of the trees but also on the level of monitoring by the forest authorities.

Two distinct groups of charcoal producers are noticeable. First, there are *farmers* who produce charcoal as part of secondary activities. The second group comprises charcoal-making *specialists* for whom charcoal making forms their primary activity. For both groups, charcoal is the most important forest-related activity. Our field survey revealed that 28% of the households are involved in charcoal making. Charcoal production often takes place in protected forests, and thus many people are reluctant to discuss their charcoal-making activities. Both categories of respondents spend around 220 days a year in the forest making charcoal. The specialists, however, are a lot more efficient. On average, they produce 267 bags a year (1.22 bags a day), while the farmers do not produce more than 171 bags (0.85 bags a day).

To calculate the total charcoal supply from the study area, our sample was extrapolated to the selected villages and subsequently to the whole Pugu/Kazimzumbwi forest area. Based on the 2002 census data, we come to 14 000 households, after population growth correction. Assuming the same share of these households is producing charcoal as in our sample (28%), we get 3920 charcoal-producing households, amounting to a total of 1 million bags of charcoal, weighing 30 kg each.

8.2.2 Economic domain

To calculate cash income of the village households, the household survey inquired about the sale of farm goods, forest products, charcoal, timber and income from a household enterprise, cash transfers from relatives or income from employment. Furthermore, we derived the value of the subsistence part of farming and forest-product collection. We expected this subsistence part to form a major share of household income. However, due to our focus on charcoal makers, the surveyed households are probably more involved in the cash economy than the average villager.

Table 8.1 shows variations in income for various subgroups. If all households are considered, a relatively high income is estimated. This is caused entirely by a limited number of timber extractors, and by one very large timber extractor in particular. For all groups the average income is significantly higher than the median income, reflecting the unequal income distribution.

If we look just at the cash income, charcoal becomes even more important. Forest products become almost insignificant, making up only 3% of cash income in the whole sample. Apparently, forest products are mainly collected for home consumption. The cash income earned from charcoal production is usually spent on other family needs such as paying school fees, medical care and investment for agriculture.

Table 8.1 *Average and median income (in TShs/household), and income shares (percentage)*

	All (n = 360)	Charcoal maker (n = 185)	Farmer (n = 138)
Total income	979 225	890 725	663 524
Median of total income	678 100	808 000	437 471
– Share charcoal	58%	80%	38%
– Share crops	14%	7%	25%
– Share livestock	1%	0%	3%
– Share forest products	17%	12%	26%
Cash income	884 219	800 646	549 649
Median of cash income	540 000	666 000	308 100
– Share cash	81%	87%	71%
– Share charcoal	69%	91%	50%
– Share crops	16%	6%	33%
– Share livestock	1%	0%	3%
– Share forest products	3%	2%	5%

Note: Groupings are based on the stated main activity of the household head.

The household survey revealed that landownership is very common (90%+), although the plots are usually small. The predominant land type of the holdings is cropland, but the land is often not actively harvested. Partly this is due to low agricultural prices. If someone has access to another form of livelihood (i.e. charcoal), the landowner tends to leave his or her land idle.

8.2.3 Environmental domain

Deforestation is one of the major environmental problems of Tanzania, mainly attributed to felling of trees for charcoal production. This has serious impacts: the forests in Tanzania have reduced from 44.3 million ha in 1961 to 33.5 million ha in 1998. The Forestry and Beekeeping Division of Tanzania estimates an annual forest reduction between 130 000 to 500 000 ha, against only 25 000 ha planted annually. According to Burgess and Muir (1994), Tanzania has just 799 km^2 of coastal forests remaining.

The felling of trees for charcoal also has significant impacts on vegetation, soil and watersheds. Our study examined the conditions of stream flow behaviours in relation to land use and land-cover conditions of Kizinga and Mzinga rivers that originate in the two forests of Pugu and Kazimzumbwi. Flow regime has changed over the years in both rivers. The land-use change study shows that plantation forest and cultivation with tree crops has increased considerably, mainly at the expense of natural forest and bush lands. This means that vegetation density has decreased and that evapotranspiration has diminished, resulting in less water use by vegetation and increased runoff.

Forest dwellers are highly aware of detrimental effects of deforestation. Almost all respondents in the producers' survey (95%) observed the disappearance of forests, and 80% expected there to be very little left in 10 years time. The respondents also confirmed that the production of charcoal is the main cause of deforestation. Respondents saw a structural move away from charcoal as the main fuel used in urban areas as the most promising measure to reduce deforestation (Figure 8.2). Stricter control by forest officials is least popular.

This strong awareness about the unsustainable status of charcoal production among producers increases the chances of successfully adopting alternative livelihood options. In the case study area several options exist:

- *Increasing agricultural productivity*: Agriculture scores second as a source of income in the area. Two crops show significant potential: cashew nut and cassava.

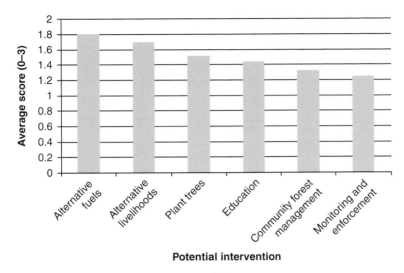

Figure 8.2 Possible measures to reduce deforestation

- *Improve marketing of cash crops:* Improving access to markets, especially for cashew nuts and honey would provide a large incentive to shift.
- *Promote agro-forestry:* Development of agro-forestry can provide a wide range of benefits to local people for subsistence, cash income, asset building etc.
- *Introduce clean development mechanism (CDM) projects:* An area with high potential for cash revenue is the development of CDM projects under the Kyoto protocol. Several possible projects have been investigated by the Tanzanian government.

A choice experiment (CE) was conducted to determine the willingness of the charcoal producers to switch to alternative income-generating activities. The choice model was run for both sets of respondents, farmers and charcoal makers. The CE shows that forest dwellers consider charcoal making as a means of last resort. Therefore, if reliable options for alternative livelihood are provided, forest dwellers in the coastal forest of Tanzania proved to be willing to switch to these other types of employment.

The results show that compensation measures differ for the two groups, implying that it will be necessary to approach local people in different ways, depending on their level of specialization. The specialized charcoal makers seem to prefer compensation in the form of jobs in forest management and loans to set up a business. Local people who are more involved in farming on the other hand would rather receive training to develop other activities. For both groups combinations of the different types of compensation are preferred over single measures.

8.2.4 Legal and institutional domain

Tanzanian natural resource management relies on an extensive regulatory, command-and-control approach, carried out largely under the direction of central government agencies. The system is characterized by a high level of bureaucracy and consequently high transaction costs. Lack of enforcement of existing laws and regulations is one of the main problems in Tanzanian environmental policy. It undermines the effectiveness of the entire regulatory and institutional framework, thus jeopardizing the sustainable management of forests. This lack of enforcement is strongly linked to the centralized administrative structure.

The Forest Ordinance, Cap.389, in particular, is supposed to regulate the use of forests but its enforcement is highly inefficient. In fact, the Forest Ordinance does not directly impact the production of charcoal because it is qualified as a secondary product obtained from harvested or gathered timber. Therefore, the only way the government can control and get revenues from the charcoal business is through a tax system. Taxes are imposed only on the charcoal that is for commercial use.

8.3 Trade

Most charcoal produced in rural areas is transported to the main Tanzanian cities by trucks or bicycles. Very few producers, less than 40%, ferry their own charcoal to the cities (Napendaeli 2004). The majority is ferried to urban centres by charcoal dealers. Only a low percentage of this flow is recorded at the official checkpoints. It has been estimated that only 10–20% of the total amount of charcoal entering the city of Dar es Salaam is officially recorded (Napendaeli 2004). Once charcoal enters the cities, it is sold at dealer-run outlets, in shops or directly to the end-users.

8.3.1 Economic and environmental domain

The assessment amongst dealers showed that start up capital to enter the charcoal business was generally low, around 50 000–800 000 TShs (US$ 50–800). Officially, dealers need an annual trading licence of 50 000 TShs. However, considering the informal structure of the business, few traders actually comply. Transport costs between 1000 and 4000 TShs per bag (survey data 2005). This cost decreases if the charcoal is gathered from fewer collection points. This forces charcoal producers to cluster in the same geographic areas or produce in larger quantities in order to minimize the dealers' transport cost. Instead of making small kilns due to decreased wood stocks, charcoal producers now tend to make bigger kilns (44 bags/ kiln) to attract traders.

As nearby forest stocks are depleted over time, charcoal production sites shift further away from the main cities, increasing the cost of transport. Dar es Salaam's charcoal supplies come from within a radius of 100–200 km, from the coast and Morogoro regions. In general, the sites have shifted from 50 km in the 1970s to about 200 km in the 1990s.

Figure 8.3 shows that wholesalers retain the highest margin of profit within the entire chain. Our study provides a clear overview of the price differential and of how charcoal prices reflect the character of the business. In the forest a *gunia* (30 kg bag) fetches a price of 1000–2000 TShs. Just a few metres outside the forest, the same bag is sold for 3000–4000 TShs. This is justified by the risk of being fined and having the charcoal and bicycle confiscated one incurs by carrying charcoal out of the forest. Within the Dar es Salaam region, but outside the actual city, a *gunia* costs around 7000 TShs. This peripheral market is located just before the checkpoint where taxes are paid. In the city itself the price is 10 000 TShs.

A strong increase in dealers' profits has been recorded as a result of the ban on charcoal imposed by the Tanzanian government in February 2006. The induced scarcity and the consequent unsatisfied demand raised the price of charcoal. However, while the price at which dealers and retailers sold, the little charcoal entering the cities rose sharply, the price at the production sites remained stable, since it was difficult for the producers to find buyers directly. After the ban was lifted, trading went back to normal,

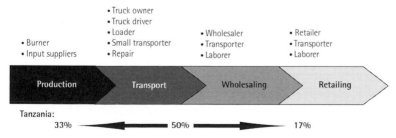

Figure 8.3 Distribution of profits along the charcoal value chain.
Source: Own calculations combined with calculations from the World Bank (2009).

but the sudden and abundant demand caused a price increase at the production sites from 3000–6000 TShs.

The highest percentage of charcoal enters Tanzania's main cities by truck. The predominantly old vehicles cause excessive emissions of CO, CO_2, NO_x and SO_2. A common explanation for using old vehicles in charcoal transportation is that the charcoal trade does not generate enough profit to pay for the upkeep and purchase of new vehicles (CHAPOSA 2002). Another issue is the dumping of charcoal dust around charcoal depots, increasing the risk of water-source pollution.

8.3.2 Legal and institutional domain

Because the Forest Ordinance does not directly impact charcoal production, the only way the government can exercise power and retrieve revenues from the charcoal business is through the tax system. Taxes are charged to traders and paid at official checkpoints. In 2005 the average tax in the Dar es Salaam region was 1000 TShs per bag, of which about 700 TShs was a central government levy while the remainder was a district levy. In 2006 the average tax was about 1600 TShs (World Bank 2009).

Tax evasion in the trade segment of the charcoal chain is a common practice. Dealers claim that in dry seasons levies can wipe out their profit entirely because of the higher prices at the production sites. Many traders take advantage of the differences in tax levels among districts paying the taxes in the cheapest districts, even if they are further from the production sites.

Monitoring is inefficient at best, causing a large loss in potential revenues. In Dar es Salaam it is estimated that only 10–20% of charcoal entering the city is recorded (Napendaeli 2004). Military loads are also deliberately unrecorded (CHAPOSA 2002). The national forest policy establishes that extraction for household use does not require a license and, consequently, no tax has to be paid on forest products obtained from this activity.

The lost revenue of the Forest and Beekeeping Division (CFBD) of Tanzania is colloquial. As shown in Table 8.2, the commercial sector and households combined, annually consume an estimated 530 million kilograms of charcoal or 17.6 million *gunia* (30 kg bag) (World Bank 2009). Each bag entering the city is taxed 1000 TShs (US$ 1) in theory leading to an annual revenue of 17.6 billion TShs (US$ 17.6 million). In 2004 the Dar es Salaam City Commission recorded the total number of bags entering the city annually at 5.6 million gunia, with corresponding annual revenue of only 5.6 billion TShs (US$ 5.6 million). This equals a loss of 12 billion TShs (US$ 12 million) in potential revenue.

Table 8.2 *Comparison between official and estimated annual flow of charcoal into Dar es Salaam*

	Official records (2004)	CHAPOSA (2002)	PREM estimates (2005)
No. of gunias	5.6 million	15 million	17.6 million
Revenues in TShs	5.6 billion (US$ 5.6 million)	n/a	17.6 billion (US$ 17.6 million)

8.4 Charcoal consumption

Charcoal in Tanzania is consumed almost exclusively in urban and peri-urban areas. In rural areas where charcoal is produced, people normally use firewood as cooking fuel. Consumers buy charcoal from various sources, such as stores, markets, kiosks, trucks and bike sellers. About 85% of the population (6.8 million) in 2009 use charcoal as their primary source of energy (World Bank 2009).

The second largest consumer of charcoal after households is the commercial sector. Nearly all entrepreneurs (99%) interviewed claimed to use charcoal in some capacity, while only 28% used electricity. As a whole they account for about 31% of the end-user demand for charcoal. This translates into a total annual consumption of 165 million kg of charcoal with an average annual expenditure of 33 billion TShs (US$ 33 million).

8.4.1 Economic and socio-cultural domain

The Tanzanian population has grown monotonically from about 10 million in the 1960s, reaching over 34 million in the last census of 2002. The resulting increase in charcoal consumption is shown in Table 8.3. The amount of charcoal reported is based on per capita consumption of charcoal reported in previous studies (see, for instance, TaTEDO 2002a). Demand has risen from almost 4 million bags in 1978 to more than 14 million bags in 2001, an almost fourfold increase. By 2005, we estimated the total consumption to have reached over 17.6 million bags.

Charcoal is deeply ingrained in Tanzanian people's daily life and is used by all income levels. To assess the demand side of the charcoal chain, a household survey was conducted in Dar es Salaam (Palmula and Beaudin 2007). The results show that the relatively low cost of charcoal ranks as the

Table 8.3 *Trends in charcoal demand for urban dwellers in Dar es Salaam*

	1978	1988	1998	2001
Total population	730 000	1 370 000	2 200 000	2 800 000
Number of households	169 767	318 605	511 628	651 163
Number of charcoal bags	3 734 874	7 009 310	11 255 816	14 325 586
Cubic meters consumed	876 000	1 644 000	2 640 000	3 360 000
Growth in demand (%)	NA	87%	61%	27%

Source: National Population Census (1978–2002), charcoal demand figures are computed using the per capita consumption estimated in the study area.

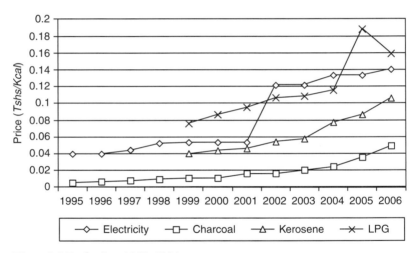

Figure 8.4 Fuel prices 1995–2006

most important reason for usage (indicated by 71% of the households). In addition, the wide availability of charcoal is seen as an important reason (52%) as well as its convenience in use (28%). As opposed to the general belief, only 10% of people use charcoal because of the effect it has on the taste of the food.

Our calculations confirm the belief of the Tanzanian people that charcoal is a cheap energy source. Looking at charcoal price trends, our analysis shows that between 1995 and 2001 the price remained at a constant level, substantially below alternative energy sources. From 2002, the price began to steadily increase, almost doubling by the end of 2006 (see Figure 8.4).

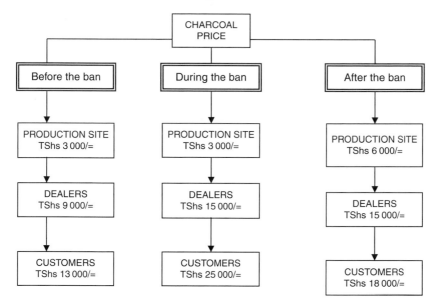

Figure 8.5 Price trends of charcoal before, during and after ban

Two facts explain this increase. First, the increasing scarcity of charcoal. Second, the ban imposed on the transport of charcoal in February 2006 had a major influence on the rising price. During the ban charcoal traders were loath to conduct business. This led to large stockpiles of charcoal at production sites and shortages in the cities. When the ban was lifted, producers began charging more for their product. Figure 8.5 shows prices before, during and after the ban as measured by the study.

8.4.2 Environmental and technical domain

Burning charcoal produces carbon monoxide, CO, with indoor usage resulting in several casualties each year. The most affected are women and children, as they are exposed to elevated concentrations three times a day. Typical CO content is between 0.3% and 0.9%, which is alarming when compared to, for instance, the European norm of 0.005%.

While there is little awareness of the risks associated with indoor charcoal usage, environmental awareness about the impacts of charcoal consumption seems quite strong. The majority of respondents (77%) stated that there are severe negative effects of charcoal burning on the environment. Deforestation was seen as the major problem, followed by drought (Figure 8.6).

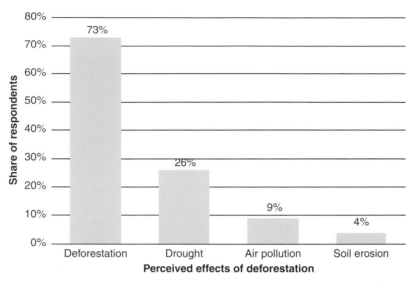

Figure 8.6 Perception of negative environmental effects related to charcoal burning

Types of stoves used to burn charcoal vary considerably in size, shape and design. The majority of household stoves are simple and homemade, are inexpensive and require almost zero maintenance. Because of the inefficiency of such stoves, the improved fuel-efficient stove, requiring less charcoal to produce the same amount of heat, has been introduced and promoted by several NGOs. According to the CHAPOSA report, 51% of the households in Dar es Salaam use conventional charcoal stoves, whereas 41% utilize improved stoves.

8.5 Conclusions and policy recommendations

This high reliance on charcoal makes Tanzanian producers, traders and consumers vulnerable to environmental problems such as deforestation. Increasing the sustainability of the charcoal chain in Tanzania calls for a comprehensive approach that accounts for a multitude of aspects (e.g. technological, economic, social and environmental). At present, the development of such a comprehensive policy is hampered by lack of information about the charcoal chain as well as the limited recognition of policymakers in Tanzania of the interdependencies among the three segments within the charcoal chain. Our calculations show that the total revenue generated by the charcoal industry for Dar es Salaam alone

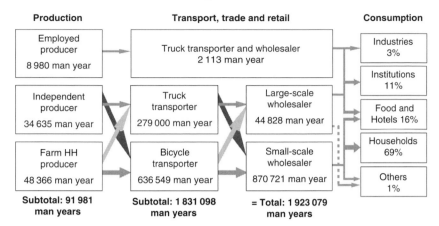

Figure 8.7 Charcoal production, trade and consumption per year in Dar es Salaam.
Source: Own calculations combined with calculations from the World Bank (2009).

amounts to 350 billion TShs (US$ 350 million) and generates employ-
ment for almost 2 million workers (see Figure 8.7). This makes the
charcoal industry one of the largest economic sectors in the Tanzanian
economy, thereby stressing the importance of developing a long-term
vision on the sustainability of the sector.

This study aims to provide a multi-dimensional overview of all three
components of the charcoal sector: production, trade and consumption.
This overview contributes to the development of a comprehensive policy
regarding the role of Tanzania's charcoal energy strategy. In this final
section policy recommendations are formulated for the improvement of
the sustainability of the charcoal chain in Tanzania.

8.5.1 Charcoal is a means of last resort to escape extreme poverty

The survey among charcoal producers reveals a number of new insights
into the characteristics of charcoal makers. First, charcoal forms the most
important source of income to different people involved in the chain
(i.e. full-time producers, subsistence farmers, employed producers).
Second, environmental awareness is high among the charcoal producers,
but their poverty leaves no alternative but to continue the profession of
charcoal making. Charcoal producers are willing to shift to alternative
livelihood options, but these are currently not available or not remunerative enough. Conditions that facilitate a shift away from charcoal include
the improvement of agricultural efficiency and the development of

markets and its accessibility to cash crops. Third, the current kiln efficiency is extremely low. This deficiency enhances the rate of deforestation. Projects supporting the improvement of kiln efficiency would greatly help local communities as well as the environment.

8.5.2 Current policies directed at the charcoal chain are inefficient

The charcoal sector has remained an informal sector in Tanzania. There is no comprehensive government policy aimed at regulating the industry. The regulations and policies that do exist are patchy in nature and are generally poorly enforced. Some typical examples of policy deficiencies include the following:

- While it is illegal to fell trees for charcoal (or for any other purpose) in Tanzania's protected forests, and licenses are required to harvest wood elsewhere, resources for patrolling forests are scant.
- Licenses given out for forest harvesting and the charcoal trade do not compare in any way with the quantities traded to Dar es Salaam. This implies that much of the activity in the charcoal chain is unregistered.
- Although charcoal that is transported to urban centres legally should be taxed, the government loses millions a year as a result of checkpoints being unstaffed.
- The 2006 ban on the transportation of charcoal to stem the flow of charcoal into urban areas and mitigate its use proved to be counter-productive and only served to increase the demand and prices.

The command and control policies dominating the approach of the current Tanzanian government need to be supplemented by market-based approaches. The command and control approach focuses too much on the symptoms of the problem, while market-based instruments are more likely to address the causes of the problem.

8.5.3 Potential improvements can be made across the charcoal chain in Tanzania

Because funds in Tanzania for interventions are limited, it is important to conduct a systematic sector-wide prioritization on the basic cost-effectiveness of the various intervention options. At present, several information gaps exist that need to be addressed before such a prioritization process can occur. Despite these gaps, a number of suggestions are made for future policy interventions. On the supply side, the need for forest resources can be partially met by further expanding production forest, plantations and agroforestry. On the demand side, targeted policies and technological innovations are needed to enhance the attractiveness of substitutes of charcoal. At the chain level, PES provide a promising route for the creation of incentives throughout the charcoal chain to move away from charcoal production to more sustainable forms of livelihood activities.

8.6 Epilogue

Soon after its completion, the study was embraced by the World Bank who extended the market assessment and added a political economy analysis, exploring opportunities and obstacles to designing and implementing reforms that would render the sector more environmentally and socially sustainable (World Bank 2010). The World Bank study took an innovative, inclusive and participatory approach to political economy analysis to collect stakeholders' views on politically and socially sensitive issues that are otherwise not put forward for open discussion. The process of conducting the study also provided a platform for mutual experience sharing, learning and awareness raising of critical (political, economic, social and environmental) issues surrounding charcoal sector reforms. In close consultation with local stakeholders, the World Bank designed an investment plan for a sustainable charcoal programme costing US\$ 40 million, following the supply chain targeting governance (70% of the budget), production (5%), trade (13%) and consumption (12%) (World Bank 2009).

References

Burgess, N. D. and Muir, C. (eds.) (1994). Coastal forests of Eastern Africa: biodiversity and conservation. Proceedings of a workshop held at the University of Dar es Salaam, August 9–11, 1993. Society for Environmental Exploration/Royal Society for the Protection of Birds, UK.

CHAPOSA (2002). Final report CHAPOSA research project-Tanzania. Swedish Environment Institute (SEI), Stockholm.

Napendaeli, S. (2004). Supply-demand chain analysis of charcoal and firewood in Dar es Salaam and Coast Region. Tanzania Traditional Energy Development and Environment Organization (TaTEDO), Dar es Salaam, Tanzania.

Palmula, S. and Beaudin, M. (2007). Greening the charcoal chain – substituting for charcoal as a household cooking fuel in Dar es Salaam. Poverty reduction and environmental management (PREM), IVM Institute for Environmental Studies, VU University Amsterdam, the Netherlands.

Sawe, E. (2004). An overview of charcoal industry in Tanzania – issues and challenges; prepared for the national R&D committee on industry and energy. Tanzania Traditional Energy Development and Environment Organization (TaTEDO), Dar es Salaam, Tanzania.

TaTEDO (2002a). Charcoal industry in Tanzania. Report, Tanzania Traditional Energy Development and Environment Organization (TaTEDO), Dar es Salaam, Tanzania.

TaTEDO (2002b). The true cost of charcoal. Report, Tanzania Traditional Energy Development and Environment Organization (TaTEDO), Dar es Salaam, Tanzania.

World Bank (2009). Environmental crisis or sustainable development opportunity? Transforming the charcoal sector in Tanzania: A Policy Note # 50207. World Bank, Washington DC.

World Bank (2010). Enabling reforms: a stakeholder-based analysis of the political economy of Tanzania's charcoal sector and the poverty and social impacts of proposed reforms. #55140. World Bank, Washington DC.

9 · Payments for environmental services in the protected areas of the Philippines

SEBASTIAAN M. HESS, EUGENIA C.
BENNAGEN, ANABETH INDAB-SAN
GREGORIO, JANET A. R. AMPONIN AND
PIETER J. H. VAN BEUKERING

9.1 Introduction

Few fathers will hesitate to chop down a tree if this helps to pay for the education of their children, even though they know that this might hurt them in the long run. Neither is it much of a choice when existing plots of farmland become unproductive, and a farmer can either let his family go hungry, or create a new plot by burning down a patch of forest. Considering these 'choices', it is no surprise that the forest and farming practices of the poor forest and mountain communities have been the main causes of forest depletion in the Philippines.

The Philippine forests are home to a large, marginalized sector of society composed of both migrant and indigenous dwellers. They constitute about 20 million or 25% of the total population and are generally considered the poorest of the poor. The relationship between poverty and forest degradation is not always the same and there are some poor forest communities that invest considerable time and resources in sustainable forest management practices while trying to meet their basic needs. In many cases, however, these communities' dependence on the forests results in the degradation of the resources they need for their livelihood and survival.

Sixty years ago, more than half of the country's 30 million hectares were covered with forests. There were 13 ha of forest for each Filipino. By the turn of the century, this had dwindled to 0.1 ha per Filipino, due to a rising population and an annual deforestation rate of 100 000 ha.

Nature's Wealth: The Economics of Ecosystem Services and Poverty, ed. P. J. H. van Beukering,
E. Papyrakis, J. Bouma and R. Brouwer. Published by Cambridge University Press,
© Cambridge University Press 2013.

The forest not only provides its poor dwellers with a livelihood, it also provides other important functions: it regulates river flows, stores carbon, provides a habitat for wildlife and offers recreational possibilities. Therefore, it is not only the forest dwellers that are dependent on a well-maintained forest. Downstream there are communities who irrigate their land with the water from the river, and rely on it as a source of drinking water. Further away, the global community needs to reduce carbon dioxide in the atmosphere and wishes to preserve biodiversity, not just because they like to see furry animals and colourful birds but also because the forest is a huge gene bank, holding agricultural crop variants and medicinal plants.

These forest services are currently enjoyed free of charge by the communities outside the forest. This has meant that those living inside the forest were only offered the paltry choices described above. The concept of PES was developed to incorporate the other values of the forest in the decisions of its managers, in this case the poor forest dwellers.

The central principle behind PES is that those who provide environmental services should be compensated for doing so and those who enjoy the services should be made to pay for their provision (Pagiola *et al.* 2005). Ensuring on top of this that the poor benefit from the PES approach is neither simple nor automatic. In fact, earlier prescriptions claimed that poverty reduction should not be a primary objective of PES programmes (Wunder 2005). More recently, however, as lessons are being drawn from ongoing PES programmes, there is an increasing interest and optimism about the potential of these programmes to help improve the plight of the poor resource-dependent communities in developing countries (Grieg-Gran and Bishop 2004, Pagiola *et al.*, 2005).

In this study we describe the potential for a pro-poor PES programme in a study site in northern Luzon, the Peñablanca Protected Landscape and Seascape (PPLS). The study takes a three-pronged approach to this. It looks at (1) science, e.g. the linkages between land use and streamflow; (2) economics, e.g. valuation of downstream benefits; and (3) the institutional environment, e.g. the set of governing rules and policies needed to support PES.

The remainder of this chapter is structured as follows: first, a description of the watershed and its dwellers is provided in Section 9.2. Section 9.3 describes the methods used in the study. Section 9.4 presents the results on the three main aspects mentioned above, and in Section 9.5 we conclude and draw policy recommendations from the findings of this study.

9.2 Study site

9.2.1 Physical description

The Pinacanauan Watershed is part of the newly proclaimed PPLS in the municipality of Peñablanca, Cagayan Province. It is located in northern Luzon to the east of the City of Tuguegarao. The Pinacanauan Watershed area covers 65 099 ha and encompasses 18 villages. In these villages, more than 25 000 people try to make a living. Their main activity is (subsistence) farming.

Based on satellite images from 2002, at least 50% of the watershed is still covered with forests. Generally, the remaining forests in the watershed are left over from intensive commercial logging activities during the 1970s and 1980s. Around 30% of the watershed is covered by grass and brush. Land use is dominated by agriculture covering 18% of the watershed land area.

The majority of upstream forest dwellers are extremely poor, living on less than a dollar a day per capita. On average, the upstream dwellers earn an annual household income of PHP 43 403 (US$ 870). Farming is the main occupation providing about half of this household income. The dependence on farming is even higher for the poorest groups (providing 68% of their income). Depending on the accessibility of the different villages, the subsistence share of production ranges from around 30% to 50%. Other important income sources are wages and remittances from relatives. Wage income is earned from work on farms and seasonal work in construction or carpentry, but also from work in government and private offices.

The poverty of the upstream dwellers also shows itself in the deprivation of assets that enhance human welfare, such as human, financial and physical capital. Education levels are low – on average, the household heads have only completed primary schooling – their health is generally poor, they have no access to official bank credit, and their farmlands are small. Moreover, they often lack property rights to their land. Only one-third of the farmers have private titles to their farm lands, and the rest are caretakers of other people's land, including government land, or squat illegally on government land, although the latter is a minority. For an extensive poverty profile of the upland dwellers see Bennagen et al. (2007).

Even though the upland dwellers should almost all be considered poor, it is not a uniform group. There are significant differences in both income levels and access to assets and opportunities. For instance, poverty incidence is significantly higher and land assets are more unequally distributed in one of the three villages surveyed. These and other differences should be taken into consideration in designing PES interventions in the project site.

9.3 Data and methodology

The three-pronged approach to this study starts with the science behind PES, providing the evidence of the link between land use and environmental services to downstream and global communities. The existing case studies on PES reveal that too little attention has been given to this aspect particularly with respect to PES in watersheds (Pagiola *et al.* 2002). This observation is particularly relevant given the ongoing debate among forestry and watershed specialists about the role of forests in hydrology, with some experts challenging the conventional wisdom that forests protect water supplies and stream flows in all cases (FAO and CIFOR 2005, Hayward 2005).

The second aspect concerns the economics behind PES. The economic basis of PES is public goods theory. The forest services that are enjoyed by downstream beneficiaries are positive externalities of sustainable forest management by the upland forest dwellers. As a market-based instrument, PES aims to internalize these external benefits by capturing their values and to channel these to the upland communities as an incentive to pursue watershed conservation and protection practices.

Finally, implementation of a PES programme will involve institutional reforms, e.g. changes in the existing legal and regulatory framework that may affect various stakeholders with different interests. For instance, it is important to have well-defined property rights to facilitate market creation (Pagiola *et al.* 2002). It is similarly important to examine the role and interests of the different organizations that constitute the actors in PES such as the people's organizations and NGOs, government organizations (at all levels of government), water districts, tour operators, etc., to formulate effective and equitable management interventions.

9.3.1 Science

Land-use and land-management practices of upland farmers affect the annual run-off and seasonal distribution of surface water and ground water recharge. Moreover, these practices and other domestic activities also affect water quality through the amount of erosion, sediment load, nutrients, organic matter and pesticides in the water. Some of these impacts can also be brought about by natural processes, including natural disasters such as earthquakes. The evaluation of land-use and water linkages for purposes of establishing watershed protection services can therefore be an extremely difficult task, and requires long-run data availability.

In the Pinacanauan watershed, the hydrological functions were assessed by analysing the historical hydro-meteorological data coupled with data on land-cover/land-use changes over time. The analysis was supplemented by relevant information elicited by key informant interviews, focus group discussions and community surveys, which were used to fill in data gaps and affirm the results of the analytical processes. Available historical stream-flow and rainfall data for both the dry and wet seasons were analysed to examine any correlation using statistical trending techniques.

The estimation of the sediment yield of the stream-flow was based on the potential surface soil erosion in the watershed. In order to get some indication of the influence of land cover and land use on the quality of the stream flow, the Universal Soil Loss Equation (USLE) was used to estimate the rate of surface soil erosion under different land-cover and land-use types. The USLE predicts the average annual soil loss per unit area as influenced by factors such as rainfall and runoff, slope, steepness, land cover and others.

Besides hydrological services, the study also looked at the potential for carbon uptake in the watershed. Land-cover data were analysed to ascertain whether suitable areas for reforestation exist. Interviews with local stakeholders supplemented the available data on current land use. Because no biomass or carbon storage data were available for the Pinacanauan watershed, a literature search was conducted to make a rough estimate on carbon storage in comparable Philippine forests.

9.3.2 Economics

In this part of the study we determined the most important marketable values of a sustainably managed watershed. The Pinacanauan River watershed supplies water to Peñablanca and to adjacent Tuguegarao City for irrigation and domestic use. These water users benefit from a sustained wet and dry season flow. The watershed also provides recreational benefits to local tourists (e.g. swimming and picnicking) and adventure tourists from Metro Manila (e.g. white-water rafting, kayaking, swimming and bat watching). The famous multi-chambered caves of Callao draw local as well as some foreign tourists. Although the caves are not directly influenced by the functioning of the watershed, the unspoilt forest surrounding the caves are important for the experience, and tourists visiting the caves also partake in other activities in the park.

The hydro-related benefits were valued using the contingent valuation method (CVM). Four sets of CVM surveys were conducted among different groups of beneficiaries to collect socio-economic data and an estimate of their willingness to pay for watershed protection. The beneficiaries include domestic water users, rice farmers with irrigated land and tourists.

The potential for a carbon sequestration project in the PPLS was estimated primarily through literature research. Cost data for Philippine forest plantations were combined with literature on monitoring and verification costs and personal communication with key informants. Costs were calculated per ton of carbon and per hectare per year to verify whether the carbon rights could be generated and sold profitably.

9.3.3 Institutions

The institutional assessment involved various elements. First, the relevant institutional framework for PES was assessed in terms of the legal and regulatory environment, property rights, cooperative mechanisms and the role of government. Second, given the multi-stakeholder nature of PES, a stakeholder analysis was implemented to identify major stakeholders and their interests and roles in a possible payment scheme. Lastly, consultation workshops were conducted with the key stakeholders on the findings of the project and the proposed institutional structure for PES. An in depth poverty survey of the upland villages in the headwaters of the Pinacanauan River was also held to gather data on the situation of the potential service providers.

9.4 Results and discussion

9.4.1 Science

Rainfall in the watershed is more or less evenly distributed throughout the year, but it is relatively dry from December to April and wet during the rest of the year. The area is characterized by hilly to rolling terrain and over half of the watershed has a slope of more than 25%.

Between 1990 and 2002 the average forest cover loss was 167 ha per year, with the rate picking up significantly between 1998 and 2002 to 240 ha per year (see Table 9.1). As forest cover decreased, agricultural land and grasslands expanded. From 1990 to 2002, agricultural areas expanded at an annual rate of 140 ha per year and grassland at a rate of 260 ha. Besides forests, brushland was also converted on a large scale. It is likely that these forests and brushland were either used first for intensive farming until the areas became submarginal or were opened up directly for grazing animals that are commonly held by upland communities.

As expected, the land cover changes have measurable had a effect on the functioning of the watershed. The rapid hydrologic assessment reveals an increasing variability in the mean annual stream-flow between 1950 and 2002. Moreover, the dry season stream-flow shows a declining trend, while the wet season flow follows an increasing trend. The declining dry season flow is generally attributed to insufficient groundwater recharge during the wet season, due to a reduction in the infiltration capacity of the watershed.

Table 9.1 also shows that the estimated sediment yield that was based on the potential soil erosion of the watershed increased from 1990 to 2002. Areas with low potential soil erosion, between 0 and 12 tons $ha^{-1}y^{-1}$, decreased while areas with higher potential soil erosion grew.

To halt a further increase in the volatility of stream-flows and erosion, the current land use trend of conversion from forest to agricultural land should be changed, especially on the steep slopes. Since the population in the protected area is still increasing and existing plots lose their fertility, the productivity of the land has to be improved. This can be done by implementing land conservation measures that also increase water

Table 9.1 *Changes in land cover and sediment yield between 1990 and 2002*

Land use / cover	Area (ha)			Annual change (ha)
	1990	1998	2002	1990–2002
Forest	34 403	33 353	32 394	−167
Brushland	9 879	7 888	7 311	−214
Riverbank	536	595	601	5
Agriculture	9 966	10 979	11 641	140
Water	887	740	608	−23
Grass	9 423	11 538	12 540	260
Total	65 094	65 093	65 095	
Annual Sediment Yield	3 427	4 029	4 658	36%

infiltration, and by the sensible use of fertilizers. A partial shift from maize cultivation to mixed cropping with agroforestry would also reduce erosion and increase infiltration.

Another option would be to reforest parts of the grassland and brushland. This would lead to extra carbon storage, which can be sold as carbon credits. For the first commitment period of the Kyoto Protocol, Certified Emission Reductions (CERs) can only be generated if the area to be reforested has been without forest since at least December 1989. In the Pinacanauan watershed this was the case for 17 321 ha. In order to minimize leakage, we assume reforestation of only 15 000 ha. Data from two other carbon projects in the Philippines (i.e. Mt. Makiling and Leyte Geothermal Reservation) show that between 255 and 393 tons of carbon ha^{-1} can be stored. For the Pinacanauan watershed this would correspond to between 3.8 and 5.9 million tons of carbon on 15 000 ha stored in the project's lifespan. Comparison with other projects worldwide shows that the lower range of this estimate is more realistic. It is therefore conservatively assumed that an amount of 4 million tons of carbon can be sequestered over 50 years in a potential project in Peñablanca.

While the linkage between deforestation and the water regime is fairly clear, the effect that reforesting would have on dry and wet season streamflow is less certain. In the short run total water flows will diminish as growing trees use more water than grass and forests and have a higher evapotranspiration. Erosion should decrease and in the longer run the infiltration capacity of the soil could increase, leading to more groundwater recharge and higher dry season flows.

9.4.2 Economics

The main benefits from sustainable management of the Pinacanauan watershed are hydro-related. Four groups of beneficiaries were targeted to establish the values they attach to watershed protection and a regular flow of the river. The first two groups are domestic water users in Peñablanca and Tuguegarao. The remaining groups are irrigated rice farmers in Peñablanca, and the local and (inter)national tourists.

The CV survey among domestic water users held in Peñablanca in 2004 showed that only 6% of the residents were serviced by the Peñablanca Water District, although the share is much higher in the urbanized downstream villages that are of interest in this study. The 68 households in the survey that had a water connection used around 17 m^3 of water a month and paid PHP 13 per m^3 leading to an average bill of PHP 215. The survey

Table 9.2 *Willingness to pay by different beneficiary groups*

Beneficiary	# Resp.	Share with positive WTP	WTP value (PHP)	Unit	Target population	Aggregated annual WTP (PHP)
Domestic users Peñablanca	68	52%	22.7	month	838 households	118 701
Domestic users Tuguegarao	401	53%	41.51	month	20 816 households	5 495 499
Rice farmers	79	65%	206.8	ha/crop	1038 ha	139 528
Local tourists	51	81%	42	visit	9955 visits	338 669
Total						6 092 397

also showed that the connected households were generally satisfied with the service and did not experience big water supply problems. However, they were concerned about possible future problems resulting from increased variability of the stream-flow in the Pinacanauan River. A small majority (52%) was willing to pay an average of PHP 23 a month on top of their current water bill to improve the management of the watershed, corresponding to around 10% of their monthly bills. As shown in Table 9.2, we end up with a total yearly WTP of PHP 118 701 when we aggregate these results to all 838 connected households in Peñablanca.

The share of connected households is much higher in the city of Tuguagarao (i.e. 80%). A total of 17% of the city's water comes from spring and pumping stations in the Pinacanauan watershed. The water district is very concerned about the sustainability of the water extraction in the watershed, because the city's population is growing and water demand is rising. On average, households in Tuguegarao use around 25 m^3 a month for which they pay almost PHP 280 (i.e. PHP 11 per m^3). Satisfaction with the water service is much lower than in Peñablanca. A total of 57% of our 401 respondents indicated they faced irregular supply or low pressure, and more than 80% were not satisfied with the quality of the water. Not all of these problems can be attributed to the deterioration of the watershed, but they make people more inclined to consider water issues and future problems of water supply. As in Peñablanca, we find that more than half of our respondents are willing to pay an extra amount on top of their water bill for watershed management. For those who did not want to pay, budget restrictions and the belief that the government should take care of the watershed were given as the main reasons. The mean household WTP was PHP 42 a month, almost 15% of the monthly water bill. The main reasons for wanting to pay were the desire to have a reliable water supply, now and in the future, and the wish to protect the watershed. Aggregation over the target population leads to a total annual WTP of PHP 5.5 million.

Good management of the PPLS is also vital for Peñablanca's rice farmers who irrigate their fields with water from the river. The irrigation is managed by the Peñablanca River Irrigation System and it services 15 out of the 24 villages in Peñablanca. Most of these farmers are not well-off either. The average farm size is less than 1 ha, and they are already faced with water supply problems, especially at the end of the irrigation system. However, this has more to do with the management of the irrigation system than with a lack of supply. A total of 65% of the 79 interviewed farmers indicated a positive WTP for watershed protection in order to have adequate and dependable irrigation water. Their stated WTP was PHP 207 ha^{-1} per cropping. Considering there is both a dry and wet season cropping each year, and aggregating over the total irrigated farmland, we find a yearly WTP of almost PHP 140 000.

The final group of beneficiaries studied are the tourists. A distinction is made between local tourists who visit the caves and go swimming in the river, and adventure tourists who participate in organized whitewater rafting or kayaking trips. In total 54 local tourists were interviewed. The contingent valuation survey results show a positive WTP for watershed protection of PHP 42 per person per visit. Aggregating this over the yearly numbers of visitors, we get a total yearly WTP of almost PHP 340 000. The number of adventure tourists (i.e. 14) was too low to run a statistical analysis and their WTP is therefore not reported here, but overall responses from this group were positive and more than half of them were willing to pay to protect the watershed and the river.

As shown in Table 9.2, the aggregation of the different WTP amounts from the four beneficiary groups results in a total annual WTP of around PHP 6 million, or US$ 120 000.

The benefits of the proposed carbon project depend on the price of carbon credits and the costs of setting up such a project. Total costs were assumed to consist of project implementation costs, transaction costs and costs for monitoring and verification. Here we aim at generating CERs under the Kyoto protocol. This means monitoring requirements and costs will be higher than for an 'informal' scheme, but so will the price of the carbon credits. The costs of a Kyoto compliant reforestation project consist of two main categories. The initial costs of the actual replanting and the cost of monitoring and verifying the carbon uptake that continues throughout the project's lifespan. The reforestation costs are based on calculations made for the Philippines by Lasco (2002). Per hectare these costs amount to approximately US$ 500. Assuming a reforestation period of 6 years, and applying these unit costs to the 15 000 ha identified in the Peñablanca region, yields total reforestation costs of US$ 6.6 million (discount rate 4%). The yearly monitoring and verification costs are estimated between US$ 1 and US$ 5 per hectare, or between US$ 15 000 and US$ 75 000 for the whole area. This culminates in US$ 0.3 million to US$ 1.6 million for the 50-year project lifespan. Total costs will therefore lie in the range of US$ 6.9 million to US$ 8.2 million. Assuming the forest matures in 50 years after which there will be no more net carbon uptake, a carbon price of US$ 5.20 would make the project break even on carbon benefits alone.

9.4.3 Institutions

The brief policy review conducted for this study shows that there are several existing legal and regulatory enactments that would support a PES programme in the study site although some realities in the watershed can also pose constraints to PES implementation.

The overriding legislation that governs the Peñablanca Protected Area is the National Integrated Protected Area System (NIPAS) law that placed all protected areas under the system in 1992. The NIPAS provides several opportunities to support PES, such as allowing both development of livelihood activities and land tenure for qualified dwellers in designated areas (i.e. multiple-use zones) in the protected area, setting up a multi-sectoral Protected Area Management Board (PAMB), and enabling the Integrated Protected Area Fund (IPAF) to sustain financing of the protected area. One of the sources of the IPAF are the fees generated from the management of the protected area. Such a fund could serve as a basis for payments to upland dwellers.

Besides NIPAS, a relatively recent executive order (EO 318, 2004) promotes the sustainable management of forests in watersheds as national government policy. The executive order advocates the development of mechanisms for proper valuation and fair and comprehensive pricing of forest products and services. Even more importantly, the policy allows ploughback mechanisms for the proceeds from environmental, watershed and forest services, such as power generation, domestic and irrigation water and ecotourism.

Other recent legislation that supports the establishment and implementation of PES are the Clean Water Act (2004), the Electric Power Industry Reform Act (2001) and the Local Government Code (1990). The Clean Water Act promotes the use of appropriate economic instruments for the protection of the country's water resources, while the Power Act imposes an environmental charge per kilowatt-hour of sales accruing to an environmental fund that will be used for watershed rehabilitation and management. The Local Government Code can be invoked by local government units to collect fees from resource access and use that can be used for watershed protection and management.

The stakeholder analysis identified the most relevant groups that would or could be involved in a PES scheme. In many cases it is convenient to cooperate with existing trustworthy organizations in setting up a PES scheme. These can be organizations that represent the buyers' and sellers' groups and that can act as intermediaries in fund collection, transfers and monitoring.

For the upstream providers there are not many organizations to work with. During the survey held in the upstream villages in 2006 we found only two villages with active farmer cooperative organizations that were accredited by local government. An alternative to these cooperatives might be to work with more informal community organizations if they exist or to mobilize the village councils to serve on behalf of the upstream service providers. However, in some villages the village councils are themselves accused of involvement in illegal timber extraction. If so, they are more likely to hinder rather than facilitate the establishment of a PES scheme. Conservation International, an organization already involved in projects in the upstream area, can also facilitate mobilization of the upstream service providers and link them with the downstream water beneficiaries.

In downstream Peñablanca there are various organizations that can represent their members' interests, such as the irrigators' association for the rice farmers, the adventure tour operator for the tourists and the village councils for the domestic water users. Our consultations with the stakeholders showed there is a preference to let the existing institutions, such as the irrigation authority and the water district, manage the collection of PES funds.

For the oversight of compensation arrangements and monitoring of sustainable practices a new private entity is preferred. Due to the central role of the PAMB in the management of the protected area it was deemed desirable that its staff would be part of the new entity. Due to a lack of trust, the environmental agency was the least preferred institution to perform this task by both the service providers and users. A newly formed private group composed of private citizens from Tuguegarao, the *DANUM ti Umili* Association, which aims to establish and manage a conservation trust fund for the rehabilitation of the PPLS watershed, could potentially form the basis of the new intermediary entity.

The most important conclusion from the in-depth poverty survey of the upstream farmers is that they are not a uniform group. There are both differences between and within villages. The difference in accessibility between villages, for instance, leads to different levels of market orientation for farm products, with more subsistence farming in the more secluded villages. There is also a large difference in the level of poverty within villages, and this difference is apparent in various indicators, such as income, education, health, access to financial and social services and landownership. This has implications for the design of the PES scheme, as these groups have different preferences for compensation and should be approached differently. It also means that it could be difficult to find upland organizations that represent all dwellers equally well.

9.5 Conclusions and policy recommendations

The science component of this study has shown there is a problem of deforestation and unsustainable land management in the PPLS. While the deforestation rates are not staggering, they have increased over the last 15 years, and, with more people still moving into the PPLS, this trend is likely to continue. The people living in the uplands already belong to the poorest groups in the Philippines, and, being dependent on the forest for building materials, fuelwood, fruit and other NTFPs, the further degradation of the forest will make them even worse off.

The study has also made a connection between deforestation and a change in the stream-flow regime of the Pinacanauan River. At the same time, through our economic analysis it has become clear that those who depend on and benefit from a stable stream-flow regime are willing to pay for conservation of the watershed to prevent further degradation and loss of watershed services. Furthermore, there is potential in the watershed for a reforestation project that would be eligible to sell carbon credits under the Kyoto Protocol. While the costs of such a project are significant, it can break even at an approximate price of US$ 5 per sequestered ton of carbon. Such a project offers jobs for the local people, providing an alternative source of income, and, considering current carbon prices, will also make a profit. Together with the funds from the watershed services, this can be used to compensate the upland dwellers to engage in sustainable farming and forest management. In summary, two important premises for a PES scheme to work therefore exist in this area: the link between forest conditions and services, and a demand for these services.

From an institutional side, important requirements for PES are also in place, at least on the level of national policy. PES fits within the legal structure of protected areas, and the laws and regulations actually stimulate the use of financial measures and incentives to manage these areas. Furthermore, land rights, whether they are communal or private, are very important for PES as they facilitate the specification of agreements and control of compliance. The NIPAS law allows for land rights for those people who have lived in the protected area for more than 5 years, however, only in designated, multi-use zones within the areas.

Looking more closely, things are more complicated. The vast majority of upland dwellers do not have secure land rights at this time. Moreover, many settlers in the forest have been there for less than 5 years, and new settlers keep arriving. These people are not eligible to receive land rights even in the designated zones. Dealing with them will not be easy, but these groups cannot be ignored if a PES scheme is to work.

Another challenge is the low presence of people's organizations in the uplands. Such organizations are important since dealing with every upland farmer individually is unpractical, and some village councils are suspected to be involved in the illegal timber trade themselves. Although a large majority of the forest dwellers acknowledge the problem and are aware of the negative impacts of forest degradation on their own livelihoods, they currently have few options other than expanding their agricultural lands and participating in the illegal timber trade to make a living. The strong awareness of the problem and the large willingness to participate in a PES scheme contribute to the feasibility of such a scheme.

Taking all these matters into consideration, we believe a PES scheme is viable in the Pinacanauan watershed, and it should be given priority by the relevant local institutions governing the protected area (DENR and PAMB) as a strategy to halt the ongoing degradation and to improve the fate and prospects of the upland dwellers.

In pursuing this strategy a number of activities have to be undertaken. First, monitoring the influx of new people into the protected area is vital. Second, the property rights of the upland people should be improved. The issue of new entrants should be addressed simultaneously, in consultation with the existing communities. Third, capacity building should take place in the upland villages to support community organizations. Downstream, information campaigns will also help increase understanding of the problem and acceptability of payments for conservation.

Given the poverty situation of the upstream farmers, a PES programme must aim to provide opportunities to increase their income through diversification of livelihoods together with capacity development, such as skills development to engage in off-farm employment. Because of the large diversity in the upstream population, the different groups should be targeted differently, although care should be taken not to create social disturbance by benefiting one group above another. Especially if the group benefiting less is the more powerful, it could undermine the whole scheme.

Furthermore, compensation in cash should also be considered as this seems to be the preferred form of compensation by the upland dwellers. Since the poor upstream service providers are dependent on the forest for part of their subsistence, some livelihood opportunities offered by a PES programme would have to be forest-based. Most of the poorest people reported gathering timber and NTFPs while the number is much lower among the less poor.

An interesting revelation of the poverty survey is the advantage of women over their male counterparts in terms of educational attainment. Considering this, a PES programme should not discriminate against women. Capacity-building and livelihood-development initiatives should include activities that are geared towards women's skills, such as handicraft making, livestock raising and retailing.

9.6 Epilogue

After the presentation of the final results of the study, Conservation International initiated conservation projects within the PPLS in partnership with local NGOs, DENR and the Local Government Unit of Peñablanca. The project's goal is to raise the community's human well-being and to reduce their dependence on the unsustainable use of their natural resources. Interested in the carbon sequestration potential of the PPLS, the Toyota Motor Corporation provided a donation and thereby created an opportunity to expand this initiative started by local stakeholders. For the long-term sustainability of the benefits the project impacts after the initial funding ends, the project will create a reforestation fund. This fund also serves the micro-financing and emergency funding needs of the project participants. The Forest Protection and Awareness

Campaign component is focused on multi-sectoral patrolling activities and information and education activities about the project and its conservation values.

References

Bennagen, E. C., Arcenas, A., Amponin, J. A., Cruz, J. D. d. and Hess, S. M. (2007). An asset-based profile of poor upstream watershed service providers in the Peñablanca Protected Landscape and Seascape. PREM report Philippines.

FAO and CIFOR (2005). *Forest and Floods: Drowning in Fiction or Thriving on Facts?* Vol. 2. Bogor, Indonesia: CIFOR and FAO Regional Office for Asia and the Pacific.

Grieg-Gran, M. and Bishop, J. (2004). How can markets for ecosystem services benefit the poor? In D. Roe (ed.), *The Millennium Development Goals and Conservation: Managing Nature's Wealth for Society's Health.* London: International Institute for Environment and Development, pp. 55–72.

Hayward, B. (2005). *From the Mountain to the Tap: How Land Use and Water Management can Work for the Rural Poor.* Hayle, UK: Natural Resources International.

Lasco, R. D. (2002). Carbon generation through multiple land use activities in the Sierra Madre. Quirino, Philippines. A feasibility analysis for development of a creditable carbon offsets project. Conservation International, Philippines.

Pagiola, S., Arcenas, A. and Platais, G. (2005). Can payments for environmental services help reduce poverty? An exploration of the issues and the evidence to date from Latin America. *World Development*, **33**(2): 237–253.

Pagiola, S., Bishop, J., and Landell-Mills, N. (2002). *Selling Forest Environmental Services: Market-based Mechanisms for Conservation and Development.* London: Earthscan.

Wunder, S. (2005). Payments for environmental services: some nuts and bolts. CIFOR Occasional Paper No. 42, Jakarta, Indonesia.

10 · The copper curse and forest degradation in Zambia

ELISSAIOS PAPYRAKIS, MUYEYE CHAMBWERA, SEBASTIAAN M. HESS AND PIETER J. H. VAN BEUKERING

10.1 Introduction

Amongst the most striking empirical relationships uncovered over the last ten to fifteen years has been the negative causal link between several measures of resource dependence and economic development. Resource-based economic development has not proven to be a great success amongst developing nations; as a matter of fact, poor nations deprived of minerals, fertile land and forests performed better compared to their resource-rich counterparts. Recent empirical evidence and theoretical work provide strong support to a resource curse hypothesis; i.e. natural wealth tends to retard rather than promote economic growth (Auty 1994, Gylfason 2000, 2001a,b, Leite and Weidmann 1999, Papyrakis and Gerlagh 2004, Rodriguez and Sachs 1999, Sachs and Warner 1995 1997, 1999a,b, 2001). The most prominent example is certainly the Organization of the Petroleum Exporting Countries (OPEC) with a disappointing annual growth rate of −1.3% on average between 1965 and 1998 despite the significant injections of petrodollars into their local economies from the oil extractive industries (Gylfason 2001a). The expectations of many early development economists (Nurkse 1953, Rostow 1960, Watkins 1963) that resource endowments could potentially support economic expansion by attracting funds from foreign creditors, channelling the primary sector rents into productive investment and escaping 'poverty traps' proved to be unrealistic. Similarly, any positive linkages between resource abundance and economic prosperity observed during the origins of the industrial revolution in Great Britain,

Nature's Wealth: The Economics of Ecosystem Services and Poverty, ed. P. J. H. van Beukering, E. Papyrakis, J. Bouma and R. Brouwer. Published by Cambridge University Press, © Cambridge University Press 2013.

Germany and the United States or more recently in countries such as Botswana, Norway and Iceland appear to be exceptional cases rather than belonging to a general applicable rule (Sachs and Warner 1995, Wright 1990).

Our analysis attempts to shed some additional light on this paradoxical relationship between resource affluence, economic performance and environmental degradation and we purposely explore the structure of the Copperbelt economy in Zambia in order to obtain additional insights into this issue. Our results can be extended to hold for a number of under-performing African nations of a similar economic structure, dominated by an extensive mining sector, as is the case for Nigeria, the Democratic Republic of the Congo or the Central African Republic. As Figures 10.1 and 10.2 depict, Zambia's economy contracted at an annual rate of 4% for a 21-year period between 1975 and 1996. An average Zambian has almost half the income level he or she relished back in the 1960s. Figures 10.1 and 10.2 also portray the extent of dependence of the Zambian economy on the primary sector and specifically on copper. Zambia's past economic reliance on the resource-based sector (and respective lack of diversification) hints that Zambia's disappointing economic performance over the last four decades is attributed to a large extent to a 'resource-curse' type development failure.

Zambia's copper production has been remarkable in terms of size. It reached above 9% of world production during World War II and above 15% after the 1960s. In 1996 Zambia was among the world's top

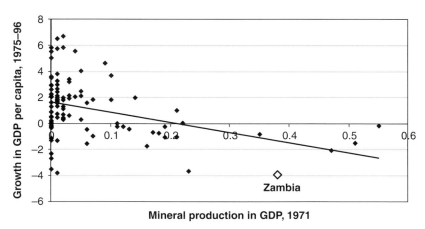

Figure 10.1 Mineral production and economic growth

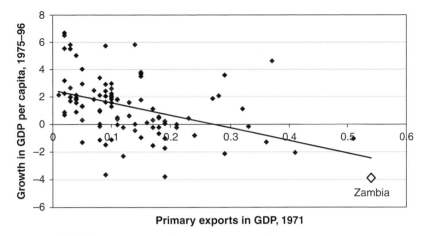

Figure 10.2 Primary exports and economic growth

ten copper producers (Dzioubinski and Chipman 1999). Since the establishment of the first mines back in the 1920s, however, the dependence
of the Zambian economy on the mining sector proved to be troublesome.
As expected, an industry of this magnitude dominated development and
all other industries and sectors depended heavily on the mines. As Elena
Berger (1974) puts it: 'administrators of such territories prayed for the
windfall of a mineral discovery to help finance development or simply pay
for current expenses'. The presence of the extensive mining sector has
been less than harmonious; mine workers often disturbed the production
process, having knowledge of their vital role in the Zambian economy
(Meebelo 1971). Such incidences already started back in the 1930s and
continued in the 1960s and thereafter, as mine workers demanded wage
increases. As Berger (1974) points out, a dangerous pattern of rising wages,
falling output and labour efficiency, soaring labour costs, declining
employment and volatile world markets exerted altogether a substantial
burden on the Zambian economy. In the meantime, while copper price
fluctuations should signal the governments to reduce dependence on
copper, more wasteful resources were pumped into the mining industry
rather than being directed into alternative sectors of production (Roberts
1976, p. 233).

Parallel to the mining sector, an informal sector started to thrive in the
Copperbelt region over the last few decades comprising activities related
to charcoal production. As wood is needed for charcoal production, it is
estimated that close to 50 000 ha of well-stocked forests are cleared

annually for charcoal production around the town of Kitwe alone (Chidumayo 1997). Zambia is the biggest charcoal consumer in the region with more than 41 000 people engaged in charcoal production on a full-time basis and 45 000 workers dealing with its transportation and distribution (Sibande 1996). Undoubtedly, high levels of charcoal production result in excessive rates of deforestation and it is projected that towards the end of the decade Zambia may experience fuelwood deficits (Kalumiana 1997).

This chapter explores the links between mining, charcoal production and economic development. With slumping and volatile copper prices and with the high-grade ores already exploited, large numbers of workers have been laid off from copper mines. In 2002 Anglo American – the global mining giant – pulled out of Zambia due to the crumbling of its mines, amid World Bank advice to the Zambian government to diversify its economy and reduce dependence on copper (*New York Times* 2002). With a lack of other formal economic activities to absorb this excessive unemployment, ex-miners are forced to make their living out of the area's natural resources; namely to engage in the informal charcoal sector, exerting increasing pressure on the local environment. Such an expansion of informal activities (as looting common natural resources) with the parallel contraction of formal taxable sectors constrain the capacity of local authorities to collect public revenues and direct them in investment projects that raise labour productivity.

In Section 10.2 we explore in depth the causalities between mining, charcoal production, and economic development and their economic foundations. In Section 10.3 we move to the description of our database and elaborate on empirical results related to socio-economic characteristics of households in the Copperbelt region. Section 10.4 concludes and draws policy recommendations.

10.2 Data and methodology

10.2.1 A profile of the Copperbelt

The Copperbelt province is situated in the northern part of Zambia between latitudes 12° 20' and 13° 50' south and longitudes 26° 40' and 29° 15' east. It covers approximately 31 014 km^2 which is 4.2% of the total land area of Zambia. The population of the province is 1 657 646 inhabitants and is divided between ten administrative districts; namely Chililabombwe, Chingola, Kitwe, Mufulira, Kalulushi (including

Chambeshi), Ndola, Luanshya, Masaiti, Lufwanyama and Mpongwe. The province is bordered to the north by the Democratic Republic of the Congo, to the west by the North-western Zambian province and to the south by the Central Zambian province.

There are 34 national forests and 12 local forests in the Copperbelt province. The national forests have a total area of approximately 475 397 ha. Local forests occupy a total area of about 44 077 ha. The vegetation covering 90% of the Copperbelt forest areas is single storey, deciduous, closed-canopy woodland known as miombo. Miombo is a mixed wood-land of *Brachystegia*, *Isoberlinia*, *Julbernardia* and *Marquesia* species with certain trends in purity of stand and species combination.

The province's economic activities are dominated by mining of copper, cobalt, silver, gold and precious and semi-precious stones. Copper is the main mineral extracted followed by cobalt. Mineral revenue from this province accounts for about 90% of total revenue for Zambia. As a result of this, most other economic activities are auxiliary to the mining sector, such as mineral refining, engineering and service provision. In parallel, there is an extensive charcoal sector, dominated by informal and often illegal activities. The charcoal market chain is made up of three stages. The production stage is made up of individuals who identify charcoal produc-tion sites and carry out the activity. These are mostly based in the forest. The second stage is made up of traders who source charcoal on site and transport it to urban areas and sell to retailers. Last, retailers interact daily with consumers.

10.2.2 A copper curse

The Copperbelt province currently has the largest share of the poor (18%) in Zambia (IMF 2002). This is higher than the share of any other Zambian province. Furthermore, Zambia as a whole had one of the highest nega-tive rates of economic growth over the last 40 years, making the country one of the famous development failures in the world. Its extensive copper reserves did not guarantee a path of sustained economic growth. This raises immediate questions on what went so wrong; a question that needs to be addressed urgently in order to avoid mistakes of the past and alleviate extreme poverty in the region.

In order to reach interesting insights into the poverty–environment nexus for the Copperbelt and Zambian economy as a whole, we model economic interactions and analyse their causal relations. We construct a Ramsey model where a formal and informal sector coexist; namely mining and charcoal production. The reason for such a choice is that it bears close resemblance to the economic reality that the majority of workers face in the Copperbelt region. The formal sector in the Copperbelt is dominated by the mining industry, mainly copper and cobalt (and to a lesser extent lead and zinc). Informally, though, workers can alternatively derive their livelihood by harvesting the Copperbelt's wood for charcoal production.

Switching between a formal and informal sector has no negative repercussions for welfare levels and economic development as such; workers simply move to the sector that brings higher earnings. In order to assess the negative implications of an extensive informal sector such as charcoal production, we have to go a step further and reflect on the endogenous character of economic development and its determinants; namely attempt to answer what drives economic development and raises living standards over time. The Zambian administration needs funds in order to finance investment that raises productivity levels; namely activities related to improved infrastructure and education. Such public investment is financed by taxes on the formal economy's output, the production of copper. The mining sector in Zambia has performed rather badly over the last three decades with falling copper prices, decreased mining rents, the closure of several mines and a large number of miners laid off from copper mines. A stagnant mining sector – which used to be the previous economic engine of the region – diverted labour to the informal charcoal sector and restricted the government's potential to finance development-promoting activities.

We assume that the local population in the Copperbelt is split between mining activities and informal activities, such as charcoal burning. Mining Y depends positively on the world price for copper P, public infrastructure A and the share $(1 - \gamma)$ of workers N employed in mining; namely $Y = f(P, A, (1 - \gamma)N)$. At the same time, individuals can alternatively direct their work effort towards informal charcoal burning. We follow Elíasson and Turnovsky (2004) assuming that charcoal production Q (and wood harvest) from applying a given work effort is independent of the stock size of the resource base. Elíasson and Turnovsky (2004) assert that this is a plausible assumption for forests where the location of the resource is easily ascertained. Charcoal production depends on the labour allocated to such activities γN (in other words the level of informal labour), as well as the level of public infrastructure A (as a side-effect). In other words, as the government invests more in infrastructure (e.g. road construction or school building) charcoal burners benefit as well. They will also be able to transfer charcoal easier or adopt more sophisticated techniques. Charcoal production also depends on a charcoal-specific productivity measure, captured by B. This can reflect an unanticipated fire that reduces the forest's biomass or an institutional change that facilitates informal activities (e.g. a relaxation of penalization schemes). The price of charcoal produced is normalized to unity for simplicity. Therefore, charcoal-making is given by $Q = q(A, \gamma N, B)$.

10.2.3 Financing public investment

The local administration collects taxes levied on mining income and utilizes all tax payments to finance improvements in labour productivity A; namely towards public investment. An overview of the Zambian economy showed that in 2002, 65% of the country's foreign earnings came from the export of copper and cobalt, even during a period in which copper production had declined by 38% between 1990 and 2000 (FAO 2002). Productivity advancements can be thought of as improvements in labour skills, educational standards, infrastructure or technologies in use. Tax payments improve work-related knowledge, finance research, import technologies or develop reliable transportation systems. Therefore investment in infrastructure A will be equal to public revenues τY, where τ is a constant tax rate; $A(t + 1) - A(t) = \tau Y(t)$.

10.2.4 Labour shifts

A decrease in the world price of copper (or mineral production in general) as captured by P results in a decrease of employment in the mining sector (or consecutively to an increase in charcoal involvement). As pointed out earlier, the mining industry in the Copperbelt has experienced a sharp and prolonged fall in world copper prices since the mid-1970s. A reduced level of copper prices deprives the mines of the potential to sustain the same level of employment. A reduced output price essentially suggests that inputs (in that case formal labour) are valued less by companies in monetary terms. When a mining firm receives reduced revenues for the same level of copper production, this bears an impact on the level of wages and employment. Necessarily, this implies that a number of employees in the mining sector are forced to seek employment in alternative activities in order to derive their livelihood (till wages equalize among sectors). In the Copperbelt case, such a scenario induced a shift of labour towards charcoal burning. In other words, charcoal burning as an income-supporting activity became more profitable in relative terms. We state this finding as a proposition:

Proposition 10.1   The share of employment in mining $(1 - \gamma)$ is decreasing in correspondence to falling copper prices. Subsequently employment shifts to charcoal burning and the informal sector expands.

10.2.5 Economic development

The Copperbelt mining industry is the main supplier of revenues to the Zambian administration. A sophisticated system of taxation (through royalties and export or income taxes) provides a large share of the value of mining production to the government (Harvey 1972). An extensive informal sector obviously deprives the government of such income. Charcoal burning takes place informally, since it is on the whole an unmonitored activity and as such not subject to taxation.

In order to achieve improvements in living standards over time, the local administration supports projects that invest in education, health, telecommunication systems, electrification and other kinds of infrastructure development that enhance labour productivity. Such initiatives involve high set-up costs and benefits that accrue to broad layers of the society. Therefore, private enterprises have a low incentive to undertake such projects, since they do not reap directly such social benefits. Especially in the developing world (where private entrepreneurship plays a minor role) the role of public investment is vital in financing such initiatives through taxation. In that respect, development projects funded through public revenues are decisive in targeting poverty and achieving welfare progress.

Saying that, there is an extensive literature on how governments in resource-abundant countries misuse resource revenues (Baland and Francois 2000, Ross 2001). Although issues of corruption and rent-seeking are not the focal point in our analysis, we recognize that often mismanagement is widespread. When government officials appropriate a proportion of mining revenues or public taxes, the sum of funds available for public investment is reduced.

Investment opportunities or productivity gains $(A(t + 1) - A(t))$ are financed by taxes on the formal economy's output. A decrease in the level of real world copper price P, as experienced till the end of the 1990s, limited the administration's capacity to obtain public revenues. This is an immediate consequence of the contraction of the formal sector and the successive expansion of informal charcoal burning due to labour shifts between economic activities. A direct reduction of copper production (e.g. due to depletion of reserves) obviously has an equivalent effect. We state this as our second proposition:

Proposition 10.2   Improvements in economic productivity (and economic development) are dependent on the level of copper prices. Reduced public revenues (in parallel with an expansion of non-taxable informal charcoal burning) restrain the capacity of the government to direct funds into projects that reach the productive base of the economy.

So far, we have mainly focused on the impact of copper prices on productivity growth and therefore treated the issue of economic expansion mainly through the perspective of mining. A change in the magnitude of the charcoal-specific technological parameter B may also have repercussions on economic development. An exogenous increase in the constant B facilitates the shift of labour towards charcoal burning, since it allows workers to achieve a higher income in the informal sector for the same level of effort. For instance, an increase in B may indicate reduced police patrolling in the area, a higher tolerance of local authorities in illegal deforestation, adoption of more advanced tree-cutting equipment, easier access to the forest or a higher demand for charcoal.

10.2.6 Environmental externalities

Informal production, namely deforestation for charcoal burning, is likely to generate negative externalities on the environment and possibly on other economic activities. Unsustainable timber harvesting is prone to upset ecological functions and reduce economic benefits related to them (see van Beukering *et al.* 2003 for an extensive discussion on the issue). The degradation of forests has potentially a negative impact on a number of environmental services such as water availability, flood and drought control, carbon sequestration and soil erosion. For instance, forest degradation is likely to upset the hydrological functions of relevant protected watersheds and deregulate the water supply system. Such irregularities in water supply will undoubtedly impose costs on water-dependent economic activities such as farming and hydroelectricity power generation. Economic sectors may closely depend on the environmental services of the forest. For instance, hydropower generation and sugar plantations in the Mazabuka region are largely supported by the Ithezhi-Thezhi Dam. The functioning of the dam strongly depends on the watershed services of the Kafue river (whose headwaters lie within the Copperbelt province), which are regulated to a large extent by local forest cover.

We keep in mind, though, that mineral production also has potentially damaging consequences for several ecological functions. This has already been widely observed in many parts of the world (Navine 1978) and to some extent in the Copperbelt region (Republic of Zambia 1985). Since mining, though, is a formal economic activity, it is relatively easier to undertake a thorough assessment of mining-caused environmental problems, monitor environmental degradation and adopt measures to meet a set of environmental standards. In the case of deforestation and charcoal burning, we deal with an informal activity, which is mostly unmonitored and non-controlled. In that respect, although our analysis' focal point lies in the negative externalities of deforestation, we keep in mind that mining as an economic activity is not only a substantial income generator in the economy but also an important polluter.

10.3 Results and discussion

This section provides some tentative findings on the determinants of household wealth in the Copperbelt area based on household data collected in 2005. Although historical data would be needed to determine a change in wealth patterns, the purpose of this exercise is to obtain some first insights into the complementarity between mining and charcoal as income-earning activities, as well as their relative contribution to welfare levels. For that purpose, a questionnaire format based on the Livings Standards Measurement Study (LSMS) questionnaires of the World Bank was developed in order to capture variation in several socio-economic characteristics of households in the Copperbelt province. The purpose was to deepen our knowledge of the Copperbelt economy and examine socio-economic linkages between welfare, mining and charcoal production. In total, there were 400 households interviewed in four regions; 300 in Lufwanyama and the rest equally spread between Masaiti and Kitwe.

To estimate income levels across households, data were collected for a series of agricultural products, forest products, livestock, charcoal production, household enterprises, mining income and income from other sources. Agricultural production included maize, millet, sorghum, cassava, Irish and sweet potatoes, soy beans, plain beans, groundnuts, tomatoes, cabbage, pumpkins, bananas, mangos, lemons, rape, impwa, carrots and okra. The value of forest production included mushrooms, forest fruit, local medicines, thatching grass, roots, honey, caterpillars, firewood and fibres. The main livestock kept in the Copperbelt comprised chickens and goats, followed by pigs and ducks. In order to estimate the value of charcoal sold across households, we used the assumption that a 50 kg bag of charcoal is sold at ZMK 5000 at production site, fetches ZMK 7000 at the roadside, and around ZMK 17 500 (US\$ 5.30) at an urban market. Household enterprises consisted of a wide range of businesses such as carpentry, knitting, milling, baking and beer brewing. Mining income was calculated according to the wages received by the mining employee, while income from other sources mainly consisted of family transfers and employment elsewhere. As can be seen in Figure 10.3, there is a wide distribution of income levels amongst households, with the majority of the households belonging to the lower end of the distribution.

The first column of Table 10.1 focuses on determinants of income differentials across households. It is needless to say that results have to be treated with caution, as the data were collected at a single point in time by visits of interviewers, and therefore there is much room for endogeneity and reverse causality. Inevitably, we also focused on income levels rather than income growth, as the latter would have needed a visit at another instant in time. We need to make explicit that column 1 is meant to generally explore potential reasons that clarify differences in living standards among households rather than prove any causal mechanisms between

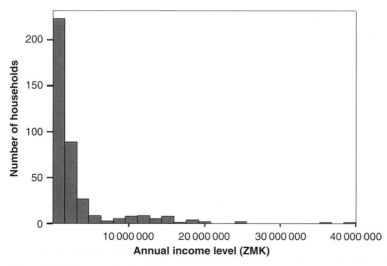

Figure 10.3 Income distribution

Table 10.1 *Income and charcoal regressions*

Dependent variable	Ln Y (1)	Charcoal (2)
Constant	14.05	0.26
Schooling (1.22)	0.06	−0.05
	(0.05)	(0.04)
Health (0.41)	−0.22	0.01
	(0.14)	(0.01)
Mining (0.25)	3.44★★★	0.04★★
	(0.32)	(0.02)
Past mining (0.18)	0.44	0.01
	(0.36)	(0.30)
Migration (0.57)	0.36★★★	0.02★
	(0.14)	(0.01)
Livestock (0.21)	0.55	
	(0.10)	
Enterprises (0.21)	1.33★★★	
	(0.31)	
Forest products (0.30)	−1.09★★★	
	(0.22)	
Other income (0.19)	1.85★★★	
	(0.33)	
Charcoal (0.09)	−2.78★★★	
	(0.66)	
Chief 1 (0.36)	−0.46★★	−0.02
	(0.17)	(0.01)
Chief 2 (0.43)	−1.31★★★	−0.03
	(0.19)	(0.02)
Ln Y (1.44)		−0.02★★★
		(0.01)
Investment (0.06)		−0.18★★★
		(0.07)
R^2 adjusted	0.45	0.06
N	386	386

Note: Standard deviations for independent variables in parentheses, based on the sample $N = 386$ of regression (1); robust standard errors for coefficients in parentheses. Superscripts ★, ★★, ★★★ correspond to a 10, 5 and 1% level of significance.

mining, charcoal production, taxation and welfare. Nevertheless, we notice that, interestingly, income levels (Ln Y: the natural logarithm of household income) are significantly higher among households earning income from the mining sector (Mining: share of mining wages in total income), running their own enterprises (Enterprises: income from enterprises in total income), and receiving income from other sources (Other income: income from other sources to total income). Migrants from other areas are also earning more (Migration: dummy variable taking the value 1 when household members come from a different town or village), while charcoal production (Charcoal: charcoal income to total income) does not seem to augment income levels. This brings some further insights. If charcoal production is harmful to economic development for the Copperbelt province as a whole (by reducing taxable income from formal activities), it seems to be detrimental to welfare levels also at a household level. The dummy variables Chief 1 and Chief 2 (having, respectively, Chimbuchinga or Mukutuma as chief in one's village) are negative and significant. The rest of the variables remain insignificant (Schooling: level of education from 1 to 6; Health: dummy variable capturing visit to a clinic during the last year; Past mining: dummy variable capturing previous employment in mining; and Livestock: the value of livestock in total income).

Column 2 of Table 10.1 is more relevant for the analysis on mining, charcoal production and the Copperbelt economy. It is interesting to notice that mining and past mining are positively correlated to charcoal production. Miners and ex-miners seem indeed to complement their income levels by engaging in charcoal burning. It also seems that wealthy households do not need to engage in charcoal burning (although it is a matter of how to interpret the causality). Investment (in terms of the value of tools and machinery owned in total income) is also negatively related to charcoal production, reflecting the fact that most tools are meant to enhance agricultural productivity (making thus charcoal income less desirable).

10.4 Conclusions and policy recommendations

10.4.1 Main lessons

The Zambian Poverty Reduction Strategy Paper (IMF 2002) explains how a series of national surveys (the Social Dimensions of Adjustment Priority Surveys) show an increase in overall poverty in Zambia during the 1990s. The majority of people suffer from weak purchasing power, homelessness and insufficient access to basic necessities such as education, health, food and clean water. There is a widespread lack of access to income-generating activities and employment opportunities. With an overall poverty rate close to 80% by the late 1990s, a life expectancy falling from 45 to 40 years just between 1996 and 1998 and a quarter of the population with no access to education, Zambia remains one of the developing countries where poverty alleviation is urgently needed.

We explore the interlinkages between mining, charcoal production and economic development, in order to explain the disappointing economic performance in the Copperbelt. A stagnated mining industry such as the Zambian copper sector in the past three decades is likely to have negative repercussions extending beyond mining that influence the livelihoods of a broad array of societal layers. A mining industry in decline, irrespective of whether this originates from slumping world prices or increased operational costs, offloads workers to the rest of the economy. In developing countries, such as Zambia, a lack of alternative formal economic activities implies that such labour surpluses are (at least partially) absorbed into informal sectors involving looting and exploiting unprotected common resources. An immediate consequence of such labour shifts is the degradation of environmental services in the surrounding areas, as well as a constraint to accumulate formal public revenues that can be reinvested into improving labour productivity.

So far, the production of copper and charcoal dominates the local Copperbelt economy, providing few alternatives for employment. A plan to create more opportunities for the establishment of local enterprises and livestock farms could help diversify the economy and offer increased employment security. As long as the copper industry dominates the formal economy, any contraction of the mining sector will reduce public revenues and investment. Increased police patrolling in the Copperbelt province and a lower tolerance of local authorities in illegal deforestation could also discourage charcoal burning, and force households to seek formal activities as a means to alleviate poverty.

10.4.2 Policy recommendations

The poverty–environment link in the Copperbelt exists in a way that degrades the environment that supports local livelihoods as well as national interests. This is largely driven by the shift of labour from mining to charcoal production to supplement household incomes (Proposition 10.1). Since charcoal is the next best alternative to mining, it is recommended that revenues from mining should be used to diversify human skills to enable individuals to earn sustainable livelihoods outside charcoal. The promotion of business enterprises based on sustainable use of local resources and value addition to NTFP is another option for diversifying Copperbelt livelihoods away from charcoal.

The charcoal business is driven by high urban demand for this source of energy amongst a range of several possible options. Demand-side management may include the promotion of alternative sources of energy such as briquets, solar and gas, and making those and the requisite appliances readily accessible to urban energy consumers. Supply-driven strategies for managing charcoal production will complement such demand-driven initiatives. In order to regulate the supply side of charcoal (and thus control the extent of deforestation), a stricter allocation of charcoal licences and adoption of a charcoal tax (as considered in Tanzania and Uganda) would enable the local authorities to collect revenues and reinvest them. Under such an initiative, the profits for charcoal burners and the incentive to engage in charcoal production would decrease.

Corporate responsibility needs to be built into the obligations of mining companies in order to ensure that they contribute to the diversification of skills of Copperbelt residents through education and training. Public pressure from interest groups, new mining codes and relevant guidelines by multilateral financial institutions already enforce several mining companies to meet corporate social responsibility targets that divert resources from resource rents towards environmental protection, educational development and anti-poverty measures (for an excellent overview, see Yakovleva 2005). This will enable inhabitants (not necessarily the mine workers alone, but also their dependants) to participate in other sectors of the economy, thereby, diversifying household income sources.

Zambia should not view copper as the only mainstay of the economy, but as a stepping stone into broad-based development. Revenues from copper, as long as the mines are operational, should be used to develop infrastructure that supports alternative sectors of the economy. This was the successful development strategy of Botswana, which used its diamond rents to diversify the economy and now exports textiles, meat and a range of agricultural commodities to other African countries.

References

Auty, R. M. (1994). Industrial policy reform in six large newly industrializing countries: the resource curse thesis. *World Development*, **22**: 11–26.

Baland, J.–M. and Francois, P. (2000). Rent-seeking and resource booms. *Journal of Development Economics*, **61**: 527–542.

Berger, E. L. (1974). *Labour, Race, and Colonial Rule: The Copperbelt from 1924 to Independence*. Oxford: Clarendon Press.

Chidumayo, E. N. (1997). *Management of Miombo Woodlands in Zambia Copperbelt*. Ndola, Zambia: Provincial Forestry Action Programme.

Dzioubinski, O. and Chipman, R. (1999). Trends in consumption and production: selected minerals. DESA Discussion Paper No. 5, United Nations.

Elíasson, L. and Turnovsky, S. J. (2004). Renewable resources in an endogenously growing economy: balanced growth and transitional dynamics. *Journal of Environmental Economics and Management*, **48**: 1018–1049.

FAO (2002). Zambia. Special Report, Food and Agriculture Organization of the United Nations, Economic and Social Department, June 2002.

Gylfason, T. (2000). Resources, agriculture, and economic growth in economies in transition. *Kyklos*, **53**: 545–580.

Gylfason, T. (2001a). Natural resources, education, and economic development. *European Economic Review*, **45**: 847–859.

Gylfason, T. (2001b). Nature, power and growth. *Scottish Journal of Political Economy*, **48**: 558–588.

Harvey, C. (1972). Tax reform in the mining industry. In M. Bostock and C. Harvey (eds.), *Economic Independence and Zambian Copper: A Case Study of Foreign Investment*. New York: Praeger Publishers.

IMF (2002). Poverty Reduction Strategy Paper: Zambia. International Monetary Fund, Washington DC.

Kalumiana, M. (1997). *Demand and Supply of Firewood and Charcoal, Copperbelt Province*. Ndola, Zambia: Provincial Forestry Action Programme.

Leite, C. and Weidmann, J. (1999). Does Mother Nature corrupt? Natural resources, corruption and economic growth. IMF Working Paper No 99/85. International Monetary Fund, Washington DC.

Meebelo, H. S. (1971). *Reaction to Colonialism: A Prelude to the Politics of Independence in Northern Zambia 1893–1939*. Manchester, UK: Manchester University Press.

Navine, T. R. (1978). *Copper Mining and Management*. Tuscon, AZ: University of Arizona Press.

New York Times (2002). World briefing, Africa. Zambia: Too many eggs in copper basket. *New York Times*, 14 February.

Nurkse, R. (1953). *Problems of Capital Formation in Underdeveloped Countries*. New York, NY: Cambridge University Press.

Papyrakis, E. and Gerlagh, R. (2004). The resource curse hypothesis and its transmission channels. *Journal of Comparative Economics*, **32**: 181–193.

Republic of Zambia (1985). *National Conservation Strategy*. Lusaka, Zambia: Government Printer,.

Roberts, A. (1976). *A History of Zambia*. London: Africana Publishing.

Rodriguez, F. and Sachs, J. D. (1999). Why do resource-abundant economies grow more slowly? *Journal of Economic Growth*, **4**: 277–303.

Ross, M. L (2001). Extractive sectors and the poor. Oxfam America Report, Boston, MA.

Rostow, W. W. (1960). *The Stages of Economic Growth: A Non-Communist Manifesto*. Cambridge: Cambridge University Press.

Sachs, J. D. and Warner, A. M. (1995). Natural resource abundance and economic growth. NBER Working Paper No 5398. National Bureau of Economic Research, Cambridge, MA.

Sachs, J. D. and Warner, A. M. (1997). Fundamental sources of long-run growth. *American Economic Review*, **87**: 184–188.

Sachs, J. D. and Warner, A. M. (1999a). The big push, natural resource booms and growth. *Journal of Development Economics*, **59**: 43–76.

Sachs, J. D., and Warner, A. M. (1999b). Natural resource intensity and economic growth. In J. Mayer, B. Chambers and F. Ayisha (eds.), *Development Policies in Natural Resource Economics*. Cheltenham, UK: Edward Elgar Publishing Ltd.

Sachs, J. D. and, Warner, A. M. (2001). Natural resources and economic development: the curse of natural resources. *European Economic Review*, **45**: 827–838.

Sibande, S. (1996). *Forest Resource Management*. Lusaka, Zambia: Zambia Forestry Action Programme.

van Beukering, P. J. H., Cesar, H. S. J. and Janssen, M. A. (2003). Economic valuation of the Leuser National park on Sumatra, Indonesia. *Ecological Economics*, **44**: 43–62.

Watkins, M. H. (1963). A staple theory of economic growth. *Canadian Journal of Economics and Political Science*, **29**: 142–158.

Wright, G. (1990). The origins of American industrial success, 1789–1940. *American Economic Review*, **80**: 651–668.

Yakovleva, N. (2005). *Corporate Social Responsibility in the Mining Industries*. Hampshire, UK: Ashgate.

11 · *Institutions and forest management in the Swat region of Pakistan*

GIDEON KRUSEMAN AND
LORENZO PELLEGRINI

11.1 Introduction

Deforestation in the north-western part of Pakistan is a long standing problem. Forest cover was 5% in 1996 compared to 20–25% in 1850 (Sungi 1996). In the district of Swat where the research was focused there is a gap (regrowth minus timbercuttings) of over 300 000 m^3 for fuelwood alone (Sungi 1996). Historical accounts tell us that forests have always been an important source of income within the livelihood strategies of local populations (Barth 1985, Sultan-i-Rome 2005). Where in the past the vast expanses of forests could support these demands, high levels of population growth (doubling of the population every 15–20 years) coupled by deforestation have drastically altered the picture. Today, the local population faces ever-decreasing forest stocks with increasing threats to both the local communities (landslides, erosion) and downstream beneficiaries of the watersheds as well as increasing threats to biodiversity.

In a broader perspective, Angelsen and Wunder (2003) distinguish five dimensions in which poor people benefit from forests and forest products: the type of beneficiaries, types of forest products and services provided, the role of forest benefits within the households strategy (subsistence versus commercial use), type of natural resource management (ranging from pure natural forest extraction to (re)planted forests) and high or low return products.

Management of forest resources, especially sustainable management thereof, seems to be suffering from severe shortcomings. In Pakistan the

Nature's Wealth: The Economics of Ecosystem Services and Poverty, ed. P. J. H. van Beukering, E. Papyrakis, J. Bouma and R. Brouwer. Published by Cambridge University Press, © Cambridge University Press 2013.

management of forest resources is generally considered the exclusive domain of the Forestry Department, especially since the state is the formal owner of all forest resources. However, the actual situation reveals an amalgam of perceived property rights concerning the forest and its resources, and a situation where institutions are ill-defined.

Travelling through Swat we see lush green hills and mountains with snow-capped peaks in the distance. The impression is of paradise. On second sight we notice that most of the lush green areas are not the virgin forests that once covered the hillsides in a not so distant past. There are only isolated spots of forest left. On the road there is a steady stream of trucks, pick-up trucks, donkeys and human beings transporting timber and fuelwood from the forests to the lower, more densely populated areas.

A large landowner living in the valley indicates that he has attempted to manage the forested areas of his land holding by devolving the management to a local community living there and paying them in forestry extraction rights, but to no avail, 'the smugglers are stronger and the local communities are not dependable'. In one such community on the forest edge an impoverished villager indicates that for them it is cut trees or die. Forest royalties from official lumber extraction pass them by – 60% is for the large landowners in the valley and 40% for the Forestry Department. Protecting the forest against local people is feasible but it is impossible to protect the forest from influential outsiders in collusion with the Forestry Department.

The Forestry Department has the sole mandate for the management of the forests, which cannot be left to the local population as indicated by senior Forestry Department officials because they are the root cause of the deforestation. The Forestry Department is undergoing a reform process including the devolution of control to lower administrative levels. The problem is that the Forestry Department is under-funded and under-staffed. Donors have pulled out of the reform process after the 9/11 terrorist attacks. A donor representative on the other hand put forward that a lack of real commitment to the reform process was the reason for the drying up of funds from the donor community. NGOs indicate that the people's perspective is lacking in forest affairs. A lawyer in Swat adds that the judiciary plays havoc in the process because there are no possibilities of redress and corruption is rampant.

Devolution is fine, a manager of joint forestry management projects notes, but there are political fetters: policy and practice are not in line. A historian adds that social institutions have eroded since the accession of Swat in the Pakistan state in 1969, 'Finding no legal redress and no redress in social orders leads to illicit use of natural resources that has become so widespread that it is now the norm.'

The Forestry Department as formal managers of the forest resources has been undergoing a long reform process aimed at improving its performance. The effects of this reform process have not resulted in less deforestation. From the policy perspective this has been leading to stated intentions to further reform the Forestry Department; the question is whether organizational reform is the answer. We think there are more limiting bottlenecks to sustainable forest management in Pakistan.

Our research question is therefore to determine the main bottlenecks to halting the current deforestation trends in north-west Pakistan taking an institutional economics perspective. We believe that such a perspective gives valuable insight into the mechanisms of resource degradation and provides a background for evaluating policies aimed at sustainable forest management.

In this chapter we explore the mechanisms behind the deforestation and try to uncover mechanisms to reverse the process. Although our conclusions are not very optimistic, we provide a framework for determining the bottlenecks in the management of common resources from the perspective of institutions.

The structure of the chapter is as follows. In Section 11.2 we provide the theoretical background used in the analysis focusing on property rights, and the historical perspective and the issue of transaction costs. In Section 11.3 we highlight the results of our analyses. These analyses are mostly based on qualitative research. The results are related to high transaction costs, corruption, vested interests and the reform process in the Forestry Department. Our results paint a very grim picture of the institutional settings surrounding the management of forest resources. Although the evidence is thin and mostly anecdotal there is some indication that there are possibilities for sustainable forest management despite the odds. This evidence is also provided in Section 11.3. In Section 11.4 we provide some conclusions regarding the institutional aspects of forest management.

11.2 Theory and methodology

In discussions on environmental degradation the issue of property rights plays an important role. The central thesis is that inappropriate property rights regimes impede sustainable management of natural resources. A strong message from the seminal paper 'The tragedy of the commons' by Hardin (1968) has echoed through the literature. The message that common-pool resources cannot be managed in a sustainable manner led the push towards private or state ownership of resources. Actually, what Hardin describes is not common property but lack of any property rights regime. This notion has become common understanding with the publication of seminal papers on this issue (Bromley 1991, Schlager and Ostrom 1992). We distinguish between categorical and concretized rights. Categorical rights are 'typified legal concepts that construct a general relationship of rights and options between categories of persons or groups with respect to categories of resources' (Bromley 1991, p. 299). Examples are constructs such as ownership and inheritance rights. In concretized rights we are dealing with the rights relationship of actual persons or groups and a resource.

The rules that define property rights – requiring, prohibiting or permitting certain actions – are effective when they are generally recognized and respected by the other economic agents. The rules governing property rights can be divided into operational level and collective choice. The former relates to the possibility to access and withdraw from the resource; the latter relates to the possibility to change management rules, exclusion rights and alienation rights (Schlager and Ostrom 1992).

11.2.1 Type and characterization of rights

We follow up on this discussion with a brief presentation of the idea of bundles of property rights, adding onto this idea the notion of bundles of property duties. The concept of bundles of property rights has been very clearly addressed by Schlager and Ostrom (1992). In their paper they distinguish between five *types of rights* that are bundled in a property rights regime, namely:

1. Access rights: the right to enter a defined physical property;
2. Withdrawal rights: the right to obtain the 'products' of a resource, both in terms of goods and (environmental) services;
3. Management rights: the right to regulate the internal pattern of usage and the transformation of the resource;
4. Exclusion rights: the right to determine who will and who will not have access, withdrawal and management rights, and how those rights can and cannot be transferred;
5. Alienation rights: the rights to sell, lease, give away or bequeath any or all of the above.

The first two components are the basic operational level rights. The last one is what is often seen as property rights in a very narrow sense. In the case of common-pool resources the last three can be considered collective choice property rights. This way of presenting property rights has proven very powerful to disentangle the complexity of common property regimes. Applying this theoretical concept to forestry resources provides us with the following picture in the case of north-west Pakistan.

At the lowest level we have access rights, meaning that a person or group is allowed physical access to a resource. The next right is the right of usufruct, implying the right to extract certain or all proceeds from the resource. Most forests can be accessed freely, but the right to withdraw products (timber and NTFPs) is limited by statutory and customary law. We distinguish management rights from usufruct rights. Management rights relate to the long-term management of the resource itself not to the extraction of the proceeds. Often it is impossible to distinguish between the two, since extraction technology and land-management practices are interwoven (even in terms of semantics). However, there are cases in which a third party has a say in the way the land is managed. At present, the management of forests is the exclusive domain of the Forestry Department despite efforts over the past decade to devolve some of the management tasks to local communities. Exclusion rights refer to the possibility of determining who is and who is not allowed access, extraction rights and management rights. According to statutory law, the forests of Pakistan are state owned, hence exclusion and alienation rights are officially in the hands of the Forestry Department. As we will argue in the next section, this current state of affairs is quite different from the historical context of forestry management in north-west Pakistan.

In northern Pakistan two sets of *de jure* institutions exist next to each other. The first is statutory law, the second customary law including the customary means of enforcement as is the case with Pukhtun Society (Barth 1985). An important aspect of the relationship between property rights and the way the land resource is utilized is the time horizon of the property holders. The time horizon depends on the subjective time preference of the rights holders and on the perception of the duration of the rights at stake. We are now moving into the *characterization of the rights*. The rights besides their content are characterized by duration and assurance. Duration of rights is the time horizon over which the rights are defined. This can be a finite period or an indefinite time span. The assurance refers to the perception of how certain it is that the rights holder can ascertain those rights during the period over which the rights are defined.

11.2.2 Institutions

These property rights are embedded in *institutions*. Since the central concept of this chapter is institutions, and especially economic institutions, it is necessary to provide a clear definition of this concept at this point. We define institutions as the formal rules and informal constraints guiding human behaviour and the means to ensure adherence to these rules of the game, in order to reduce uncertainty in the transfer of goods and services between individuals and/or groups (Ahmed 1980, North 1990). Formal rules are usually defined as those rules laid down in statutory law, also known as *de jure* institutions. However, if well-defined customary law exists including the customary means of enforcement as is the case with Pukhtun Society (Ahmed 1980) this too should be considered as *de jure* institutions.

In northern Pakistan two sets of *de jure* institutions (i.e. statutory and customary law) exist next to each other. To some extent the rules these two sets of institutions lay down complement each other, but more often they contradict each other. In the undefined space created by the competing *de jure* institutions, de facto property rights regimes embedded in informal institutions can exist.

Informal de facto institutions also entail culturally accepted norms and modes of behaviour. The institutions themselves shape the process of updating the expectations with regard to the rules of the game according to actual experience. Hence, the natural dynamics of institutions is a slow process of gradual adaptation to changing circumstances. This process follows different speeds. Some institutional arrangements are deeply embedded in culture and tend to change only very slowly, while others evolve more quickly.

The common hypothesis is that institutions matter when talking about poverty reduction and resource conservation (Barrett and Swallow 2006). If there is uncertainty or lack of information about property rights and the way they are enforced, this can give rise to conflicts over those resources that are detrimental for their conservation, (see for instance Amman and Duraiappah 2004, Angelsen and Kaimowitz 1999, Kabubo-Mariara 2004). The key issue is that it does not really matter which rules are adopted by a community or country, rather it does matters how well they are embedded and how well they are enforced (Barrett and Swallow 2006).

Following Williamson (2000) embeddedness refers to the spontaneous, non-calculative informal institutions anchored in customs, traditions, norms and religion. These institutions evolve only very slowly. Slightly quicker is the evolution of the formal institutional environment. These are the formal rules of the game especially related to formal property rights and anchored in policy, bureaucracy and the judiciary system. Whilst the institutional environment tends to remain stable over decades, the play of the game itself is prone to more rapid change. We are referring to the actual governance structures that shape the actual marginal conditions. These marginal conditions shape resource allocation and day-to-day management of common-pool resources.

11.2.3 Transaction costs

Management and controlled extraction of products from natural resources not only requires clearly defined property rights but also some system for monitoring and evaluation of the state of the resource. Even when property rights are well defined, enforcement of the property rights may well entail substantial costs. These costs are part of a larger set of costs commonly known as *transaction costs*.[1] In short, transaction costs include contact, contract and control (North 1990, pp. 28–33).

This entails the cost of measuring the valuable attributes of what is being managed. This may be difficult, because of asymmetric information: resource managers on one side of the market have much better information than those on the other side. Moreover, there are the costs of policing and enforcing agreements. Enforcement poses no problems when it is in the interests of the other party to live up to agreements. But without institutional constraints, self-interested behaviour will exclude complex exchange because of the uncertainty that the other party will find it in his or her interest to live up to the agreement. This problem is particularly relevant for agreements in which there are conflicting interests.

Policing and enforcing agreements (or rules, laws, etc.) may involve substantial costs. Successful institutions can be characterized as those institutions where, on the one hand, the rules and regulations are strongly embedded in the mind-set of the stakeholders to keep the transaction costs linked to the enforcement to a minimum, while on the other hand, the institution offers a locally optimal solution to complex coordination issues that without the institution would lead to insurmountable transaction costs in various fields.

---

[1] A transaction can be defined as something that occurs when a good or service is transferred across a technologically separable interface. One stage of activity ends and another begins (Williamson 1985). Transaction costs entail the transfer of resources, either physically or legally.

The notion that transactions are costly has important theoretical ramifications. It is a reasonable assumption that individuals or organizations have to use time and resources to secure information and have limited ability to process data and formulate appropriate action.[2]

11.2.4 Duties and obligations

The rules and regulations governing the interactions between individuals and/or groups where it concerns property are usually presented in terms of property rights. However, besides the notion of rights, there are also obligations or duties concerning the property. These duties can be placed alongside the rights as the flip side of a coin. This implies that we can now also distinguish the following duties:

1. Management duties: the obligation to regulate the internal pattern of usage and/or the transformation of the resource in such a way that certain preconditions are met, including basic notions about sustainability;
2. Exclusion duties: the obligation to exclude or grant rights to certain groups or individuals under certain circumstances;
3. Alienation duties: the obligation to pass on any of the other rights and/or duties to another party under certain circumstances.

Obviously, with common-pool resources, those exploiting the resource are often individuals or households, while those managing the resource are communities. In principle, each right and/or duty can be attributed to one or more stakeholders. If there are more stakeholders, there are grounds for conflict. If there is conflict there is a need for means to mediate in the conflict, either in a formal (courts) or informal (village councils) sense.

11.2.5 Monitoring and control

There is another issue that plays an important role when more than one stakeholder is concerned. This is the issue of *monitoring and control*. A stakeholder may have certain rights and obligations vis-à-vis other stakeholders, but if there are only limited possibilities to have checks and

---

[2] From a theoretical point of view, the fact that decision-makers have neither perfect knowledge, nor foresight, and may make errors implies that they will always function inefficiently compared to the hypothetical decision-makers in mainstream neoclassical economic theory.

balances in place, the rights and obligations will have little meaning. Extraction rights may be limited to a few selected stakeholders, but if there is no control possible over 'illegal' extraction, those extraction rights are of limited value especially if rights are combined with restricting obligations that do not hold for those who extract outside the framework of the property rights regime.

An important concept in relation to monitoring and control is that of transaction costs. The transaction costs are the costs that have to be incurred to effectuate the monitoring and control of the property-rights regime. Very often failure of property-rights regimes to deliver the desired results regarding common-pool resources can be attributed to high transaction costs for monitoring and control when many stakeholders exist.

Transaction costs are very important in understanding institutional aspects of economic behaviour (Furubotn and Richter 1997) since they relate not only to the issues of monitoring and control, but also to the cost of information in general. When property-rights regimes are unclear, the costs of action and inaction can be substantial.

Especially when poor households in affected rural communities depend critically on the natural resources for their livelihoods, well-defined rules and regulations governing the use of those natural resources, whether de facto or *de jure*, are indispensable. This is a necessary but not sufficient precondition as we have just argued. It is also important to have the means to enforce the rights and obligations related to the property-rights regime and to have a system for monitoring and control with transaction costs that are not prohibitive.

Following the notions put forward by Williamson (1985), we argue that lower transaction costs in complex settings are associated with well-embedded institutions. Well-embedded institutions imply that stakeholders will be less willing to infract the rules, and other stakeholders will be more willing to monitor and control the adherence to the rules. Strong customary law can be very instrumental.

11.3 Results and discussion

11.3.1 History of Swat

History matters for entitlements because it is at the root of the status quo and, as such, it shapes expectations (claims and rights). The history of Swat has a bearing on current land tenure and forest management. Historical developments are at the origin of the right holding of heirs of the most important families allied to the local dictator. In this section we will outline the historical evolution leading to the modern management regime.

Swat has a distinct history in relation to Pakistan and the' rest of the Indian subcontinent. While the rest of the area was colonized by the British crown, Swat succeeded in conserving its autonomy and, once Pakistan became independent in 1947, it did not fully access Pakistan until 1969.

11.3.1.1 Pre-accession

The earlier history of Swat (from the end of the nineteenth century) was marked by the emergence of a political structure centred on an autocrat. The dictator, called the *Wali*, was leading a rebellious country and needed the support of allies from the powerful families of Swat in order to defend the Princely State of Swat (as the state was officially called before the accession to Pakistan). Challenges to the state came from external powers (the British and the neighbouring Kalam state) and from internal ones (other local leaders trying to seize power) (Sultan-i-Rome 2005). As a result, the rulers were under continuous threat and built alliances with wealthy local lords.

The history of the *Wali* is especially important for the management of the forest, because natural resources were valuable assets used for building and keeping alliances, thus their exploitation was crucial to the survival of the regime. The management system compounded harsh punishments to those who illegally encroached the forest, while allowing favouritism and smuggling to the *Wali*'s – and his allies' – benefit. The decision power, with respect to logging and forest exploitation, was centralized and rested in the hands of the *Wali*. For a detailed historical account of the autocratic rule of the *Wali* of Swat, the strictness of the application of the law to the commoners, the importance that forestry had under his rule, and the illegal cutting and favours for his allies (the *Khans*) see Sultan-i-Rome (2005). For an altogether different version of the *Wali's* rule, see his autobiography (Barth 1985).

11.3.1.2 Post-accession

The accession to Pakistan of the Princely State of Swat brought a break-down of the existing institutional setting: the national state, formally, took under its control the management and exploitation of forests. The law provided only for a minute disbursement of royalties (10% of the total net revenues) to be paid to the legal right holders and established some extraction rights for local communities (to meet consumption needs of the residents).

The Pakistani state's management, from the moment it took over the forest sector in Swat, was characterized by a command-and-control approach whose effective enforcement was beyond the state's capability. The post-accession regime implied that most of the forests were declared reserved or protected, i.e. the management regimes strictly constrained exploitation rights.[3] In both systems property and management rested in the hands of the state, through the Forestry Department, but locals and former owners' rights are different under the two management regimes. In the reserve forest, no cutting is allowed and members of the local community have limited rights for extracting dry wood and NTFP. In protected forest, the Forestry Department is in charge of the management plan and at times when there are cuttings, the forest development corporation is in charge of the cutting of marked timber and shares of the net sales revenues are given to right holders (Steimann 2004). Moreover, the 'local quota' of timber is reserved for local residents for their needs and there is an 'emergency quota' that can be used under special circumstances (e.g. to rebuild a house after a fire).

The shares of revenues from wood sales directed to right holders have been increasing over time, from 10% at the time of accession, to 60% nowadays. Those entitled to shares of the royalties of the forest are the landowners (mostly *Khans* families – the former allies of the *Wali*) that reside far away from the forest, or local communities, depending on the property situation of the forest at the time it was taken over by the state. As a result, even today, contention over property rights is a source of endless litigations that are dealt with by Pakistan's corrupt and inefficient judiciary system. The incapability of the state to implement its policies, together with the dismantlement of the previous regime of property rights, gave rise to an open-access regime.

Additionally, the gap between *de jure* and de facto management of the forest created multiple bases for claiming rights on the forest: the statutory law, custom and the de facto regime. In such a situation, some agents chose to refer to competing legal structures to justify their claims

---

[3] A third category of forests is the guzara forests. They are those whose property was left to the communities, even though relevant management responsibilities were taken over by the Forestry Department. Guzara forest is an institutional setup that characterized almost no forest in our study area.

according to their interests and other agents preferred to operate altogether outside of the legal framework (Meinzen-Dick and Pradhan 2002).

The open-access regime on the forest implied that landowners and local
residents were exploiting the forest restrained only by local power relations. Since the Forestry Department started operations in Swat (along
with the rest of the Pakistani public administration), the gap between the
regulatory framework and the actual management practices was not
bridged. Encroachments and illegal cuttings continued. Furthermore,
legal disputes over the cuttings' royalties began and court decisions settling
them were long awaited (at times for decades) and, in many cases, court
decisions are still pending.

With respect to the Forestry Department, we collected information
through focus group discussions and key informant interviews during the
fieldwork. It appeared that corruption widely affects all the operations of
the Forestry Department and of the Forestry Development Corporation
(who is in charge of exploitative operations). Forest guards are paid salaries
that cannot satisfy the needs of families (Steimann 2004) and they are
lacking the support of the other enforcement agencies (which are themselves inefficient and plagued by corruption). The overall enforcement
system seems more geared to make ends meet for forest officials, rather
than protect the forest. Also, the manning of numerous check-posts along
the roads going down the Swat Valley, serve little purpose apart from
extracting bribes in exchange for turning a blind eye to timber smugglers.
This was confirmed by numerous interviews and by our own witnessing
of the ease of movement of truckloads of illegal logs through the valley.

11.3.2 Forest management in Swat

The present situation in Swat can be seen as the result of historic processes
described in the previous section. While the overall goal of the Forestry
Department and the local communities may not differ that much: sustainable management of the forest resources to ensure a stable stream of
revenues and services, the outcome of the strategic actions of the actual
stakeholders is quite the opposite.

Let us start by disentangling the different stakeholders to then relate
them to the different aspects of the current property-rights regimes. The
main stakeholders are: the Forestry Department, the local communities on
the fringe of the forest, large landowners who are not themselves part of
the aforementioned local communities. Besides these main stakeholders
there are also those who have a vested interest in a steady supply of timber

from the forests of North West Frontier province (NWFP), often referred to as the Timber Mafia.

Extraction rights and obligations related to those rights are handled somewhat differently under statutory and customary law. Under *statutory law*, the Forestry Department is the sole entity with extraction and management rights related to timber. The ban on logging that has been in effect over the past decade or so implies a very strong restriction on extraction. Extraction rights are limited to local needs for construction and firewood purposes. The Forestry Department has the duty to pay royalties to the rightful stakeholders in the forest. Under *customary law*, the owner of the forest has the disposal of the extraction rights. He can lease out those rights to individual loggers. Under the rule of the *Wali*, commercial logging did not take place on a large scale, and extraction was mainly for domestic purposes. There was no royalty system.

Over time, overall land tenure systems have changed. The traditional system of rotating landownership amongst large landowners with whole communities as tenants has given way to a mixed system of large land-holdings alongside small-holdings of individuals and communities. With the mixed system regarding agricultural land an unclear situation regarding rights to forestry royalties has arisen. It is unclear to which owners of agricultural land certain forestry lots are related.

11.3.3 High transaction costs of sustainable forestry

Management of the forest resources used to be the domain of the local landowners or their tenants. With state ownership of forest resources, management is the exclusive domain of the Forestry Department. However, monitoring and control of the forests and the activities therein is costly and hence transgression of formal rules is usually not punished.

What we need to keep in mind is that there are a number of conflicting goals and aspirations of different stakeholders vis-à-vis the natural resource base. Where local communities will often regard the forest as a source of revenues especially in the absence of other sources of income, the state will often have a more general perspective by viewing the forest as a continuing source of ecological service provision, in terms of bio-diversity, watershed management, etc. The benefits of these ecological services do not accrue to the local communities but to beneficiaries else-where. Sustainable management of forest resources implies that the con-flicting goals are somehow resolved.

Besides the definition of *clear property rights*, sustainable management of natural resources requires a number of criteria to be met that are currently not met in NWFP. One refers to the realization that the management costs and the benefits of the ecological services are not linked to the same stakeholders. Some sort of *compensation mechanism* should be in place if non-benefiting stakeholders are supposed to provide management inputs (see the growing literature on PES).

If management of the forest resources is not effectuated through working economic instruments, the alternative is a system of command and control. This entails the definition of clear indicators for management and a costly system of *monitoring*. Even under systems of communal management (with and without outside payment) there are costly systems of monitoring necessary to ensure that the flow of ecological services is safeguarded.

The *transaction costs* involved in ensuring sustainable forest management increase with the decrease in felt environmental threats. People facing moderate, credible environmental threats are more likely to be willing to manage the forest resources in a sustainable manner than those who do not face those threats but just exercise those threats. This former group may need small economic incentives to get involved in sustainable forest management, more so than those living in relative harmony with nature. Although paying the latter may be perceived as 'fair', it does not create additional ecological service provision.

A final criterion for sustainable forest management is the presence of a stable *institutional setting*. Transaction costs for monitoring sustainable forest management also increase with disarticulated organizational structures. If the organizational structures at local level are not recognized by formal authorities or not taken seriously, it becomes difficult to set up a system of joint monitoring. Trust is an important element in transactions. With unclear property rights the level of distrust between various stakeholders tends to be high.

11.3.4 Corruption

Corruption plays an important role in illegal logging operations that take place across Swat (Pelligrini 2011: 121). Information on corruption and forestry has been accumulated during the fieldwork. The source of such information is, in most of the cases, group interviews with members of the local communities, with students of history, lawyers and employees of the Forestry Department (of different ranks).

The enforcement agencies (first of all, the Forestry Department) whose official goal is protecting the forest have in many occasions turned into the main culprits. In practice, the main purpose of many agents that are supposed to watch over the management and protection of the forest is to gain the maximum personal advantage from the power that they hold. The extraction of personal gains through the exercise of public power is the very definition of corruption (Transparency International 2004) and in the enforcement agencies of Swat it takes the shape of bribe extraction. Inducements are disbursed throughout the illegal logging process: when logs are felled and when they are transported.

Forest guards patrolling the forest are generally ready to demand a bribe for turning a blind eye on illegal activities.[4] When agreement on illegal payments cannot be made, possibly because of some incorruptible employees of the Forestry Department, confrontations among smugglers and forest guards can escalate into violence. Furthermore, forest officials artificially create occasions for bribes extraction because they create bureaucratic obstacles to members of the local community who have a right to extract wood for self-consumption (e.g. they delay the issue of extraction permits). Indeed, some of the most common complaints of locals is that they are harassed when they exercise their rights and when villagers have been involved in participatory projects they required changes to the procedure for issuing logging permits (Suleri 2002).

Corruption and the related enforcement problem in the forestry sector are well known across the country and are often reported by the media. Reform attempts do recognize the shortcoming of the present management regime and suggest changes to the enforcement mechanisms. These reform attempts are often focused on increasing the likelihood of catching those involved in illegal activities and increasing the penalties ('crime and punishment' approach), or require radical reforms at the state level. The crime and punishment approach focuses on improving the enforcement of existing legal arrangements. The problem with this approach is that it requires that the control agencies are not corrupt, while in Pakistan the judiciary and the police (that is, those who

---

[4] The situation of bribes extraction is not as clear-cut as it could seem at first. Forest guards themselves are not aware of the rules that should be applied if someone is caught breaching the law. In such a confused situation the difference between a bribe and a fine (whose proceeds are used to complement a meagre salary) becomes blurred. Steimann (2004) illustrates different perceptions – within the Forestry Department – of the *de jure* management regime with respect to enforcement.

should coerce forest officials to respect the law and resist the temptation of colluding with illegal loggers) are very corrupted agencies themselves (Transparency International 2002). Thus the controller cannot be relied upon.

National reforms, such as the ones that are requested by international NGOs that focus on fighting corruption, require major changes in the institutional framework and meet the opposition of powerful vested interests. The transition away from a status where corruption is so pervasive needs to be embedded in other interests in order to be successful and the use of textbook rules for achieving such a transition is commonly resulting in failure (Johnston 2005). This is witnessed by the fact that only seldom do countries manage to decrease sensibly the amount of corruption in a short period of time (Kaufmann *et al.* 2005). While radical reforms such as introducing guarantees for free mass media, democratizing the political process and encouraging citizens' scrutiny of politicians are initiatives that deserve encouragement and are valuable objectives (per se, beyond their implications for corruption), their achievement in the short run seems unlikely and improving management in the forestry sector cannot wait for ideal conditions.

Moreover, the successful enforcement of existing property rights, even if possible, would run the risk of cementing historical injustices: above, we highlighted the origins of skewed rights distribution towards large landowners. Historical developments with respect to land holdings are still determining skewed land distribution because land reforms' implementation have failed (see Naqvi *et al.* 1987 and Social Policy and Development Centre (SPDC) 2001). The effective enforcement of rules based on existing rights would deprive members of the local communities of large shares of their income and, ultimately, of their necessary means of survival.

Note that the poor, especially those living in the more marginal areas, heavily depend on resources that would not be available to them if the law was successfully enforced (Khan and Khan 2009). In particular, such decrease in the availability of resources would be problematic if the interpretation of the law and the resolution of disputes on property rights were biased in favour of large landowners. The latter is probable.

Finally, reforms that focus on enforcement end up giving increasing power to forest officials. When confronted with opportunities to extract bribes, these officials are predictably going to use their powers to increase their share of rents. Forest officials would find that the easiest way to extract rents is using their increased powers against those agents that are not in the condition to resist. As a result the most corrupt members of the Forestry Department are most likely going to increase their revenues at the expense of villagers who have difficulty meeting their own basic needs.

11.3.5 Vested interests

Economic agents coordinate themselves to take advantage of profit opportunities offered by the existing institutional setting. Over time, the selection of successful organizations will create wealthy interests who have a stake in the same institutions that allowed them to operate and enrich themselves (North 1990). In the forest sector of Swat the opportunities offered by the open access situation, following the breakdown of earlier institutions that characterized accession to Pakistan, were seized by a constellation of economic agents and gave rise to a new de facto institutional setting. Those who have gained the most from these institutions have large stakes in the status quo and have the resources to defend these interests. In this section we will identify the stakeholders in the management of forests in Swat and highlight the vested interests that need to be tackled in order to achieve institutional reform.

The main actors, with stakes in the forestry sector, are: federal and provincial governments, the Forestry Department, the forest development corporation, local communities (differentiated among themselves mostly because of right-holding issues), NGOs and international donors, who all have some qualified interest in forest conservation. Additionally, there are the interests involved in illegal logging for commercial purposes (the Timber Mafia), since the forest has been put under such strain by overextraction that in the medium run only logging for local consumption would be possible. Thanks to corruption, these agents are able to unduly influence some of the other actors.

The vested interests that will resist most changes from the current situation are those that profit from illegal logging. These include the smugglers, the wood processing industry and the economic agents involved in activities induced by the timber trade. All of these actors depend – to a different degree – on the continuation of unsustainable logs' extraction.

The Forestry Department, though not in its entirety, is among the vested interests opposed to change. The current regime is giving demanding tasks to the Forestry Department (beyond feasible reach), but also granting it political and economic power. Some members of the Forestry Department, at different levels from the top to the bottom, have become accustomed to receiving bribes to complement their meagre salaries. These illegal disbursements depend on the power that the department can wield. Apart from the economic incentives related to bribes it also appears that the department developed its own routines and it is unwilling to give up part of its responsibilities even though it lacks the resources to fulfil the tasks related to them. The broad extent of this attitude was confirmed by more progressive members of the Forestry Department in several meetings.

The provincial government is the recipient of 40% of royalties on commercial logging and, more importantly, members of it are also collecting bribes in relation to illegal extraction, or are linked to the illegal logging industry. The reluctance of the provincial government to give away power and part of the revenues from the forestry sector were – as we will see – some of the main reasons for the failure of previous reform attempts aiming at introducing participatory forestry in Swat.

The current institutional regime is embedded in these interests and successful reform can be achieved only when the reform process is equally embedded in a coalition of potential winners. Furthermore, once institutional change is in order, the resistance to change of vested interests can be diluted by external resources. Such resources could be used for creating a win–win solution for insiders.

11.3.6 Reform attempts

The unsustainability of de facto management practices and the environmental, social and economic shortcomings of current deforestation rates in most of the NWFP is apparent and some steps have been undertaken in order to reform the forestry sector. For the Swat district the most relevant reform effort was the Forest Sector Strategy, whose inception was supported by a loan from the Asian Development Bank (ADB) and by technical assistance to the programme provided by the Dutch Royal Embassy.

The main task of this donor-driven programme was to move the management of the forest from a state of command and control (and the related coercive practices and the social outcome thereof) towards co-management. In the intentions of the donors, the reform would have strengthened the local capabilities, and the emphasis of the operations of the Forestry Department, with respect to local communities, would have moved from punishment towards partnership.

The process is characterized by contradicting steps. On the one hand, local committees[5] have been established in several villages, so that they could be an interlocutor of the Forestry Department and, most importantly, contribute to the village land-use plan. On the other hand, the process seems to have been controlled at the local level by the most influential members of the community (so that the committees over-represent the wealthy) and the Forestry Department has been unwilling to give up part of its powers and enter in a real partnership with the locals. Eventually, due to implementation slowness and especially because of the resistance at the provincial level to undertake the necessary steps to support the reform in due time, the support of donor agencies was with-drawn in 2004. This was a major setback, especially in the few places where the reform had started to yield benefits (see Suleri 2002). Now trust in this type of intervention is lost and it will be even more difficult for new projects to build confidence.

The Forest Ordinance of 2002 confirmed that the reform process was proceeding in contradicting ways. The inception of the ordinance was admirable – geared towards setting co-management and partnership in a stable regulatory framework – but under the insistence of more conservative powers in the Forestry Department the final result was only a compromise. Even Sungi, a Pakistani grass-roots NGO working for the empowerment of the most marginal members of the society, which from the beginning was supportive of the reform process, eventually rejected the Forest Ordinance. Of special importance in determining Sungi's stance were the punitive powers that the ordinance gave to the Forestry Department and that can be used against local communities.

Since the donors' support for the reform process was stopped, the whole process has halted. The latest development was, if needed, additional proof of the resistance that there was to the reform process. The latest initiatives are towards the strengthening of the Forestry Department's control of the forest and – in accordance with the 2002 Forest Ordinance – the forest force has been instituted. Now a 500-man-strong armed force will patrol the forest of the North West Frontier Province. The Forestry Department itself estimates that at least 3000 members costing PKR 60 million (€0.8 million) would be required in order to patrol the forest of the province effectively. It appears that the

---

[5] The committees are the Village Development Committee (composed of men) and the Women Organization (composed of women).

insufficient number of men involved in the force will in the future be used as an excuse for another failure of the 'crime and punishment' approach.

Note that if the existing force was effective much fewer than 500 armed men would be necessary to control extraction activities in the area. Few roads run along the valley and functioning check-posts should be sufficient to stop a large part of the smuggling. If the guards manning the check posts are colluding with the smugglers, is arming the very same guards going to help to halt deforestation? Is having armed guards patrolling the forest, even more removed from possible detection, going to stop illegal cuttings? Such questions should be addressed if the resources devoted to the reform process have to produce the intended environmental, economic and social dividend.

11.3.7 Virtuous examples of forest management

So far we have depicted a very grim image of the forestry sector of northwest Pakistan with rampant deforestation due to unclear property rights and obligations and very few means to enforce them. However, there are a few isolated examples of initiatives that rein in deforestation. These examples do shed a light on the way property-rights regimes at a local level can be used for sustainable forest management.

Local groups of citizens at various times have organized themselves in order to defend some stretches of forest from exploitation. The most effective way that has been found to enforce extraction restrictions was to have limits on the transport of logs: community-manned check posts have been set up and timber loads have been blocked along the roads (Khan et al. 2004, Killeen and Khan 2001). These check posts are operated independently from the Forestry Department's check posts (Khan and Khan 2009), which is a clear indication of the level of mistrust between local communities and the state organizations.

Other efforts to protect the forest involved local communities in concert with the Forestry Department. The ones that were initiated under the donor-sponsored initiatives include community-manned check posts, patrols from the local community guarding the forests and joint patrols of community and Forestry Department members, and have been associated with improvements in forest cover (Steimann 2004, Suleri 2002).

In NWFP a forest sector reform was initiated with the help of the ADB and bilateral foreign donors. As part of this initiative people belonging to the forested areas of NWFP set up a platform, named Sarhad Awami Forestry Ittehad (SAFI) to raise their voices and make the process of reforms more democratic, transparent and people friendly. Although there is a voiced opposition to the way the forestry reform is progressing, which is not promising for joint management by state and local communities, there is at least the beginning of focused organization.

Donor-funded pilot projects of community-managed forests in neighbouring Dir District were quite promising for a while but encountered severe difficulties after funding stopped. Joint forest management seems to have scope for success but depends critically on the willingness to endorse such a process by the Forestry Department (Sungi 1996).

11.4 Conclusions and policy recommendation

Institutional change is costly. Let us examine two types of institutional change. The first type is institutional change that originates from within the confines of the stakeholder group that adheres to the original institutional arrangement. The second is institutional change that is enforced on stakeholders by an 'overlord'. Obviously actual institutional change takes place on a spectrum between the two extremes.

Institutional change that originates from within can be characterized as an adaptation to changing circumstances. The changing circumstances can be autonomous processes, such as population growth (demographic pressure), integration into the outside world (market access), changes in the resource base (agro-ecological potential) and the knowledge base (technological innovation). If the change in circumstances facing the decision-makers is gradual, then institutions can evolve alongside in order to deal with the new situations. Information is costly; hence institutional change from within reflects the trial-and-error adaptation and may well not coincide with the theoretically most efficient solution. However, institutions that evolve from within tend to be robust.

Costs associated with collective action tend to be lower if groups are small and homogenous (Baland and Platteau 1997, 1998). This implies that small special interest groups have disproportionate power compared to large amorphous groups.

Institutional change that originates from outside the original stakeholder confines may depart from traditional rules in order to address issues that local institutions cannot deal with because the changing outside circumstances have become too overwhelming. The solutions offered by this exogenous institutional change may be theoretically more efficient (especially in the long run); however, the institutions are less (or even not) embedded in the local communities, which implies that the enforcement costs of the institutional arrangements can become very high.

When the institutional arrangements are not firmly embedded in the local community the principal stakeholder implementing the institutional arrangements is faced with a series of managerial and political transaction costs. These include costs for setting up, maintaining or changing the organizational design; costs (information, physical transfer) of running the organization; and the costs of running the polity.

In summary we can say that in circumstances where institutional change is necessary we are faced with a trade-off between the transaction costs related to the enforcement of 'improved' institutional arrangements and the transaction costs improving enforceable institutional arrangements. Incurring these transaction costs only makes sense if the benefits from improved institutional arrangements outweigh the transition costs.

When we relate this dilemma to the management regime of the forest in Swat, we identify at the one end of the spectrum the ideal forest management system; at the other end we see the spontaneous evolution of self organization. Different stakeholders in Swat have their own ideal management system to put forward:

1. Private property rights for the local large landowners.
2. State ownership (effectively enforced by the Forestry Department) as the goal of the government.
3. Management by local communities (guaranteeing socially and environmentally sustainable outcomes) as common-pool resources is identified by the local communities characterized by higher level of trust and most NGOs and donors.[6]

Obviously the division of ideal management regimes will not relate so clearly to each class of actors; most notably the most progressive elements of the Forestry Department will also identify community-based management as an ideal. The current situation is in between an incoherent set of external interventions and strategic reactions by different agents in the local communities. The emergent system of management is the one producing the present dismal outcome.

---

[6] During some public meetings, in communities marked by social conflict, several participants claimed that their desire was not community-based management, but improved management by the Forestry Department.

To be effective, the role of donors and external interventions should not go in the opposite direction with respect to embedded evolutionary processes. The role of these interventions could be complementary to local process and contribute to guarantee that their evolution will respect social and environmental objectives. From this perspective, external actors could provide the additional funds to orient locally evolved institutional change to produce desired outcomes (low transaction costs, equity and sustainable management of the natural resource base) and provide incentives so that no spoilers' coalition arises that stops the process. In this framework, external agencies would be directing some of the final goals, but leaving the organizational issues to local actors who can behave according to their preferential knowledge and produce locally embedded institutional change.

Joint forest management seems to have possibilities given the scattered positive signals highlighted earlier. However, we argue that the preconditions in terms of well-defined institutional arrangements regarding property rights, control and financial remuneration for management activities should be addressed prior to embracing any scheme that is considered some sort of silver bullet.

References

Ahmed, A. S. (1980). *Pukhtun Economy and Society: Traditional Structure and Economic Development in a Tribal Society*. London: Boston, Routledge & Kegan Paul.

Amman, H. and Duraiappah, A. (2004). Land tenure and conflict resolution: a game theoretic approach in the Narok district in Kenya. *Environment and Development Economics*, **9**: 383–407.

Angelsen, A. and Kaimowitz, D. (1999). Rethinking the causes of deforestation: lessons from economic models. *The World Bank Research Observer*, **14**(1): 73–98.

Angelsen, A. and Wunder, S. (2003). Exploring the forest-poverty link. Key concepts, issues and research implications. CIFOR Working Paper. Bogor, Indonesia.

Baland, J. M. and Platteau, J. P. (1997). Coordination problems in local-level resource management. *Journal of Development Economics*, **53**(1): 197–210.

Baland, J. M. and Platteau, J. P. (1998). Wealth inequality and efficiency in the commons, part II: the regulated case. *Oxford Economic Papers*, **50**(1): 1–22.

Barrett, C. B., & Swallow, B. M. (2006). Fractal poverty traps. *World Development*, **34**: 1–15.

Barth, F. (1985). *The Last Wali of Swat*. Bangkok, Thailand: White Orchid Press.

Bromley, D. W. (1991). *Environment and Economy: Property Rights and Public Policy*. Oxford and Cambridge: Blackwell.

Furubotn, E. G. and Richter, R. (1997). *Institutions and Economic Theory: The Contribution of the New Institutional Economics*. Ann Arbor, MI: University of Michigan Press.

Hardin, G. (1968). Tragedy of the commons. *Science*, **162**(3859): 1243–1248.

Johnston, M. (2005). *Syndromes of Corruption: Wealth, Power, and Democracy*. Cambridge: Cambridge University Press.

Kabubo-Mariara, J. (2004), Poverty, property rights and socio-economic incentives for land conservation: the case for Kenya. *African Journal of Economic Policy*, **11**(1): 35–68.

Kaufmann, D., Kraay, A. and Mastruzzi, M. (2005). Governance matters IV: governance indicators for 1996–2004. World Bank Policy Research Working Paper Series. World Bank. Washington DC.

Khan, S. R., Bokhari, S. and Cheema, M. A. (2004). Resource rights and sustainable livelihoods: a case study of Pakistan's Dir-Kohistan forests. International Union for the Conservation of Nature and Natural Resources (IUCN). Islamabad, Pakistan.

Khan, S. R. and Khan, S. R. (2009). Assessing poverty–deforestation links: evidence from Swat, Pakistan. *Ecological Economics*, **68**(10): 2607–2618.

Killeen, D. and Khan, S. R. (2001). Poverty and environment. Opinion: World Summit on Sustainable Development. International Institute for Environment and Development, London.

Meinzen-Dick, R. and Pradhan, R. (2002). Legal pluralism and dynamic property rights. CAPRi Working Paper, no. 22. International Food Policy Research Institute (IFPRI), Washington DC.

Naqvi, S. N. H., Khan, M. H. and Chaudhry, M. G. (1987). *Land Reforms in Pakistan: A Historical Perspective*. Islamabad, Pakistan: Pakistan Institute of Development Economics.

North, D. C. (1990). *Institutions, Institutional Change, and Economic Performance*. Cambridge: Cambridge University Press.

Pellegrini, L. (2011). *Corruption, Development and the Environment*. Dordrecht, the Netherlands: Springer.

Schlager, E. and Ostrom, E. (1992). Property-rights regimes and natural-resources: a conceptual analysis. *Land Economics*, **68**(3): 249–262.

Social Policy and Development Centre (SPDC) (2001). *Social Development in Pakistan: Annual Review 2000*. Oxford: Oxford University Press.

Steimann, B. (2004). Decentralization and participation in the forestry sector of NWFP, Pakistan: the role of the state. IP6 Working Paper. Swiss National Centre for Competence in Research, Islamabad Pakistan.

Suleri, A. Q. (2002). Regional study on forest policy and institutional reforms: final report of the Pakistan case study. Asian Development Bank, Manila, Philippines.

Sultan-i-Rome (2005). Forestry in the princely state of Swat and Kalam (North-West Pakistan). IP6 Working Paper, Swiss National Centre for Competence in Research, Islamabad, Pakistan.

Sungi (1996). *Forests, Laws and People: A Focus on Hazara*. Peshawar, Pakistan: Sungi Development Foundation.

Transparency International (2002). *Corruption in South Asia: Insights and Benchmarks from Citizen Feedback Surveys in Five Countries*. Berlin: Transparency International.

Transparency International (2004). *Corruption Perceptions Index 2004.* Berlin: Transparency International.

Williamson, O. E. (1985). *The Economic Institutions of Capitalism: Firms, Markets, Relational Contracting.* New York and London: Free Press and Collier Macmillan.

Williamson, O. E. (2000). The new institutional economics: taking stock, looking ahead. *Journal of Economic Literature,* **38**(3): 595–613.

Part IV

Water–related ecosystem services

ROY BROUWER AND RASHID HASSAN

Water resources are essential for sustaining the life of humans and other living species on earth. Living organisms and ecosystems, however, face two distinct situations of water supply dynamics depending on the location and season. Many experience water shortage situations (at least seasonal scarcity), while others are endowed with water abundance throughout the year. Both situations often cause serious hardships to people and pressures on ecosystems supporting human life (i.e. drought and flooding) that need careful management for mitigation and prevention. Interference with natural water flows to reduce the direct effects of such undesirable phenomena, however, can disrupt ecosystem functions leading to negative consequences for human well-being, especially among the poor. The gains from interventions to regulate the hydrological cycle have often been achieved at high costs in terms of resulting ecological disruptions with major implications for human well-being. Like all other ecosystems, wetlands and freshwater systems provide multiple services to people and other life on earth (Fisher *et al.* 2008, Turner *et al.* 2000). Human actions altering natural systems to improve the provision of one or few components of the bundle of services provided by ecosystems often reduce the system's capacity to provide other services (MEA 2005). Examples of tradeoffs in human–freshwater interactions are many and sometimes quite complex. Regulation of natural water flows through construction of dams and reservoirs to increase provision of water for irrigation, domestic and industrial use, flood protection and hydropower generation, for example, is known to have many negative ecological

Nature's Wealth: The Economics of Ecosystem Services and Poverty, ed. P. J. H. van Beukering, E. Papyrakis, J. Bouma and R. Brouwer. Published by Cambridge University Press, © Cambridge University Press 2013.

consequences. Habitat and biodiversity loss and widespread pollution are typical consequences of interventions to regulate natural water flows with many negative impacts on human well-being. Among the negative impacts of these ecological disruptions are fragmentation of aquatic habitats and consequent damages to fisheries, increased chemical pollution from irrigation agriculture and effluent discharge by industrial activities leading to loss of important health services and biological diversity, and disruption of provision of sediment and nutrient inputs to floodplains and the natural habitat for fish spawning and breeding (MEA 2005). It is therefore necessary to consider such important tradeoffs in ecosystem services for proper evaluation of the true costs and benefits to human well-being of water management strategies and policy interventions (Brouwer et al. 2003).

Significant gains have been achieved from human interventions to regulate the hydrological cycle contributing to improved well-being of millions of people through various pathways. To manage situations of spatial and temporal (seasonal) water supply fluctuations, huge investments have been made in engineering large water management infrastructures such as dams, reservoirs and interbasin transfer schemes (Hirji 1998) that currently intercept about 40% of all surface water runoff and store water stocks, which are three to six times the size of those held by natural rivers (Vorosomarty et al. 2003). Major gains have been realized in terms of economic development through these engineering interventions to regulate global river runoff and make more water available for irrigation, industrial, domestic and recreational use (Rosegrant et al. 2002). Over the past four decades total water use by human society has more than doubled. Irrigation was the main source of this expansion achieving 100% growth in total area under irrigation between 1961 and 1999 (UNWWAP 2003). Irrigation agriculture is currently contributing 40% of the world crop production (Gleick 2002), which represents major progress towards improved food security. There is some evidence that irrigation has contributed to reduced poverty through the pathways of increased farm productivity and incomes, rural employment and lower food prices for all. These positive impacts, however, are believed to be much weaker among the landless and less endowed small farmers in the absence of appropriate complementary pro-poor institutional and technological interventions (Hussein and Hanjra 2003, Lipton et al. 2003, Rijsberman 2003).

In addition to irrigation benefits, more than 5 billion people are now connected to a regular supply of clean water and more than 3 billion have access to sanitation through these investments in water infrastructure (UN-HABITAT 2003, WHO/UNICEF 2004). Large public health benefits from reduced exposure to water-related diseases have been realized as a result, contributing significantly to improved well-being (e.g. Pryer 1993). Nevertheless, about 1.1 billion people remain with no access to clean water supply and more than 2.6 billion still lack access to improved sanitation (WHO/UNICEF, 2004). Water-related diseases (diarrhoea, cholera, typhoid, malaria) inflict huge well-being losses estimated at between 2 million to 12 million deaths per year in developing countries mostly among the poor and particularly children (Gleick 2002). Achieving the Millennium Development Goals (MDG) for water is therefore a big challenge to human society. Studies have estimated gains of US$ 3–34 in health benefits from every US$ 1 invested in improved water supply and sanitation with 10% global reduction in diarrhoea episodes. Benefits from meeting MDG targets in drinking water and sanitation are estimated at US$ 84 billion per year in terms of reduced costs of health care, days of reduced illness, averted death, time savings from proximity to drinking water and sanitation facilities (Hutton and Haller 2004). On top of the above outlined contributions to human well-being through improvements in food security and health gains, investments in water infrastructure also provided important other benefits such as protection against floods and electrification through regulation of the hydrological cycle.

While gains in human well-being from broad improvements in economic efficiency in water use and reduced pollution through the pathways of reduced water scarcity and improved public health discussed above are relatively clear, linkages of some demand management measures such as charging a water price to poverty reduction are less obvious and may appear to work in opposite directions. Water users around the world are typically highly subsidized and hence tend to overuse and waste water. On the other hand, the financial viability of water supply agencies is seriously affected from not recovering actual costs of delivering water and maintenance of its supply infrastructures (Tarfasa and Brouwer 2013). Public sources of funding to bear the burdens of almost free water supply have dwindled and are no longer available at low costs. Naturally, the consequences are decline in water service quality and reliability, higher rates of transmission losses from decaying infrastructures and inability to expand water services to unconnected users, who are primarily the poor, forcing them to resort to more expensive and unclean alternative sources. Pricing water for cost recovery therefore has a number of indirect benefits to poor and marginalized communities and enables most countries to use cross-subsidization in pricing schemes to support provision of cheap water for basic human needs and other targeted economic purposes (Vorosomarty et al., 2005). Also, 90–95% of all sewage and 70% of industrial waste in developing countries are dumped directly into surface water causing huge risks to people and ecosystems' health downstream which usually affect the poor whose livelihoods are highly dependent on services of natural ecosystems (Vorosomarty et al. 2005). Taxing polluters is accordingly a necessary measure and incentive for reducing environmental hazards and health risks, conserving water and reducing scarcity as well as generating financial resources to improve and expand sanitation services to the poor and marginalized in the developing world. Other market-based policy instruments such as establishment of water markets and tradable pollution permits have proven to work well for improved water use efficiency in a number of countries (MEA 2005).

Organization of Part IV

This part of the book documents analyses of case study experiences attempting various strategies for managing situations of water scarcity and abundance to promote economic development and protect livelihoods of poor people in the developing world. Three of the studies carried out in Mali, South Africa and Ethiopia analyse strategies for managing water scarcity. The Mali and Ethiopia studies are concerned with supply intervention options, whereas the study in South Africa evaluates water-demand management solutions. While the Ethiopia study assessed the impacts of micro-level supply solutions (ponds and wells) on household poverty, the costs (mainly ecological disturbances) and benefits (hydropower, irrigation and tourism) of macro-level supply interventions (dams) are analysed in Mali. Both the Mali and Ethiopia studies employ a cost–benefit analysis framework to evaluate the direct and indirect costs and benefits of engineering supply solutions (damming and small-scale harvesting of water). Market as well as non-market valuation methods have been used to measure the value of ecosystem services affected by water supply interventions in Mali and Ethiopia. The South Africa study on the other hand employs a general equilibrium approach to trace the economy-wide implications of a water tax combined with recycling of tax revenues to support the poor. In addition to achieving efficiency gains in water use from economic pricing the study explores the double dividend in poverty reduction from the revenue transfer policy regime. In contrast with the above three, the study in Bangladesh addresses the issue of water abundance evaluating engineering solutions to control flooding. A cost–benefit analysis framework is also employed in the Bangladesh study to investigate direct and indirect consequences of embankments to control floods for the welfare of competing water uses and ecosystem services of flooding. Benefits of flood protection measured include reduced damages to crop agriculture, livestock and fish farming (aquaculture), property and health. The costs of flood prevention on the other hand include declined soil fertility due to reduced nutrient deposition services to agriculture, losses in capture fisheries from disrupting fish migration and spawning habitat and reduced water transportation services. Impacts on the livelihoods of households involved in all these economic activities (farmers, fishermen, water transport workers) are evaluated accordingly.

References

Brouwer, R., Turner, R. K. and Georgiou, S. (2003). Integrated assessment and sustainable water and wetland management: a review of concepts and methods. *Integrated Assessment*, **3**(4): 171–183.

Fisher, B., Turner, R. K., Zylstra, M. *et al.* (2008). Ecosystem services and economic theory: integration for policy relevant research. *Ecological Applications*, **18**(8): 2050–2067.

Gleick, P. H. (2002). Dirty water: estimated deaths from water-related diseases 2000–2002. Pacific Institute Research Report, Pacific Institute for Studies in development, Environment and Security, Oakland, California.

Hirji, K. (1998). Inter-basin water transfers: emerging trends. World Bank Report No. 226775, Environment Matters, World Bank, Washington DC.

Hussein, I. and Hanjra, M. (2003). Does irrigation water matter for rural poverty alleviation? Evidence from South and South East Asia. *Water Policy*, **5**(5): 429–442.

Hutton, G. and Haller, L. (2004). Evaluation of the costs and benefits of water and sanitation improvements at the global level. WHO/SDE/WSH/04.04, WHO, Geneva.

Lipton, M., Litchfield, J. and Faurès, J.-M. (2003). The effects of irrigation on poverty: a framework for analysis. *Water Policy*, **5**(5): 413–427.

Millennium Ecosystem Assessment (MEA) (2005). *Ecosystems and Human Well-being: Biodiversity Synthesis*. Washington DC: World Resources Institute.

Pryer, J. (1993). The impact of adult ill-health on household income and nutrition in Khulana, Bangladesh. *Environment and Organization*, **5**: 35–49.

Rijsberman, F. (2003). Can development of water resources reduce poverty? *Water Policy*, **5**(5): 399–412.

Rosegrant, M, Cai, X. and Cline, S. (2002). *World Water and Food to 2025: Dealing with Scarcity*. Washington DC: International Food Policy Research Institute.

Tarfasa, S. and R. Brouwer (2013). Estimation of the public benefits of urban water supply improvements in Ethiopia: a choice experiment. *Applied Economics*, **45**(9): 1099–1108.

Turner, R. K., van den Bergh, J. C. J. M., Soderquist, T. *et al.* (2000). Ecological-economic analysis of wetlands: scientific integration for management and policy. *Ecological Economics*, **35**: 7–23.

UN-HABITAT (2003). *Water and Sanitation in the World's Cities: Local Actions for Global Goals.* London: Earthscan.

United Nations World Water Assessment Program (UNWWAP) (2003). Water for people: water for life. UN World Water Development Report, UNESCO, Paris.

Vorosomarty, C., Meybeck, M., Fekete, B. *et al.* (2003). Anthropogenic sediment retention: major global-scale impact from the population of registered impoundments. *Global and Planetary Change*, **39**: 169–190.

Vorosomarty, C., Leveque, C., Ravenga, C. *et al.* (2005). Fresh water. In Millennium Ecosystem Assessment (MEA), *Ecosystems and Human Wellbeing. Volume 1. Current State and Trends.* Washington DC: Island Press, Chapter 7.

WHO/UNICEF (2004). Meeting the MDG drinking water and sanitation target: a mid-term assessment of progress. WHO, Geneva.

12 · Small-scale water harvesting and household poverty in northern Ethiopia

FITSUM HAGOS, EYASU YAZEW,
MEKONNEN YOHANNES,
AFEWORKI MULUGETA, GIRMAY
GEBRESAMUEL ABRAHA,
ZENEBE ABRAHA, GIDEON KRUSEMAN
AND VINCENT LINDERHOF

12.1 Introduction

The climate of Tigray, northern Ethiopia, is mainly semi-arid and most of the region experiences erratic and inadequate rainfall that remains insufficient for crop production. Climatic change over the past decades has resulted in a temperature increase of about 0.2 °C. This has resulted in a notable decrease in the amount of, and altered the distribution of, precipitation in Ethiopia (NMSA 2001). The distribution of mean annual rainfall over the country is characterized by large spatial variation and ranges from 2000 mm in some pockets in the south-west to less than 100 mm in the Afar lowlands in the north-east. Trend analysis of annual rainfall showed that rainfall levels remained more or less constant when averaged over the whole country while a declining trend has been observed over the northern half and south-western Ethiopia. The average annual minimum temperature has been increasing by about 0.25°C while average maximum temperature has been increasing by about 0.1°C every decade (NMSA 2001).

Climate change may have far reaching implications for Ethiopia for various reasons. The country's economy depends mainly on agriculture, which is very sensitive to climatic variations. A large part of the country is arid and semiarid, and is highly prone to desertification and drought.

Nature's Wealth: The Economics of Ecosystem Services and Poverty, ed. P. J. H. van Beukering, E. Papyrakis, J. Bouma and R. Brouwer. Published by Cambridge University Press, © Cambridge University Press 2013.

Ethiopia has a fragile highland ecosystem, which is currently under stress due to population pressure. Forest, water and biodiversity resources of the country are also climate sensitive. Vector-borne diseases such as malaria, which are closely associated with climatic variations, affect Ethiopia. The country has experienced environmental problems such as recurring droughts, high rates of deforestation, soil degradation and loss, overgrazing, etc., which may be exacerbated by climate change. Climate change is, therefore, a case for concern (NMSA 2001). Assessing vulnerability to climate change and exploring adaptation options should therefore become a critical element of the development programme of the country. Among the possible adaptations to recurrent drought is the promotion of small-scale irrigation agriculture through small household-managed water harvesting structures.

Since 2003, household-level water harvesting schemes have been expanding as an integral part of the Tigray regional food security and extension programmes aiming at breaking the cycle of famine. The idea is to make water available to supplement rain-fed agriculture during the critical stages of plant growth when rainfall is inadequate, and to promote home garden development. Shallow wells are expected to enable farmers to grow vegetables and fruit trees and other high value crops (e.g. spices) and render diversification of food and cash crops' production possible. These are achieved by holding and diverting rainwater or stream flow during the wet season directly onto agricultural fields or through the exploitation of shallow underground water supplies and small-scale ponds. These would provide adequate water for households for a period of 4–16 weeks in the case of ponds and on a continuous basis in the case of wells.

Given that reliable marketing opportunities and other supporting services (e.g. credit services) are in place, such measures may lead to higher income and improved welfare of farm households. Furthermore, this is expected to improve nutrition and dietary intake because of improved ability to buy (as a result of increased income) and increased supply of vegetables and fruits grown in home gardens. Besides, the overall increase in income and household welfare may lead to investment in land conservation thereby contributing positively to reversing the spiral of poverty-induced environmental degradation.

While harnessing water potential at the micro-level can mitigate some of the negative effects of climate change (water availability), it does increase the risk of malaria in areas where malaria currently thrives and expands this further to areas previously unaffected. It is important to evaluate whether the expected benefits from such interventions will offset these negative external costs to justify intended actions.

The present study investigates the following research questions:

1. What is the impact of ponds and wells on household welfare measured in terms of higher income and improved diet and nutrition through adoption of high value crops?
2. What are the most important engineering, biophysical and environmental factors that influence the performance of small-scale water harvesting structures?
3. What is the extent of expected negative health externalities (e.g. incidence of malaria) associated with water wells and ponds?
4. Does higher household income from ownership of wells and ponds translate into a higher willingness to pay (WTP) for improved malaria control?

The purpose of the present study is to find out by how much interventions such as provision of irrigation water can increase the productivity and income of households (reduce their poverty), and at the same time decrease water-related illnesses and their implications for labour supply and health expenditure. We claim that such knowledge could be very valuable to help low income countries develop better poverty reduction strategies and to achieve the MDGs.

The chapter is structured as follows. Section 12.2 presents a description of the study area, sampling techniques and methods of analysis. In 12.3 we present the major findings. Finally, in 12.4 we analyse the results and present the policy implications of the findings.

12.2 Data and methodology

The data used in this study is drawn from an integrated study on environment, health and nutrition based on household and plot surveys of 650 randomly selected farm households in 13 tabias (villages) from four zones in Tigray region, northern Ethiopia. The communities were selected on the basis of their differences in agro-ecology (low land, middle altitude and highland), presence of ponds and water wells in the villages, distance to market, and availability of secondary baseline information. Using proportional sampling, 50 households with and without ponds and water wells were randomly selected from each community. For the health studies, data was collected in four rounds in two seasons (after and during the start of the rainy season) including information on water bodies (mapped using a global positioning system (GPS)) and household characteristics (type of housing, number and kind of animals, etc.). The survey also collected data on larval abundance and density in different types of breeding sites to quantify the role of each type in the overall adult output of vector populations, level of indoor resting/visiting densities by adult mosquito vectors and prevalence of malaria infections. The sampling strategy involved selection of six villages (communities with and without ponds and wells from each of the three agro-ecologies). The level of indoor resting/visiting densities by adult mosquito vectors in the study communities was established twice monthly using light traps. Households within 100–200 m of the ponds were selected for sampling. Furthermore, we determined the prevalence of malaria infections in children under 10 years old by sampling 120 children in the study communities. Sampling started from the second week of September and was conducted fortnightly until the ponds dried up. The same process was repeated during the start of the rainy season. A follow-up contingent valuation study on 250 households was conducted to elicit households' WTP for improved health services (specifically malaria control measures) and examine whether access to ponds and wells has increased their WTP.

Finally, for the environment study, we sampled three ponds and wells from each community from the different agro-ecologies in the four zones. In total we sampled 108 water-harvesting structures. The environment study is based on household interviews, field measurement and observation, where we looked into structure design, water-holding capacity and utilization. Investigation of the water balance of the household water-harvesting schemes in Tigray can be used as a tool to assess the efficiency of the water storage and utilization. It can also be used as an input to improve the construction and operation of the schemes. The water balance of the ponds and hand-dug wells generally include inflow, evaporation loss, seepage loss and the net harvested water (Yazew *et al.* 2007).

The household ponds are generally small in size and could be quickly silted up by erosion from the catchment area. As a result, most of the ponds are provided with silt trap structures to minimize the sedimentation problem. Known volumes of runoff were collected before and after the silt trap structures at various times to assess their efficiency. The runoff samples were then oven dried to determine the amount of silt load. The silt trap efficiency was calculated using standard approaches as stipulated in Yazew *et al.* (2007).

Moreover, we looked into the effect of water utilization from ponds and wells on environmental indicators such as salinization. The causes of salinity could be natural or human induced. The natural process involves the accumulation of salts as a result of long-term processes such as weathering and one-time submergence of soils under seawater (Ghassemi *et al.* 1995). On the other hand, the two most common causes of secondary salinity, which is most severe in arid and semi-arid areas, are excessive application of irrigation water and a shallow groundwater table (Ghassemi *et al.* 1995, Ritzema *et al.* 1996, Smedema and Rycroft 1983, Umali 1993, Vagen *et al.* 2000). The effect of salinity and sodicity on crop growth can be classified into three categories, namely, dispersion, osmotic and toxicity problems (for detailed discussion of the effects see Yazew *et al.* 2007). Hence, soil and water samples collected from the various household water-harvesting schemes were used for the assessment of salinity. The electrical conductivity of the irrigation water and the soil was measured by using an electrical conductivity meter (Yazew *et al.* 2007). The water and soil analysis results were then interpreted based on the classifications of Richards (1954).

The socio-economic impacts of ponds and wells were assessed using simple and complex statistical techniques. The simple statistical techniques involved using simple mean separation tests of the impact of ponds and wells on growing of high value crops (e.g. vegetables), income from such cash crop sales and perceived changes in diet and utilization of inputs such as fertilizer. We also examined the impact of ponds and wells on household welfare, where welfare is measured as per capita household expenditure, using matching and other econometric techniques.

Matching is widely used in the estimation of the average treatment effect of a binary treatment. It uses non-parametric regression methods to construct the counterfactual under an assumption of selection on observables. We considered having access to ponds/wells as a binary treatment, expenditure per capita as an outcome and households having ponds/wells as treatment group and non-participant households as control group variables. Matching estimators aim to combine (match) treated and control group households that are similar in terms of their observable characteristics in order to estimate the effect of participation as the difference in the mean value of an outcome variable (for details on the theory and application of such techniques, see Dehejia and Wahba 2002, Heckman and Vytlacil 2007, Rosenbaum and Rubin 1983, Wooldridge 2002). In this chapter we used matching estimators using propensity scores.

The intuition behind using propensity scores is that two individual households with the same probability of adoption will show up in the treated and untreated samples in equal proportions. These methods identify for households the closest propensity score in the opposite technological status; then compute investment effect as the mean difference of households' per capita expenditure between each pair of matched households. For details of these methods we refer to Becker and Ichino (2002) who also provided the STATA software code we used in this study.

To elicit WTP for improved malaria control, we followed the so-called double-bounded dichotomous-choice format to elicit households' WTP for improved public health services (Arrow *et al.* 1993, Cameron and Quiggin 1994, Hanemann *et al.* 1991). To improve the precision of WTP estimates, we introduce follow-up questions to the dichotomous-choice payment question following the recommendation of Hanemann *et al.* (1991). It is important to note that the dichotomous-choice approach does not observe WTP directly. At best, we can infer that the respondent's WTP amount was greater than the bid value (if the respondent is in favour of the programme) or less than the bid amount (if the respondent votes against the plan), and form broad intervals around the respondent's WTP amount. Mean WTP is estimated statistically from the data of responses obtained from respondents.

Double-bounded dichotomous-choice payment questions typically require a different type of statistical analysis, based on the assumption that if the individual states she or he is willing to pay the bid amount, their WTP must be greater than the bid. If the individual declines to pay the stated amount, then their WTP must be less than the bid. In both cases, the respondent's actual WTP amount is not observed directly by the researcher.

12.3 Results and discussion

12.3.1 Water balance of the schemes

The economic importance of ponds and wells is strongly affected by whether they hold water, and if they do, how long they hold water and ultimately whether the water is used productively. Approximately 58% of the sampled ponds hold water for an average of about 3 months with standard deviation of 2 months. Almost all wells hold some water throughout the year. Part of the runoff accumulated in the ponds during the rainy season is lost as evaporation and seepage. The net harvested water, the amount that remains in the ponds at the end of the rainy season, is used for supplementary irrigation.

The results of our analysis show that the evaporation loss during the observation period is small and has little impact on the net harvested water of the ponds. The amount of the evaporation loss is also generally related to the surface area of the ponds. The net harvested water is largely affected by the seepage loss and lining of the bed and walls of the ponds; the choice of lining material plays a key role in this regard. Our results show that plastic lined ponds hold more water for longer periods. On the other hand, evaporation loss from wells is less than from ponds due to the smaller surface area of wells. Seepage loss is also small. The amount of net harvested water is generally affected by the capacity of the groundwater to recharge wells.

Results for the water balance of ponds imply the need and importance of minimizing evaporation and seepage losses. Protecting the net harvested water from evaporation and seepage loss can increase the irrigated area by as much as 95%. Proper lining of the bed and walls of the pond and shading of the surface area would improve the water balance of these water-harvesting schemes significantly. Moreover, the extent of the irrigable land by the ponds can be maximized if the reduction in evaporation and seepage loss is accompanied by better field water application efficiency. Since most of the farmers are new to irrigation, they do not have sufficient knowledge of proper irrigation scheduling. Besides, the water is generally applied to the fields by either watering cans or treadle pumps, which are of low efficiency. These might contribute to poor performance of the schemes. Sediment deposition may reduce the water-holding capacity of the ponds. Proper construction and maintenance of silt trap structures is important to minimize this problem. Our empirical results show, however, that effectiveness of silt trap structures of ponds is generally low (Yazew et al., 2007). Lack of timely removal of sediment deposited in the structures and location of silt trap structures are among the main reasons. Since the structures are located close to the ponds, they do not give much chance of sediment settlement and deposition before the runoff arrives at the pond. As far as water and soil quality are concerned, results from the analysis of 39 water samples indicate that ponds have lower electrical conductivity (less than 0.3 dS/m). Since they are recharged by the groundwater flow, the electrical conductivity of hand-dug wells is generally higher. Based on the above, quality of water from ponds can be categorized as suitable for irrigation. On the other hand, water from wells is more saline and is used for more salt-tolerant crops along with leaching. Electrical conductivity of all soil samples was found to be very low, indicating that the soils of all schemes are salt free.

12.3.2 Socio-economic benefits

Households use water from ponds and wells to grow vegetables. From the survey we found that about 40% of the households who own ponds and wells grow vegetables. To examine whether households are benefiting from wells and ponds and whether that is reflected in per capita consumption, expenditures are assessed using simple mean separation tests. Results of the tests suggest that mean differences in per capita expenditure of households with and without ponds/wells are not statistically different from zero. Cash sales of vegetables with and without ponds and/or wells indicate that the average revenue of cash vegetable sales of pond owners is Birr 2.5 while the average revenue of cash vegetable sales of well owners is Birr 197.13. This indicates that households with wells have a much higher cash revenue from vegetable sales. We also examined whether households have different input use as increased access to water may encourage increased use of purchased farm inputs such as fertilizer. The results show that households with ponds have significantly higher fertilizer use compared to households without ponds (Table 12.1). There is no significant difference in fertilizer use between those owning a well and those who do not.

The test shows that there is high variability in per capita expenditure among households owning ponds or wells. Although it needs rigorous analysis to quantify the potential gains in efficiency, one could see that farmers' income and welfare could improve significantly if water-harvesting schemes are utilized efficiently.

The matching estimates where the treated and control households are matched on the basis of their scores using nearest neighbour, kernel methods and stratification matching estimators, show that there is no significant effect on household welfare from owning a pond or a well (Table 12.2 and Table 12.3). This implies that although there is no intrinsic reason as to why ponds and wells could not be economically viable, the critical issue remains their water holding capacity and the efficient utilization of the harvested water. The estimated ATT (Average Treatment Effect) is, however, positive in almost all of the cases.

12.3.3 Determinants of poverty

In the previous section, we assessed the impact of water harvesting on household expenditure. In this section, we analyse how poverty correlates with a set of household and demographic characteristics, village level factors, policy-related variables and households' access to ponds and wells. The intention is to evaluate the effect on poverty of having access

Table 12.1 *Mean separation tests of some important variables of households with access and without access to ponds/wells*

Access to ponds/wells	Mean per capita expenditure in Birr (SE in parentheses)	Difference	t–test
Without pond or well (n = 348)	885.32 (43.50)		
		31.56 (85.96)	0.37
With pond (n = 101)	853.76 (54.54)		
Without pond or well (n = 348)	885.32 (43.50)		
		40.02 (109.55)	−0.37
With well (n = 63)	925.35 (92.36)		
Without pond with well (n = 63)	925.35 (92.36)		
		−354.65 (308.52)	−1.15
With pond and with well (n = 9)	1280.00 (515.79)		
Without well, with pond (n = 101)	853.76 (54.54)		
		−426.23 (234.78)	−1.82
With pond with well (n = 9)	1280.00 (515.79)		
Access to ponds/wells	Mean fertilizer use in kg (SE in parentheses) –	Difference	t–test
Without pond (n = 145)	61.89 (2.37)		
With pond (n = 22)	170.13 (88.36)	−108.24 (34.37)	−3.14★
Without well (n = 132)	79.58 (15.06)		
With well (n = 35)	63.21 (4.00)	16.36 (29.38)	0.56

★Significant at 1%

to ponds and wells, while correcting for the endogeneity of access to ponds/wells.[1] We define the welfare indicator W_i as:

$$W_i = Y_i/Z \tag{12.1}$$

Where Y is per capita consumption expenditure and Z is the poverty line.[2] Using X_i to denote the vector of explanatory variables, we perform the following regression:

[1] To correct for the possible endogeneity of having ponds or wells we used the estimated probability of having ponds and wells from the independent probit regressions in Equation (12.2).

[2] We used the mean per capita expenditure (807 Birr/person/year) to define the food poverty line.

Table 12.2 *Results of matching method to measure impact of ponds on household welfare (bootstrapped standard errors)*

	Kernel matching method				Nearest neighbour matching method				Stratification method			
	Treatment	Control	ATT*	t	Treatment	Control	ATT*	t	Treatment	Control	ATT*	t
	78	241	87.9 (72)	1.2	78	61	37 (109)	0.34	121	417	11.9 (71)	0.17

* Mean difference of households' per capita expenditure (in Birr) between each pair of matched households

Table 12.3 *Results of matching method to measure impact of wells on household welfare (bootstrapped standard errors)*

	Kernel matching method				Nearest neighbour				Stratification method			
	Treatment	Control	ATT*	t	Treatment	Control	ATT*	t	Treatment	Control	ATT*	t
	38	245	−30.4 (169.7)	−0.18	38	30	−130.1 (256.0)	−0.51	77	448	109.3	0.94

* Mean difference of households' per capita expenditure (in Birr) between each pair of matched households

$$LogW_i = \beta'X_i + \varepsilon_i \tag{12.2}$$

In this regression, the logarithm of per capita consumption expenditure divided by the poverty line is used as the welfare response variable. Explanatory variables on the right hand side include (1) household characteristics and demographic factors such as sex, access to education and age of the household head, (2) asset holding (oxen, adult labour disaggregated by sex and household members with primary education and access to ponds and wells); (3) access to different services (health, market); and (4) the geographical location of the household, and some community characteristics such as rainfall variability. The β coefficients in Equation (12.2) measure the degree of association between explanatory variables and levels of consumption expenditure (relative to the poverty line) and not necessarily their causal relationship. These parameter estimates could be interpreted as measures of marginal gains in welfare (or reduced poverty) from an increment in the respective explanatory attribute (Wodon 1999). We used survey regression estimation techniques to adjust the standard errors to both stratification and clustering effects (Rogers 1993, Wooldrige 2002) and thereby correcting for heteroskedasticity. We also tested for other possible misspecifications (e.g. multicollinearity). Results are shown in Table 12.4.

As can be seen from the regression results, owning a pond or a well does not significantly affect household welfare. A host of household attributes, asset-holding variables and access to services were found to be significant in explaining the welfare of households in the study areas. These results imply that ensuring a household's access to assets and employment opportunities could be important instruments in reducing poverty. From among the village level variables, distance to the nearest health centre and rainfall variability were found to be negatively correlated with household welfare. Distance to market was not significantly correlated with household welfare although it has the expected sign. Finally, households located in mid- and lowland agro-ecologies were found to be better off than households located in highlands, indicating agro-ecological differences in welfare conditions, which could be related to differences in agricultural potential and other opportunities.

12.3.4 Impact on health

Asked whether ponds and wells have side effects, about 60% of the households reported that ponds particularly provide good breeding

Table 12.4 *Determinants of per capita expenditure relative to the poverty line*

Dependent var: LOG (per capita expenditure/poverty line)			
Variable name	Variable description	Coefficient	SE
Constant	Intercept	0.607	0.266★
ownwell2	Own well (dummy)	−0.158	0.397
ownpond2	Own pond (dummy)	−0.404	0.403
sexhh1	Sex of household head (Male=1)	−0.229	0.090★★
hh_age	Age of household head (in years)	0.004	0.003
adult_f	Number of female adults in the household	−0.311	0.084★★
adult_m	Number of male adults in the household	0.181	0.062★★
CWratio	Consumer:worker ratio (family size/ male and female adult)	−0.112	0.026★★
pcoxen96	Per capita oxen 2004 (oxen units)	0.466	0.115★★
primary_s	Number of people with primary education	0.002	0.003
health_c	Presence of a health centre (dummy)	−0.003	0.001★★
rainvar	Rainfall variability (coefficient of variation in rainfall)	−0.548	0.256★
weredam	Distance to wereda (district) market (in minutes)	−0.0002	0.001
agroecodmy2	Low land (dummy highland = base)	0.351	0.115★★
agroecodmy3	Mid land (dummy highland = base))	0.254	0.094★
Ce_well	Estimated probability of owning a well	−0.045	0.231
Ce_pond	Estimated probability of owning a pond	−0.284	0.232

Number of obs = 261
F(16, 235) = 13.19
Prob > F = 0.0000
R-squared = 0.414

★★ and ★ significant at 1 and 5, respectively.
± *Tabia* dummies were found to be collinear with many regressors and hence were excluded.

grounds for mosquitoes. The results of the malaria prevalence study conducted in a subsample of the study communities are shown in Table 12.5. The results show that malaria is becoming a major public health problem, especially in the lowland zones, where prevalence rates exceed 30%. Moreover, prevalence rates are significantly higher in sites where there are ponds and wells in contrast to the control sites.

To complement this, during the household survey 35% of households reported that a member(s) was ill from malaria in contrast to only 3% reporting other illnesses such as diarrhoea, skin and other infectious

Table 12.5 *Malaria prevalence at intervention and control sites (% of children under 5 years sampled)*

Zone/intervention/control[*] sites	Nov. 2004	Dec. 2004	March 2005	May 2005
Higher altitude				
Modoge/Sofoho	0.9 (1.9)	4.5 (0.0)	1 (2.3)	0 (0.0)
Gergera/Hiwilwal	18 (0.0)	10.9 (1.0)	10.7 (1.0)	1.98 (0.0)
Midland				
Mai Daero/ Mai Beja	9.1 (0.0)	8.5 (0.0)	3.6 (0.0)	2.1 (0.0)
Zongi/ Adi Tegemes	2.1 (0.0)	3.7 (0.0)	3.1 (0.0)	3.0 (0.0)
Lowland				
Hashia/ Rarhe	35 (10.1)	32.6 (30.5)	37.0 (4.9)	33.5 (7.4)

[*]Prevalence of control sites in parentheses.

diseases. Hence, malaria seems to be the most dominant disease in the sample villages. The consequences of having malaria are reported to be serious (drawing on meagre household income to pay for treatment, the opportunity cost of missed working days and income and costs incurred to avert the illness) (Yazew *et al.* 2007). Dominant aversive strategies reported are to regularly disturb the water body (habitat of mosquitoes) (92%) and use mosquito nets (9.4%) while repellents or sprays are least used.

12.3.5 Household's WTP for improved malaria control

The next section presents the results of the WTP elicitation exercise by describing how respondents generally responded to the contingent valuation method (CVM) exercise and exploring further the most important variables explaining people's difference in WTP. The intention here is to show the external costs (e.g. increased incidence of malaria) associated with water harvesting while the preceding sections outlined the estimated benefits of using harvested water from pond and wells.

With respect to their responses, only about 13% of the households accepted the initial bid, while 28% of those who did not approve the initial bid accepted the second lower bid. The results indicate that there is a high level of disagreement with the hypothesized benefits of an improved public health service programme. The stated mean maximum WTP for the public health programme is Birr 3.8/month. The expressed high discontent with improved provision of health services may be because surveyed communities are relatively satisfied with existing health services or due to the fact that they do not perceive malaria to be a major health problem. However, 52% of the households rated the available health service as poor, 29% consider it as satisfactory and 19% as good. On average, households need to travel for about 42.7 minutes to receive health service from the nearest health post or centre. About 25% of the respondents reporting zero WTP believed that malaria is not a serious problem. Nearly 58% believe that they are too poor to afford it, while about 4.5% believed that they are too old and it is the government's responsibility to provide such services. Clearly, poverty is a major reason for households' low WTP, a point to which we turn to below.

Results of the regression analysis of determinants of WTP are reported in Table 12.6. Given the limited number of respondents (39% of the sample) who were willing to pay (either the amount of the first or second

Table 12.6 *Determinants of willingness to pay for a public health programme*

Variable name	Coefficient	SE
Highland (dummy)	0.57	1.71
Midland (dummy)	−2.02	1.76
Female-headed household (dummy)	2.79	0.98★★
Literate head (dummy)	0.17	1.17
Own pond (dummy)	0.97	2.34
Family size	0.52	0.24★
Own well (dummy)	−1.06	1.89
Credit access	3.22	1.08★★
Per capita expenditure	0.001	0.001★
pcoxen96	2.82	0.97★★
Malaria illness (dummy)	3.33	1.01★★
Other disease (dummy)	−2.77	1.26★
Satisfactory health service (dummy)	−2.47	1.80
Good health service (dummy)	−.677	1.96
_cons	−1.87	2.45
Sigma	3.46	0.385

Number of obs. = 91	0 left-censored observations
Wald chi2(12) = 53.91	0 uncensored observations
Log pseudo likelihood = −102.1274	5 right-censored observation
P > chi^2 = 0.0000	86 interval observations

★★ and ★ significant at 1 and 5%, respectively.
± *Tabia* dummies were found to be collinear with many regressors and hence were excluded.

bid), the regression results may have limited scope for extrapolation. Pond and well ownership does not seem to increase a household's WTP for improved health services. This suggests that the effect of ponds and wells on household welfare remains very low. Access to institutional credit positively affects a household's WTP. Likewise, asset wealth (oxen holding) and income wealth (per capita expenditure) positively affects WTP suggesting again, the importance of poverty in determining households' WTP.

Agro-ecological location does not seem to have a significant effect on WTP even though we know that malaria is more serious in the lowlands compared to the mid- and highlands. However, the presence of a malaria-sick household member increases a household's WTP. The perception of the households of existing health services does not significantly influence households' WTP, although the sign of the coefficients is consistently negative with better existing services. Finally, household factors such as sex of the household head (in this case being female-headed) and having larger family size seem to influence WTP positively.

12.4 Conclusions and recommendations

12.4.1 Improving the design of water harvesting structures

The results of this study show that evaporation loss during the rainy season has a small impact on net harvested water and depends on the extent of the surface area of the ponds. On the other hand, the impact of the seepage loss is very high unless there is proper lining of the bed and walls of the ponds. It has to be noted that the value of evaporation and seepage loss indicated in this report is mainly for the rainy season. The amount would be higher after the rainy season, where there is only outflow without inflow. This suggests that the irrigated area can be increased considerably if proper measures to reduce water loss are implemented. Some of the measures that may improve water harvesting and utilization efficiency include shading of the ponds with local materials such as straw and bushes to minimize evaporation; reducing the open surface area of the ponds to minimize both evaporation and seepage losses; proper lining and subsequent maintenance of the bed and walls of the ponds; and introduction of better field irrigation methods and mechanisms (e.g. drip irrigation). In addition, farmers need to be trained in effective water harvesting, storage and management practices.

Current silt trap structures are not effective in minimizing sediment deposition, which calls for improvements in the design, construction and maintenance of these structures to reduce sediment deposition and increase the water-holding capacity of ponds. The use of extensive ponds and hand-dug wells for supplementary irrigation is a very new experience in Tigray. As a result, the soils of almost all schemes are currently salt free, which suggests that salinity from using ponds for supplementary irrigation may not be a threat to farmers at this stage. However, the water quality of wells is poor. Besides, since they are continuously recharged by groundwater, most of the wells irrigate for longer periods than the ponds. Farmers using wells would have to implement necessary measures to minimize salinity. Since this study was carried out only during one rainy season, the results must be used with caution. A more comprehensive follow-up study of water balances, sediment deposition and salinity aspects is therefore crucial. In addition, crop water requirements, field water use efficiency, areas suitable for irrigation from individual household water harvesting schemes and consequent yield gains need to be adequately investigated for proper evaluation of the net economic benefits from these interventions.

12.4.2 Socio-economic impacts

The results show that household access to ponds and wells has no significant impact on welfare. Regression analyses results also confirm that households with ponds and wells are not better-off compared to households without ponds/wells. A host of other variables proved to be much more important in explaining disparities in welfare of households in the study areas. The main reason behind these results lies in the fact that ponds and wells have not yet been fully exploited to show impacts on people's livelihoods. Earlier results also indicate that there is still a lot to be desired in the design and management of existing water harvesting schemes to increase the effectiveness of ponds and water wells. One could easily understand that the effect of ponds and wells on household welfare may not be significant under the current circumstances. Moreover, irrigation agriculture is a relatively new concept many rural households in northern Ethiopia. Households learn by doing and it may take some time until water is used in its most efficient way. These data also show that there is a lot of variability in income and revenue from sales, which is indicative of the potential gains in efficiency. This also calls for strong extension support to help households introduce new and more profitable crops. There is no doubt that water harvesting will reduce the hazards of climate variability. The main challenge is to come up with technologies that increase water use efficiency and crops that have higher returns.

12.4.3 Health impacts

Malaria is becoming a major public health problem with serious welfare and economic consequences; unfortunately these are external costs of the presence of ponds and wells. While malaria is a serious health problem in the study villages, particularly in the lowlands, the respondents' WTP for a hypothetical improved malaria control service is very low. This may have a lot to do with the level of poverty in the study communities which is reinforced by the regression results that indicated that the asset holding and income wealth have significant effects on households' WTP for improved health services. Cash constraints may also play an important role in influencing households' WTP. The implications of these results are that: (1) it poses a question on the validity of using CVM in cash-constrained and poor economies; (2) using the estimated mean WTP to measure the external health cost of wells and ponds may underestimate the magnitude of the problem. Furthermore, if household ponds and wells fail to yield their full economic potential, as the evidence seems to suggest, then they pose high external costs to the economy.

References

Arrow, K., Solow, R., Portney, P. R., Leamer, E. and Radner, R. *et al.* (1993). Report of the NOAA panel on contingent valuation. *Federal Register*, **58**: 4601–4614.

Becker, O. S. and Ichino, A. (2002). Estimation of average treatment effects based on propensity scores. *The Stata Journal*, **2**(4): 358–377.

Cameron, A. T. and Quiggin, J. (1994) Estimation using contingent valuation data from a 'dichotomous choice with follow up' questionnaire. *Journal of Environmental Economics and Management*, **27**: 218–234.

Dehejia, R. H. and Wahba, S. (2002). Propensity score matching for non-experimental causal studies. *Review of Economics and Statistics*, **84**(1): 151–161.

Ghassemi, F., Jakeman, A. J. and Nix, H. A. (1995). *Salinization of Land and Water Resources: Human Causes, Extent, Management and Case Studies*. Sydney, Australia: University of New South Wales Press, Ltd.

Hanemann, N., Loomis, J. and Kanninen, B. (1991). Statistical efficiency of double-bounded dichotomous choice contingent valuation. *American Journal of Agricultural Economics*, **73**(4): 1255–1263.

Heckman, J. J. and Vytlacil, E. J. (2007). Econometric evaluation of social programs, part I: causal models, structural models and econometric policy evaluation. *Handbook of Econometrics*, Vol. **6B**. Elsevier: Amsterdam, pp. 4780–4874.

NMSA (National Metrological Services Agency) (2001). *Initial National Communication of Ethiopia to the United Nations Framework Convention on Climate Change.* Addis Ababa, Ethiopia: National Metrological Services Agency.

Richards, L. A. (1954). *Diagnosis and Improvement of Saline and Alkali Soils, US Department of Agriculture Handbook*, Vol. **60**, Washington DC: US Department of Agriculture, p.160.

Ritzema, H. P., Kselik, R. A. L. and Chanduri, F. (1996). Drainage of irrigated lands: irrigation water management training manual. Food and Agriculture Organization of the United Nations, 9, Rome, Italy.

Rogers, W. H. (1993). Regression standard errors in clustered samples. *Stata Technical Bulletin*, **13**: 19–23.

Rosenbaum, P. R. and Rubin, D. B. (1983). The central role of propensity scores in observational studies for causal effects. *Biometrika*, **70**(1): 41–55.

Smedema, L. K. and Rycroft, D. W. (1983). *Land Drainage: Planning and Design of Agricultural Drainage Systems.* London: Batsford Academic and Educational, Ltd.

Umali, D. L. (1993). Irrigation-induced salinity: a growing problem for development and the environment. World Bank technical paper. The World Bank. Washington DC.

Vagen, T. G., Eyasu, Y., Kibrom, G. and Pandey, R. (2000). *Studies of the Tsalet River Diversion Scheme and Mayshum Micro-dam Project in Tigray, Ethiopia.* Ås, Norway: Jordforsk.

Wodon, Q. T. (1999). Micro determinants of consumption, poverty and growth and inequality in Bangladesh. Policy Research Working Paper, 2076, The World Bank, Washington DC.

Wooldridge, M. J. (2002). *Econometric Analysis of Cross-section and Panel Data.* Cambridge, MA: MIT Press.

Yazew, E., Abraha, G. G., Hagos, F. *et al.* (2007). Water harvesting for poverty reduction and sustainable resource use: environmental and technical issues. PREM Working Paper 07/02. IVM, VU University Amsterdam, the Netherlands.

13 · Water services, dam management and poverty in the Inner Niger Delta in Mali

PIETER J. H. VAN BEUKERING,
BAKARY KONE AND LEO ZWARTS

13.1 Introduction

Mali's Poverty Reduction Strategy Paper (PRSP) constitutes the sole framework for Mali's development policies and poverty reduction strategies (GoM 2002). This influential document highlights the need to exploit the country's hydro-electric and hydro-agricultural potential in the order of 5000 GWh/annum and 2 million hectares, respectively.

A review of the PRSP by the International Development Association (IDA) and the International Monetary Fund (IMF) confirms this, and states, 'further development of Mali's untapped hydrological potential is a critical need, as it directly addresses one of Mali's core vulnerabilities, that of the temporal and spatial variability in rainfall, as well as the uncertainty of climatic conditions' (IDA and IMF 2003). Although Mali's hydro-electric and hydro-agricultural potential has yet to be fully realized, it is widely questioned whether the costs and benefits of such mega-investments are properly estimated. Besides the economic feasibility (i.e. direct costs and benefits) of additional dams, it is still unclear what the indirect effects of hydro-electric and hydro-agricultural schemes are on downstream beneficiaries of rivers.

The overall aim of this chapter is to support decision-making at a basin level with regard to the management and construction of dams and irrigation schemes in the Upper Niger, and how this may affect food security and ecological conditions downstream (in the Inner Niger Delta). This is achieved by conducting an extended cost–benefit analysis (CBA) for the

Nature's Wealth: The Economics of Ecosystem Services and Poverty, ed. P. J. H. van Beukering, E. Papyrakis, J. Bouma and R. Brouwer. Published by Cambridge University Press, © Cambridge University Press 2013.

main economic sectors: agriculture, fisheries, cattle breeding, transport and electricity.

The chapter is structured as follows: The methodology underlying the CBA of dams and irrigation schemes in the Niger River basin is explained in Section 13.2. The valuation of the direct costs and benefits of three policy scenarios for the Office du Niger, Sélingué and the Fomi dams and one baseline scenario (scenario Ø) with no dams are described in Section 13.3. The indirect costs and benefits of the four scenarios are also estimated. The sectors affected indirectly include agriculture, fisheries, livestock, transport and biodiversity. The extended CBA is then conducted. Conclusions are drawn in Section 13.4.

13.2 Data and methodology

This study takes a novel approach to the estimation of the costs and benefits associated with dams in the Niger River basin. CBA is an indispensable economic tool in any large infrastructure project. Dams are no exception. Traditionally, a CBA would be performed using a limited set of parameters. In most cases the costs were restricted to the direct capital investment, construction costs and operational costs. Likewise, only direct (measurable) benefits, such as power generation, irrigation benefits and tourism were taken into account. Nowadays, social and environmental effects are increasingly considered in the planning of dams, through the application of an extended CBA. This analysis requires economic valuation of indirect costs and benefits (Aylward *et al.* 2000).

Like any other large infrastructure project, dams require substantial investments in the planning and construction phase. Investments take the form of financial capital as well as technology and human resources. In comparison with initial investment costs, operation and maintenance costs for dams are relatively low. Besides initial investments and operational costs, large dam projects often have significant impacts on society and the natural environment. These can represent an additional cost of the project. The best example of social impacts caused by large dam projects is the displacement and resettlement of inhabitants of the flooded area. Whereas in the past, resettlement used to be overseen in the planning phase, at present, resettlement costs are increasingly budgeted in project planning. Environmental impacts associated with dams include reduction in wetland habitat and restricted fish migration. As with social impacts, the costs of mitigating environmental impacts are more likely to be included in project planning than in the past.

Direct costs for dam construction projects vary significantly as a result of site characteristics. The World Commission on Dams (WCD 2001) conducted a worldwide survey on the costs of dams and their results have been used to estimate operation and maintenance costs. The direct costs of dams can be divided into four main categories: (1) construction costs (major component of the total project); (2) resettlement costs (zero to 25%, depending on the local demographic situation); (3) environmental mitigation costs (fish migration systems, habitat restoration and artificial flooding of wetlands); and (4) operation and maintenance costs (1–3%).

Indirect costs and benefits have been estimated using the 'impact pathway approach' (Figure 13.1) on the basis of a series of methodological steps, including (1) definition of the boundaries of the study (the four scenarios): evaluate different water management scenarios along the Niger River, with a special emphasis on the Inner Niger Delta. The temporal boundaries of the project are 2005 to 2030; (2) identifying significant impacts: fishery, agriculture, livestock, transport and biodiversity as economic activities and services in the Inner Niger Delta; (3) physically quantifying these impacts: development of a model that approximates the main effects of each scenario on the various benefit categories and evaluates the changes for the various districts (i.e. upstream and downstream). To calculate these impacts, simplifying assumptions were adopted, such as for climatic and hydrological conditions and future economic activities; and (4) finally calculating the monetary values and conducting a sensitivity analysis.

Another important assumption underlying the model relates to annual climate variations. The annual rainfall in the catchments area of the Upper Niger varies between 1100 and 1900 mm, with an average of 1500 mm.

13.3 Results and discussion

For the period 2005–30, a simulated flooding area scenario for the Niger Delta has been applied to all sectors (Figure 13.2), in which the projected climate change has been taken into account.

13.3.1 Costs

The CBA of the three artificial dam structures in the Upper Niger is somewhat unusual because it compares the Office du Niger Irrigation Zone and the Sélingué Dam (which were established a long time ago) with the Fomi Dam (which has yet to be built). To make a meaningful comparison, we consider a future time period of 2005 to 2030, in which

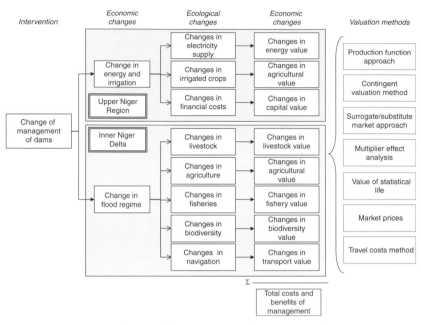

Figure 13.1 Impact pathway of the economic evaluation procedure

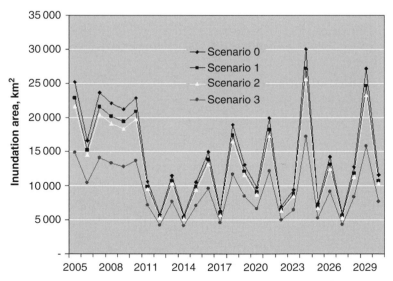

Figure 13.2 Simulated flooding area for the four scenarios (in km²)

we assume all dams are active and subsequently generate benefits. In valuing the capital costs, the following assumptions have been made. First, depreciation of the capital stock is set to 0.5% per year. Of the rehabilitation costs from the past, 25% is assumed to be additional investments in fixed capital (e.g. roads, canals, turbines). Moreover, in the early stages of operation of the dam and the irrigation scheme, the operation and maintenance (O&M) costs are assumed to be 2% of the value of the capital stock (WCD 2001). To account for ageing of the infrastructure, this fraction increases by 1.25% per year. Therefore, the more recently the dams and irrigation schemes have been established, the lower the O&M costs. International funding agencies and national donors have covered most of the investments in dams and irrigation schemes in Mali. The opportunity cost of capital has been set to 8% of the actual capital stock. Various sources have been consulted to estimate capital and rehabilitation costs for the two existing dams, including the Office du Niger area (Schreyger, personal communication, 2002, Slob, personal communication, 2002), and for the proposed Fomi Dam (Agence Canadienne pour le Développement International 1999).

Limited information is available on the financial costs of the Sélingué Dam. Table 13.1 shows the initial investments and rehabilitation costs made for the Sélingué Dam for the period 1980–2002 and the Office du Niger. Note that costs for the 80-year period by the Office du Niger include the costs of the construction and rehabilitation of the dam itself, as well as the development of the irrigation area, which presently measures around 70 000 ha. The irrigation area is expected to expand by another 40 000 ha by 2030. The cost of the expansion of the irrigation area is estimated at €2300 per hectare.

The construction of the Fomi Dam was initially considered several years ago. As such, most of the background information originates from the late

Table 13.1 *Financial costs of Sélingué Dam for the period 1980–2002 and the Office du Niger for the period 1919–96 (in €)*

Year	Initial investment	Rehabilitation	Total costs
Sélingué Dam (1980–2002)	53 361 793	49 010 000	92 371 793
Office du Niger (1919–1996)	163 700 000	181 170 000	385 870 000

Source: Schreyger, personal communication, 2002; Slob, personal communication, 2002.

1990s. Still, limited financial information is available. The 42-m high dam is expected to produce 374 GWh per month and is scheduled to provide irrigation to almost 30 000 ha of cultivable land (Zwarts *et al.* 2005). The construction period of the Fomi Dam itself will take 44 months. The costs are expected to be around US$300 million (Agence Canadienne pour le Développement International 1999).

13.3.2 Benefits

Benefits for each sector have been estimated directly (in euros) from the available production information, including fisheries, livestock, agriculture and transport, as well as indirect benefits of biodiversity.

13.3.2.1 Electricity

Theoretically, the installed capacity of the Sélingué Hydropower Plant is 47.6 MW. This means that the plant could produce 34.8 GWh per month if all four turbines are available and the reservoir is full. In reality, the maximum generated energy is around 25 GWh per month, which is around 70% of the theoretical value. The value added of one kilowatt-hour is FCFA 75. (The exchange rate applied for the Francs Communauté Financière Africaine (FCFA) against the euro is 660). Because the Fomi Hydropower Plant is scheduled to have a maximum installed capacity of 90 MW at full head, we assume that the power production is twice as big as that of Sélingué: 26 GWh per month. The same value of one kilowatt-hour is assumed.

13.3.2.2 Agriculture

The agricultural sector in and around the Inner Delta can be subdivided into irrigated agriculture and flood-related agriculture, and separate production functions have been applied in the simulation model. For reasons of simplicity, the value added of rice and other crops has been set to FCFA 95 000 and 75 000 per metric tonne, respectively. The main contribution to agricultural production in Mali comes from the Office du Niger of which the area is assumed to expand by 1500 ha per year. The other important source of rice, sorghum and other crops in the region is expected to be the Fomi Dam. Parallel to the implementation of the hydropower capacity, the irrigation area will be developed over a period of 15 years, at 2000 ha per year.

13.3.2.3 Fisheries

Fisheries are heavily affected by changes in the inundated area (Figure 13.3). The short-term fluctuations are caused by the standard

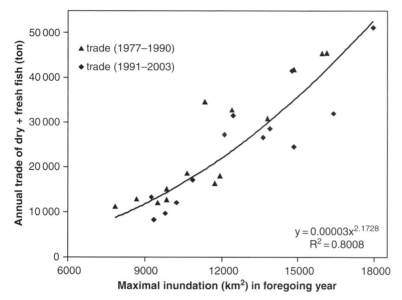

Figure 13.3 Annual trade (ton fresh fish equivalents) of dry and fresh fish in the Inner Niger Delta as a function of the maximal inundation in the previous year

variation in climate conditions. Scenario 0 generates the highest benefits, while each additional dam in operation leads to a further reduction in the fisheries industry. The differences in fish catch are particularly large during wet years.

13.3.2.4 Livestock

Livestock is valued on the basis of meat production (Figure 13.4). It is assumed that on average 2% and 8% of the sheep and goats, and cattle, respectively, is marketed each year. The year-to-year fluctuations are smaller than for fisheries due to the mobility of the herds. However, livestock is vulnerable to long-term droughts, as demonstrated by its collapse in the period 2010 to 2013, which is modelled as extremely dry. Another lesson from Figure 13.4 is that in extremely wet years (i.e. 2005 to 2010) the presence of dams can actually benefit cattle, sheep and goats. This is due to the fact that livestock heavily depend on the availability of bourgou (i.e. hippo grass). If the water level is too high, bourgou is negatively affected, and so are the cattle. By dampening the extreme peak flows and thus creating a more optimal bourgou habitat in extremely wet years, scenario 3 performs well in periods with abundant rain. By reducing the

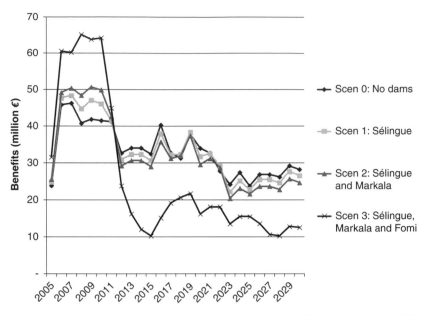

Figure 13.4 Benefits in the livestock sector over time for the four scenarios (in million €/year)

peak flow far beyond optimal levels in extremely dry years, scenario 3 performs poorly during years with exceptionally little rain.

13.3.2.5 Transport

The seasonal variations in depth of the waterways of the Niger greatly affect the transport potential of the river. The scenarios that perform best are the Sélingué Dam and Office du Niger. These dams secure sufficient water in the dry season for the smaller boats, without causing too much damage in the wet season for the larger boats. Depending on whether the year is relatively wet or not, scenario 0 (no dams) and scenario 3 (Fomi) switch positions. In extremely dry years the system with the Fomi Dam performs better in transport terms, while in wet years the absence of dams is preferred.

13.3.2.6 Biodiversity

The Inner Niger Delta is a Ramsar Wetland Site of International Importance, designated under seven of the possible eight criteria, thus it is important for its biodiversity values. For example, it accommodates two of the largest known breeding colonies of large wading birds in Africa and in addition, is a vital part of the eco-regional network supporting up to 3–4 million staging waterbirds, residents and migrants from all over Europe and Asia (Zwarts *et al.* 2005). The hydrological and related ecological conditions in the Inner Delta largely determine the population size of these waterbird species. Behaviour of migratory water birds reveals the interrelations between different wetland ecosystems, thousands of miles apart. An example is the direct relationship between breeding population size of purple herons in Europe and water levels in the Inner Niger Delta, where they reside outside their breeding season. This international connection represents an economic value. To capture this value, a survey was carried out in the Netherlands in which Dutch citizens were asked about their willingness to support protection of birds in the Netherlands and in sub-Saharan Africa (Brouwer *et al.* 2008, Sultanian and van Beukering 2008). The average willingness to pay has been estimated at around €15 per household per year. If extrapolated across Europe, the fund available for migratory bird protection would be more than €2 billion. It is assumed that 1% of this amount is available for bird protection in Mali in 2005. Birds in the Inner Niger Delta depend heavily on bourgou, which does not grow well in extremely deep water, so that scenario 2 scores somewhat better than scenario 0 in extremely wet years. However, across the full period, a situation without dams generates the highest biodiversity value. Scenario 3 leads to an extremely low value of biodiversity. The reduction in flooded area that results from the Fomi Dam forces the water birds to concentrate in limited areas, which not only restricts the availability of food, but also makes them more vulnerable for human exploitation.

13.3.3 Cost–benefit analysis

The present value, calculated at a discount rate of 5%, of the overall net benefits of the four scenarios was both aggregated over the full period and expressed as annual values. These values represent the total net economic value of each scenario. As shown in Table 13.2, scenario 2 generates the highest net benefits, while scenario 3 generates the lowest. This implies that construction of the Fomi Dam will have a negative impact on the overall economy.

To analyse incremental economic impact of the three combinations of dams, the difference of the dam scenarios – 3 with scenario 0 (no dams) were considered. These additional net benefits of the three dam scenarios have been calculated by subtracting the overall net benefits of scenario 0 from those of scenarios 1, 2 and 3. The difference between

Table 13.2 *The present value (PV) of the net benefits of the four dam scenarios*

	Overall		Marginal	
Scenario	PV of net benefits (in million €)	PV of annualized net benefits (in million € per year)	PV of net benefits (in million €)	PV of annualized net benefits (in million € per year)
Scenario 0	1903	132	–	–
Scenario 1	1971	137	68.5	4.8
Scenario 2	2283	159	380.2	26.4
Scenario 3	1781	124	−121.8	−8.5

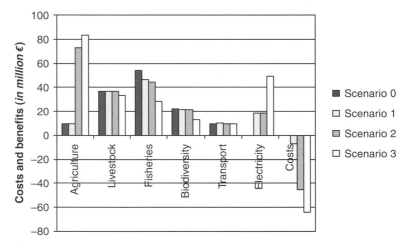

Figure 13.5 Overview of the sectoral benefits and costs of the four scenarios (PV = present value; million €/year)

scenarios 2 and 3 represents the additional net benefit of the Fomi Dam compared to the present situation (Markala and Sélingué). By building the Fomi Dam, society at large will lose more than €500 million (i.e. €121 million + €380 million), which implies an annual loss of €35 million (i.e. €8.5 million + €26.4 million) as shown in Table 13.2. The Sélingué Dam generates additional net benefits of €68.5 million until 2030. The Markala dam is economically the most attractive dam of the three by generating aggregated net benefits of €312 million (i.e. €380 million − €69 million), which is equal to almost €22 million per year (i.e. €26.4 million − €4.8 million).

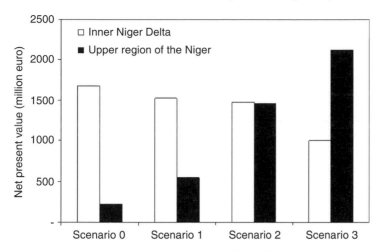

Figure 13.6 Spatial distribution of the overall benefits divided between the Inner Niger Delta and the Upper Niger region, which includes Mali and Guinea

An important dimension of the study is the spatial distribution of the benefits under the different scenarios. Besides changes in the absolute level of welfare, dams are likely to cause transfers of benefits from one region to another. Figure 13.6 shows the allocation of overall benefits between the Inner Niger Delta and the Upper Niger region. The Upper Niger region includes all those districts in Mali and Guinea in which dams generate economic activities such as irrigated agriculture and hydropower. In Mali these districts are Segou, Macina, Niono and Yanfolila. The pattern in Figure 13.6 clearly shows that with each additional dam, benefits are transferred from the Inner Niger Delta to the Upper Niger region. This transfer is especially significant in scenario 3. This implies that the Fomi Dam will substantially benefit Guinea at the expense of Mali.

13.4 Conclusions and policy recommendations

After combining information provided by hydrologists, ecologists, engineers, fishery experts and agriculturalists, the economic analysis was the final step in a long series of scientific exercises in this study. Despite the fact that simplifying assumptions were made, several conclusions can be drawn from the results of this study.

The economic value of dams in the Niger River depends predominantly on the amount of water diverted from the river. The Sélingué and the Markala Dams appear to be economically feasible. They jointly generate

€26.4 million of benefits per year to society at large. The addition of the Fomi Dam would reduce economic prosperity by €35 million per year.

The benefits accrue to various sectors and vary widely depending on the level of water diversion from the Niger River. The additional financial costs of the Fomi Dam are only partly compensated by additional electricity and agricultural benefits. Moreover, the indirect loss in fisheries, livestock and biodiversity downstream override these direct revenues. These negative downstream effects are less pronounced in the case of the Office du Niger Irrigation Zone and the Sélingué Dam.

Besides changes in the absolute level of welfare, dams are likely to cause transfers of benefits from one region to the other. The results clearly show that with each additional dam, benefits are transferred from the Inner Niger Delta to the Upper Niger region. This transfer is especially significant in the case of the addition of the Fomi Dam, which substantially benefits Guinea at the expense of Mali.

Dams in the Niger have mixed effects on poverty. The population of the Inner Delta experiences a significant decline in per capita income with an increase in the number of dams. The per capita economic benefits of the Upper Niger population increase with additional dams. The average river-related benefit per person increases with each additional dam from €44 (no dams), to €48 (Sélingué) to €68 (Sélingué and Markala). The Fomi Dam is expected to reduce the welfare of the Malian population from €68 to €52 per capita.

Finally, the sensitivity analysis of climatic conditions reveals that the Inner Delta (and, to a lesser degree, the Upper Niger region) suffer from increased drought. The vulnerability of the Inner Niger Delta increases substantially by the construction of the Fomi Dam.

References

Agence Canadienne pour le Développement International (1999). *Etudes de Réactualisation du Dossier de Faisabilité du Barrage de Fomi: Rapport de Faisabilité.* Montréal: SNC-Lavalin International.

Aylward, B., Berkhoff, J., Green, C. *et al.* (2000). Financial, economic and distributional analysis, Thematic review III.1. Prepared as an input to the World Commission on Dams.

Brouwer, R., van Beukering, P. J. H. and Sultanian, E. (2008). The impact of the bird flu on public willingness to pay for the protection of migratory birds. *Ecological Economics*, **64**: 575–585.

Government of Mali (GoM) (2002). Poverty reduction strategy paper. Document prepared and adopted by the Government of Mali, Bamako.

International Development Association (IDA) and the International Monetary Fund (IMF) (2003). Joint Staff Assessment of the Poverty Reduction Strategy Paper of the Republic of Mali. International Development Association (IDA) and the International Monetary Fund (IMF), Washington DC.

Sultanian, E. and van Beukering, P. J. H. (2008). Economics of migratory birds: market creation for the protection of migratory birds in the Inner Niger Delta (Mali). *Human Dimensions of Wildlife*, **13**: 3–15.

World Commission on Dams (2001). Orange River Development Project, South Africa. Case Study prepared as an input to the World Commission on Dams, Cape Town.

Zwarts, L., van Beukering, P. J. H., Kone, B. and Wymenga, E. (eds.) (2005). *The Niger, A Lifeline: Effective Water Management in the Upper Niger Basin*. Mali / The Netherlands: RIZA, Lelystad / Wetlands International, Sévaré / Institute for Environmental Studies (IVM), Amsterdam / A&W Ecological Consultants, Veenwouden.

14 · The environmental and social impacts of flood defences in rural Bangladesh

A. K. ENAMUL HAQUE, LUKE BRANDER, ROY BROUWER, SONIA AKTER AND SAKIB MAHMUD

14.1 Introduction

Flood mitigation is clearly a very important issue for Bangladesh. It is highly related to other high priority policy goals such as food security. As a result, protection against flooding has often been combined with efforts to intensify agricultural production. Most of the flood control projects in Bangladesh are so-called 'flood control, drainage, and irrigation' projects. These projects have recovered thousands of hectares of land from the floodplain through the construction of dykes or embankments. However, this measure has come under criticism because: (1) it provides benefits in terms of increased agricultural productivity and is therefore biased towards landowners; (2) it has resulted in a long run decline in soil fertility due to the cessation of sediment deposition on land during flood; (3) it prohibits the migration and spawning of fish, and thereby reduces overall fish stocks and the livelihood of fishers; (4) flood control embankments also hinder water transportation because they prevent free water flows between rivers within and outside embankments. Consequently, flood control projects often reduce the incomes of water transportation workers and force some to leave this sector.

Thus the construction of flood protection embankments has a complex pattern of positive and negative impacts for different sections and occupational groups within the affected population. Some of these impacts are through environmental channels. As mentioned above, fish stocks are negatively affected by the presence of embankments, which affects the livelihoods of those engaged in capture fisheries. This is a big problem because fisher communities in Bangladesh are generally one of the poorest

Nature's Wealth: The Economics of Ecosystem Services and Poverty, ed. P. J. H. van Beukering, E. Papyrakis, J. Bouma and R. Brouwer. Published by Cambridge University Press, © Cambridge University Press 2013.

occupation groups. Similarly, negatively affected people engaged in water transportation also tend to be from the poorest strata in the community.

Most of these impacts are long-term impacts, which are difficult to ascertain before the initiation of the project. The Government of Bangladesh has recently proposed its poverty reduction strategy plan where the key objective is to develop a *pro-poor development policy* implying that development projects must be poor-friendly in terms of their impacts. The Poverty Reduction Strategy Paper (PRSP) also envisages the provision of social security measures for the poor and vulnerable groups in the country, the majority of whom live in the floodplains. Considering this thrust in public policy along with the commitment by the multilateral and bilateral donor organizations and countries, it is important that we understand the long-term impacts of development projects using an *ex-post* framework so that future policies are relevant to Bangladesh's development priorities.

The research objective addressed in this chapter is to revisit these issues with evidence from a case study in which the long-run impacts can be assessed. Our chosen flood control project for this research is the Meghna Dhonogoda Irrigation Project (MDIP), which was completed in 1988 and for the last two decades has protected its target area from flooding. MDIP is located in the Matlab North thana,[1] about 120 km south of Dhaka with a population of nearly 299 000 (2001 census data).

The structure of this chapter is as follows: Section 14.2 describes the study and control sites used in the analysis and presents the household survey conducted for this appraisal; then the framework for the cost–benefit analysis (CBA) is aligned. Section 14.3 presents the estimations of costs and benefits for each impact category and sets out the results of the CBA and examines the distribution of impacts across occupational groups. Section 14.4 provides conclusions and policy recommendations.

14.2 Data and methodology

14.2.1 Case study areas

14.2.1.1 The study site
The MDIP project is situated in the Chadpur district in the south-east of Bangladesh. The project has a gross area of 33 220 ha of land[2] and water

---

[1] An administrative unit in Bangladesh under one police station.

[2] We have used GIS maps to determine the exact land and water area in the project. This produced a figure that is about double the estimate of 17 584 ha quoted in existing documentation of this project site.

bodies inside it. It is bounded by the Meghna River to the north and west and by the Dhonogoda River to the south and east. Total agricultural land in the area varies significantly depending on the season. Of the total land area, 24% is used as aman crop land,[3] 31% is used as boro crop land,[4] 2% is used for potato production, 15% is used for human settlement, 12% for other production and 16% are water bodies.[5]

In terms of land elevation, nearly 47% of the land is low lying (without the embankment this land would be mostly flooded during the monsoon months), 26% is medium land that floods occasionally, and 15% of land is above flooding level. Most of the highland is used for human settlement, whereas other land is used primarily for crop agriculture.

The MDIP project was completed at a cost of US$ 32.8 million to protect nearly 19 060 ha of land, and to provide irrigation facilities to 14 175 ha of crop land (CIRDAP 1987, p. 9). The project consists of 64 km of embankment, 282 km of canal system for irrigation and 125 km of drainage canals.

The MDIP has been the subject of three earlier appraisals. It is useful to review these studies briefly in order to allow a comparison with the results of the present study. The first of these appraisals was conducted by the Asian Development Bank (ADB) prior to funding the project (ADB 1977). This analysis included just the value of increased rice yield on the benefit side, and just the construction and operation and maintenance costs on the cost side. Other potential benefits were identified but not monetized, including employment opportunities, environment and foreign exchange savings. The central net present value (NPV) estimate from this study is US$ 7.6 million per year, and the estimated internal rate of return is 17.9%. This appraisal has been complemented by a second project completion report by the ADB, which details the time and cost overruns experienced by the project (4 years and 36%, respectively), and the breaching of the embankment in 1987 and 1988 (ADB 1990).

An *ex-post* appraisal of the MDIP was conducted in 1992 for the Bangladesh Ministry of Irrigation, Water Development and Flood Control (Hunting Technical Services Ltd. 1992). This third study finds a highly negative outcome of the project. In addition to the high cost overrun, the benefits to agriculture were found to be overestimated in the

---

[3] A rice season in Bangladesh where rice is produced from mid–March to mid–November.

[4] A rice season in Bangladesh where rice is produced from mid–March to early August.

[5] The Centre for Environmental Geographic Information System (CEGIS), Dhaka provided GIS data for this research.

original feasibility study and appraisal, and a devastating impact on capture fisheries is identified. The NPV was now estimated to be US$ -8.7 million and the internal rate of return (IRR) to be 6.7%.

The appraisal described in this chapter extends further than these existing studies in terms of the long-run assessment of impacts from the embankment and also in the use of production functions to estimate the net impacts of the embankment on agriculture and capture fisheries.

14.2.1.2 The control sites

Since we produced a partially *ex-post* CBA for understanding the overall impact of the MDIP, we needed to develop suitable control areas that would meaningfully represent the 'without project' scenario for the project. After studying the locality, we chose two separate thanas that might effectively give us a 'without project' scenario. Matlab South was part of the Matlab thana until 1991, after which the original Matlab thana was divided into two. Matlab North is the MDIP project area and Matlab South is the area outside the MDIP project. However, there are a number of reasons why Matlab South may not represent a perfect 'without project' scenario: (1) Matlab South is much more urbanized than Matlab North; (2) Matlab South is better linked with the rest of the country by road and hence its economic activities might have been significantly influenced by the factors outside the locality; and (3) Matlab South is less prone to flooding than Matlab North because it is located on the south-eastern bank of the river Dhonogoda and not on the larger Meghna river. Considering this, we also used a slightly distant location, Homna thana, as another control area. Homna is located on the south-eastern bank of Meghna river, is a flood-prone area, and is currently without embankments.

14.2.2 Analytical framework

The analytical framework used in this study to assess the economic, environmental and social implications of the MDIP is CBA. CBA is an evaluation method in which all the costs and benefits associated with a project are expressed and compared in monetary terms. Through the calculation of net present values (the discounted stream of future benefits minus the discounted stream of future costs), CBA provides an indication of how much an investment contributes to social welfare.

CBA is essentially a 'with and without' analysis, i.e. it involves a comparison of the value of economic activities and environmental services in an observed scenario (with embankment) against their values in a counter-factual scenario (without embankment). The elaboration of a counter-factual scenario for the MDIP is assisted by the use of 'control sites' that have similar characteristics to the study site but do not have a flood control embankment. In this study we use two control sites (see Section 14.2.1.2).

As previously mentioned, the construction of flood protection embankments results in a complex set of both positive and negative impacts that are distributed over the lifetime of the project. The categories of impacts that we include in this appraisal are: (1) agricultural crop production; (2) capture fisheries; (3) aquaculture; (4) livestock and poultry; (5) fruit trees; (6) housing; (7) human health; and (8) construction and maintenance costs.

The presence of the embankment results in avoided damage costs of flooding to crops, culture fisheries, property and health. These benefits of the embankment are estimated by comparing the damage costs per household within the project area with the damage costs in the control sites. The difference is then extrapolated over the relevant population and the lifetime of the project. The presence of the embankment, however, is observed to cause increased water-logging during the rainy season, which results in damage across the various impact categories. Although the embankment allows cropping all year round and prevents flood damage, it also prevents the deposition of nutrients by flood waters and results in declining soil fertility over time. These costs of the embankment are estimated in a similar manner and included in the appraisal. In the case of crop agriculture, the impact of the embankment is more complex and is estimated using a production function approach.[6] By estimating a production function for crop agriculture, these opposing effects can be taken into account. Similarly, a production function is estimated for capture fisheries. These estimations are explained in more detail in Section 14.3.1.

In addition to the direct impacts of the embankment on damage costs, the presence of the embankment is also observed to influence land use patterns and population characteristics – and these also need to be taken

---

[6] A production function describes the relationship between inputs and outputs in production. For example, the production of rice may be described as a function of labour, fertilizer, pesticides, soil fertility, etc. A change in quantity or quality of an input may result in both a change in total output and a change in the use of other inputs (and therefore the cost of production). For example, a reduction in soil fertility results in both a reduction in the harvest of rice and an increase in the quantity of fertilizer required to produce a given quantity.

into account in the appraisal. To accommodate changes in land use (i.e. a higher proportion of land being used for crop agriculture and aquaculture under the project scenario), the land-use pattern of one of the control sites is used as the land-use pattern under the 'without project' scenario. Population growth, density and overall pattern of occupation are also likely to follow a different path following construction of the embankment. We use data from four separate points in time to extrapolate the changes in population and its distribution. The first two sets of point estimates on the population and its occupational distribution came from CIRDAP (1987) for 1978 and 1986, the third point estimate came from MIWDFC (1992) – popularly known as FAP-12, and the final estimate is derived from our household survey for 2005. Based on these estimates, linear interpolations were made to estimate the population for intermittent years. After 2005 we assumed no further changes to provide a more conservative estimate of changes.

As mentioned above, the various costs and benefits of the MDIP project are distributed over time. The costs of construction are borne in the initial phase of the project, whereas the benefits (e.g. increased agricultural output) and maintenance costs accrue after completion of the embankment construction and until it ceases to function. It is therefore necessary to select an appropriate time horizon over which to estimate the costs and benefits associated with the project. We selected a 50 year time horizon, starting in 1978 (the year in which construction started) and ending in 2038. The analysis in this study is therefore partially *ex post* (i.e. for the period 1978–2005) and partially *ex ante* (for the period 2006–38). Although it is possible that the embankment may continue to function beyond 2038, any discounted costs and benefits accruing after this point will be negligible given any realistic discount rate. We do not consider the case that the embankment is effective for a shorter time horizon.

14.2.3 Survey design and the sample

This study uses data from an extensive rural household survey which investigated agricultural production, aquacultural production, fisheries (capture fisheries), nutrition, recent damage due to flooding in the control area, recent damage due to water-logging in the project area, other health-related information, willingness to pay (WTP) for flood protection, agricultural and fish production, and general demographic, socio-economic characteristics of residents.

A total of 589 households from the project area and 672 households from Homna and Matlab South were interviewed face-to-face in 2005 using a stratified random sampling procedure. A structured questionnaire with several modules for each subsection of information was used. The questionnaire consists of five main parts, three of which were designed for specific occupational activities (including household production and consumption patterns). Based on the pattern of occupational distribution used in the Flood Action Plan study on the MDIP project, the survey team selected six categories of households in 60 villages. Of these villages, 28 are from the project area and the rest are from control areas.

Prior to the survey, the field investigators were trained and questionnaires sufficiently tested to reduce possible errors and biases during the survey.

In addition to the household survey, 45 semi-structured key informant interviews were carried out by the research team. Whereas some quantitative information was asked from the key informants, i.e. population of the village, per capita income of villagers, water level during flood, etc., most of the information collected was qualitative in nature. Interviews were designed for individuals from different professional backgrounds and were conducted by local college teachers who were trained and briefed thoroughly about the objective of the interviews. Local primary school teachers, fishing community leaders as well as field-level agricultural extension officials, health workers and NGO workers were also interviewed for the study.

The key informant interviews were conducted from the second week of April to the second week of May 2005. On average each interview with key informants lasted for 1.5 hours. The questionnaire covered impacts of flooding on different occupational groups, coping mechanisms during and after flood, and information regarding household activities during normal and flood years.

14.3 Results and discussion

14.3.1 Benefit and cost estimation

The following impacts were estimated as input into the CBA: (1) agricultural production; (2) capture fisheries; (3) aquaculture; (4) livestock and poultry; (5) fruit trees; (6) housing; and (7) human health. The impact on agricultural production and capture fisheries are directly estimated using the production function approach while other benefits are estimated indirectly using a two step procedure. In the first stage agricultural benefits are estimated. In the second stage other impacts are estimated using a proportion of households affected and the proportion of damage as a percentage of agricultural benefits.

As mentioned above, a production function approach has been used to measure the impact of the embankment on agriculture and fisheries. Combined data for the project and control sites were used to estimate production functions for rice production and fish catch, controlling for variation in input variables and site effects. Given the following production function:

$$Q = f(I, \theta) \tag{14.1}$$

Where, Q is the output, I is the vector of inputs, θ is the site dummy with a value of 0 for the project area and 1 for the control area. $\Delta Q = f(I, 0) - f(I, 1)$ measures the impact of the embankment. A positive value means a gain and a negative value implies a loss due to the embankment. Four production functions were estimated, three for rice production (see Table 14.2) and one for capture fisheries (see Table 14.3). These production functions are described in more detail below.

For benefits and cost estimation in terms of health, poultry, livestock, agricultural equipment, housing, trees and aquaculture, this study resorted to indirect measurement using the benefit from agriculture as the benchmark. There are two reasons for using this method: First, to estimate these benefits using the production function approach, a large set of data is required, which could not be collected during the field survey (mainly because it would have increased the number of questions in the questionnaire and would potentially have caused non-sampling errors). Second, protection of agricultural output from flooding was the main reason for construction of the embankment. Therefore, these benefits were estimated as a proportion of agricultural benefits. The following formula is used to estimate the benefits from protection of flooding and costs due to waterlogging in the project area:

$$B_i = \Gamma \cdot \gamma_i \cdot \alpha_i \tag{14.2}$$

Where Γ is the mean estimate of damage in agriculture found in the no-embankment scenario, γ_i is percent of households affected by flooding and reported ith type of damages, and α_i is the proportion of ith type of damage as a percent of agricultural damage due to flooding. As a result, B_i measures the ith type of benefit due to flood control in the project area.

$$C_i = \Lambda \cdot \varphi_i \cdot \beta_i \qquad (14.3)$$

In Equation (14.3), C_i measures the cost due to water-logging. Λ is the mean estimates of damages in agriculture due to waterlogging in the project area, φ_i is the percent of households who reported ith type of damage in the project area due to waterlogging, and β_i is the percent of agricultural damage.

14.3.1.1 Agriculture

There are four major crops produced in the project area. These are: (1) aman rice, (2) aus rice, (3) boro rice and (4) potato. Table 14.1 presents the seasonal production of these crops. The table shows that the aus and aman crops are likely to be directly affected by flooding without flood protection. Hence, the embankment reduces the damage to aus and aman rice crops and benefits agricultural production. In addition, by preventing flooding there may also be an increase in the net cropping area.

Data on both output and inputs were collected for all crop types from both the project and control areas. Input data on the use of fertilizer, pesticide, labour (both home and hired labour), draught animals, tractors and irrigation technology were collected in the household survey. However, we did not have enough observations to estimate a production function for wheat and potato. Consequently, benefits in terms of increased production of wheat and potato due to flood protection have not been estimated.

A Cobb–Douglas production function was estimated for aus rice using labour, draught power, fertilizer and pesticides as explanatory variables

Table 14.1 *Cropping seasons in the MDIP area*

Crop	Jan	Feb	Mar	Apr	May	Jun	July	Aug	Sep	Oct	Nov	Dec
							Flooding months					
Boro rice												
Aus rice												
Aman rice												
Wheat												
Potato/ Vegetables												

(see Table 14.2). The estimated coefficient on each input variable measures the elasticity or responsiveness of output to a change in each input. The model suggests that labour is the most influential input, followed by pesticide and draught power. Fertilizer is not significant but it has the correct sign.

The variable of interest for this study is the site dummy, which is positive and significant. This indicates that, *ceteris paribus*, aus rice productivity is higher in the control site than in the project site. This is also true for boro and aman rice production. Productivity estimates, therefore, show that after almost 20 years of operation of the flood control project, land productivity in the project area has decreased compared to that of the control area, despite crops being protected from flooding.

In theory, the reduction of land productivity could be due to a large number of factors such as crop variety, production technology, farmers' knowledge of production system, soil quality, etc. However, during the field survey it was observed that the same variety of rice is produced in these two regions, that production technology in the two areas is very similar, and that there are no significant differences in the education level of farmers. Given this, it can be concluded that the productivity difference is mainly due to differences in soil fertility, which has been reported by many farmers during the field survey.

As a result of the embankment, the land-use pattern within the embankment area has changed. This has two opposing effects on crop production. On one hand, the increased cropping intensity has reduced the productivity on a seasonal basis. On the other hand, by growing crops year round, it has increased annual productivity per hectare. In other words, while seasonal output has declined on a per hectare basis, flood-protected land is used for more than two crops a year and hence the total output per hectare of land has gone up on an annual basis. Taking this into consideration, the present value of net benefits from agriculture were calculated to be 256 million Taka[7] (or US$ 3.7 million) for the 50-year life of the project at a 10% rate of discount. Net benefits are calculated using (1) increased total production per year due to flood control, (2) plus crop damage avoided from flooding due to construction of the embankment and (3) minus damage due to waterlogging resulting from mismanagement of the irrigation and drainage networks.

---

[7] US$ 1 = 69 taka in 2006.

Table 14.2 *Estimated aggregate production functions for aus, aman and boro rice*

(All inputs are measured in per ha of land)	Dependent variable (yield in kg per ha per season)					
	ln (aus production)		ln (aman production)		ln (boro production)	
	Coefficient	t-value	Coefficient	t-value	Coefficient	t-value
Intercept	5.620	17.63***	5.425238	5.85***	6.014	27.67***
ln(draught power)	0.174	1.87*			0.260	4.95***
ln(fertilizer)	0.095	1.22	0.1540082	1.37		
ln(pesticides)	0.270	3.42***			0.068	1.92**
ln(labour)	0.296	3.42***	0.2498513	1.10	0.174	3.43***
ln(irrigation)					0.150	3.28***
Site dummy	0.220	2.34***	0.6098547	1.98**	0.155	2.02**
Target = 0						
Control = 1						
Other regression parameters	$R^2 = 0.8171$ $N = 93$ $F = 77.73$ ***		$R^2 = 0.3784$ $N = 16$ $F = 2.44$		$R^2 = 0.3218$ $N = 383$ $F = 35.77$***	

Draught power measured in number of days per ha, fertilizer in Taka value per ha, pesticides in taka value per ha, labour is home plus hired labour per ha, and irrigation water is measured in taka per ha. (US$ 1 = 69 taka).

Note: * significant at 10%, ** significant at 5%, *** significant at 1%.

Table 14.3 *Estimated production function for large fish*

Dependent variable: ln(large fish catch measured in kg of production per year per fishing team)

	Coefficient	t-value
Intercept	−0.145	−0.08
ln(no. of fishing trips per year per team)	0.468**	1.91
Dummy for boat type	−0.975[+]	−1.61
Dummy for net type	−0.008	−0.01
Site dummy (Project = 0, Control = 1)	0.559***	−2.68
Other regression parameters	$R^2 = 0.18$	
	$N = 106$	

Note: ** significant at 5%, *** is significant at 1% and [+] significant at 11%.

Capture fisheries   Using data from the household survey, we estimated a production function for capture fisheries (see Table 14.3). The site dummy was found to be positive and significant.[8] This indicates that the project area, *ceteris paribus*, has lower harvest in large fish than the control area. The present value of the loss to capture fisheries is around 10.55 million Taka (or US$ 0.15 million) over the 50-year time horizon of the project. Prices for fish were collected from key informants and are in 2005 prices.

14.3.1.2 Culture fishing

This benefit is estimated using an indirect approach. Improved flood control activity induces aquaculture activities in the flood–protected area. We estimate that there is nearly a 501 million Taka (in present value terms at a 10% rate of discount) or US$ 7.27 million increase in production of culture fishing over the 50-year time horizon of the project. At the same time, it was also observed that due to waterlogging in the area, some of these fish farms also lost huge sums of money. For a 50-year period the present value of such a loss is nearly 40.5 million taka or US$ 0.58 million (see Table 14.4).

---

[8] We initially estimated production functions for three sizes of fish (small, medium and large) but were only able to find a significant difference in production between the study site and control sites for large fish.

Table 14.4 *Estimates of damage avoided due to non-flooding in the project area*

Type of damage avoided	Taka per household per year[**]	Total damage avoided[*] (million taka)
Crop damage	16300	502.4
Agricultural equipment damage	502	0.3
Poultry and livestock damage	2861	22.0
Fruit tree damage	6864	22.9
Homestead damage	16224	47.3
Health damage	4188	10.2
Education-related damage	13556	
Other damage	73	3.4

Notes: [*]In present value terms using a 10% rate of discount for 50 years in million Taka. US$ 1 = 69 taka (2006). [**] Field survey data.

14.3.1.3 Livestock and poultry

Flooding negatively affects livestock and poultry production. Consequently, it is observed that livestock production and poultry production increased after the embankment was constructed. It was also observed that livestock gains are not significantly different between the target and control sites. However, there is significant gain in poultry production in the flood-protected area. It was also noted that waterlogging has been responsible for damages to the poultry industry. Using the control and project area data and the estimates on population and its distribution, it is estimated that the present value of avoided damage costs to poultry is 22 million taka (US$ 0.31 million), but at the same time there is a loss of 14.8 million taka (US$ 0.21 million) from poultry and livestock due to increased waterlogging over the 50-year lifetime of the project.

14.3.1.4 Property damages

Flood protection is expected to provide a significant positive benefit to people in terms of reduction in property damages. At the same time, because of the protection provided by the embankment, many people have built homes below the flood-level and these are vulnerable to waterlogging due to mismanagement of the drainage facilities. Table 14.4 shows the gains in terms of avoided flooding damage to properties in low-lying areas of the project site and Table 14.5 shows losses due to waterlogging in the project area. It is estimated that there is a net loss (in present value terms) of 30 million taka (or US$ 0.44 million) over the 50-year lifetime of the project.

Table 14.5 *Estimates of damages due to waterlogging in the project area*

Damage type	Taka per household per year[**]	Total damage costs[*] (million taka)
Crop damage	21841	245.8
Agricultural equipment damage	66	0.003
Livestock damage	681	14.3
Aquaculture damage	7968	40.6
Poultry damage	411	0.6
Fruit tree damage	4975	10.1
Homestead damage	1994	77.3
Health cost	159	0.05
Other damages	55	0.002

*In present value terms for 50 years with a 10% rate of discount and in million taka. US$ 1 = 69 taka (2006). ** Field survey data.

For all of these estimates the population projections, its distributional changes and the damage cost information from the survey data were used.

14.3.1.5 Fruit tree benefits

Again, there are both losses and gains in terms of fruit trees. This includes losses due to waterlogging in the project area and loss avoided due to flood protection as gains. Per household loss due to waterlogging has been estimated from the responses on average losses in the project area. Here per household gain or loss avoided due to flood protection is estimated from the responses in the control area. It should, however, be noted that the two areas do not have equal vegetation. Using the indirect method of estimation it is found that a net loss is 12.1 million taka (or US$ 0.19 million) in present value terms for the 50-year lifetime of the project has been avoided due to embankment.

14.3.1.6 Health benefits

Flooding causes significant health issues for people living in the flooded areas. During a flood the provision of clean drinking water becomes a major problem and the incidence of water-borne diseases increases. By reducing the level of flooding through flood protection embankments, these health problems are much reduced. The indirect method of measurement was used to estimate this benefit. Table 14.4 shows that health benefits associated with the embankment are around 10 million taka in present value terms (or US$ 0.15 million). On the other hand, waterlogging also caused some health problems in the protected area, mostly related to skin diseases as the quality of water deteriorates in the water-logged area. The monetary value of this impact is estimated to be 0.053 million taka (or US$ 0.001 million).

14.3.1.7 Agricultural equipment benefits

The majority of economic activities in the floodplains are based upon agriculture. In the flood plains farmers are usually aware of the cost of floods in terms of damage to their equipment. Data from the control area suggested that flood damage to agricultural equipment related to the tilling of land and weeding could be avoided if flood protection is put in place. However, it has also been observed in the survey that farmers in the protected area often increase the amount of equipment used due to their highly intensive use of land while waterlogging could also cause some damage. Using an indirect approach to measurement, it was found that the present value of such net benefits accounting for a 10% rate of discount for 50 years is only 0.28 million taka (or US$ 0.004 million).

14.3.1.8 Other benefits (costs)

Households living in the floodplain reported that flooding has resulted in other costs to their families. These include loss of income to continue education, temporary displacement of families and increased cost of living in flooded areas. These could have been avoided if flooding would have been controlled. On the other hand, in the protected area, families also report similar costs due to waterlogging. Using indirect measures, the net impact is around US$ 3.41 million.

Finally, Table 14.6 presents the summary of all costs and benefits for the MDIP project. Once again, to estimate future losses for years up to 2038, a population projection was used based on secondary data from the Bangladesh Bureau of Statistics (BBS, selected years). No estimates of benefits and costs in terms of biodiversity losses (due to the embankment) were used in this calculation. However, this calculation includes partial estimates of environmental costs in terms of losses in open access fisheries due to the embankment.

14.3.2 CBA results and distribution of impacts

The net present value of the MDIP project using a 10% discount rate is −1080 million Taka (US$ -15.6 million) in 2006 constant prices. The internal rate of return of the project (the discount rate at which discounted costs and benefits would be equal) is about 5.32%. This result is

Table 14.6 *Summary of benefits and costs MDIP*

Item of benefits and costs	PV in million taka (at 10% rate of discount)	Explanatory notes
Agri-equipment – benefits	0.3	Due to flood control
Agriculture production – benefits	466.8	Due to flood protection
Agriculture production – rice – benefits	35.6	Due to increased cropping intensity
Culture fishing – benefits	542.5	Due to culture fishing
Fruit tree – benefits	18.7	Due to flood control
Health – benefits	8.8	Due to flood control
Housing – benefits	47.3	Due to flood control
Livestock – benefits	3.4	Due to flood control
Other benefits	3.4	Due to flood control
Poultry – benefits	10.2	Due to flood control
Potato and wheat – benefits	–	Not measured due to lack of data
Total benefit	1136.9	(A)
Agri-equipment – costs	−0.003	Due to waterlogging
Agricultural production – costs	−245.8	Due to waterlogging
Capture fishing – costs	−10.6	Due to flood control (environmental cost)
Culture fishing – costs	−40.6	Due to waterlogging
Fruit tree – costs	−10.1	Due to waterlogging
Health – costs	−0.05	Due to waterlogging
Housing – costs	−77.3	Due to waterlogging
Livestock – costs	−14.3	Due to waterlogging
Poultry – costs	−0.6	Due to waterlogging
Other costs	−0.002	Due to waterlogging
Biodiversity costs	–	Environmental costs
Total costs	−399.3	(B)
Total project cost	−1818.0	(C)
NPV at 10%	−1080.5	(A) + (B) + (C)

different from other studies done on this project. The MDIP project had an *ex-ante* IRR value of 17.9% in its first feasibility report (ADB 1977). In order to examine the distributional effects, we look at the impacts of the project on the income levels of different occupational groups. Figure 14.1 shows the sources of income for the six main occupational groups in the project area and we use this to deduce the distributional

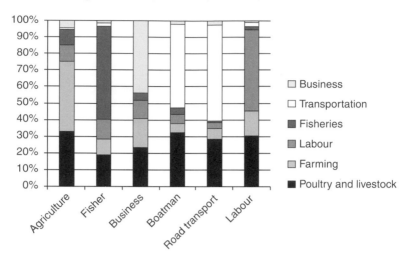

Figure 14.1 Sources of income by occupation groups

impact of the embankment. It shows that growth in aquaculture is most beneficial to landowning farmers. Similarly, growth in agriculture will mainly benefit farmers, people engaged in business activities and labourers. The benefit of the embankment to poultry and livestock production is more evenly distributed across occupational groups. The negative impacts of the embankment on capture fisheries and river transportation services are disproportionately likely to affect the fishermen and boatmen groups, respectively. Considering this, it can be inferred that the livelihood impact of the embankment is tilted against the people who are dependent on water and water resources. At the same time it is also important to note that these groups of people did not migrate to other sectors as had been predicted.

14.4 Conclusions and policy recommendations

The MDIP has come under severe criticism on environmental grounds, particularly in terms of its impact on capture fisheries. Although these impacts have not been quantified in previous appraisals of the MDIP, the Ministry of Water Resources has halted further construction of FCDI projects in Bangladesh on grounds of its potential negative impact on poverty and environment.

The analysis presented in this chapter supports this decision and shows that the MDIP resulted in a net welfare loss, i.e. the wider social costs of the project are greater than the benefits. This outcome is a consequence of a number of factors, including higher than anticipated construction costs, lower benefits to agriculture due to loss of soil fertility over time, higher waterlogging damages and the highly negative impact on capture fisheries. The combined result is a negative net present value for the project of 1080 million Taka (US$ 15.6 million) using a 10% discount rate and an associated internal rate of return of 5.32%. The unanticipated environmental impacts of constructing such an embankment have made this project an expensive mistake. The project has also had detrimental distributional consequences, which compound the already negative outcome. Although landowners as a group have gained from increased crop yields, reduced property damage and increased poultry, livestock and aquaculture production, the project has also had a significantly negative impact on two occupational groups who were dependent on water and water resources. They are fishermen and the river transport workers, both of which comprise already poor sections of society.

It is important to note that the annual undiscounted loss of around 118 million Taka (US$ 1.7 million) from waterlogging in the embankment area could have been avoided if proper management could be ensured. If waterlogging impacts could have been avoided with better management of the project, the internal rate of return would have increased from 5.35% to 7.04%. This highlights the importance of water and irrigation management during the operational phase in ensuring the success of such projects. Together with the need to address negative environmental and distributional impacts of flood protection, this is an important lesson for future flood control projects. The need for flood protection in Bangladesh is evident and, given the anticipated impacts of climate change, this need will increase in the future. Flood protection projects do, however, need to address environmental, distributional and water management issues in order to ensure that they are socially desirable.

References

ADB (1977). Appraisal of the Meghna-Dhonagoda Irrigation Project in the People's Republic of Bangladesh. Report No. BAN: Ap-18.

ADB (1990). Project completion report of the Meghna-Dhonagoda Irrigation Project in Bangladesh. PCR: BAN 21177.

CIRDAP (1987). *The Impact of Flood Control, Drainage and Irrigation (FCDI) Project in Bangladesh (CIRDAP).* Centre on Integrated Development for Asia and the Pacific, Dhaka, Bangladesh.

Hunting Technical Services Limited (1992). Project impact evaluation of Meghna-Dhonagoda irrigation project. FAP 12 FCD/I Agricultural Study.

MIWDFC (1992). Bangladesh Flood Action Plan Fap-12: Project Impact Evaluation of Meghna-Dhonogoda Irrigation Project, Ministry of Irrigation, Water Development and Flood Control, Government of Bangladesh, Dhaka.

15 · Double dividends of additional water charges in South Africa

JAN H. VAN HEERDEN, RICHARD S. J. TOL,
REYER GERLAGH, JAMES N. BLIGNANT,
SEBASTIAAN M. HESS, MARK HORRIDGE,
MARGARET MABUGU, RAMOS E. MABUGU
AND MARTINUS P. DE WIT

15.1 Introduction

The purpose of this chapter is to show how double dividends could be obtained from using market instruments to tax water use in a developing country. The double dividends are namely environmental (water conservation) on the one hand, and poverty reduction dividends on the other. We apply a water tax on selected industries in South Africa to reduce demand for water, and then transfer the revenue from this tax to the poor to achieve reduction in absolute levels of poverty.

South Africa is classified as a semi-arid country. Precipitation has been fluctuating over the years with an average of 500 mm per annum, well below the world average of about 860 mm (DWAF 2002). The total flow of all the rivers in the country combined amounts to approximately 49 200 million m^3 per year, while the National Water Resource Strategy estimated the total water requirement for the year 2000 at 13 280 million m^3 per year, excluding environmental requirements. In addition, South Africa is poorly endowed in groundwater as most of the country is underlain by hard rock formations that do not contain any major groundwater aquifers (DWAF 2002).

While currently only about 24% of rural people have access to water on site, additional sources of water supply are environmentally, financially and politically hard to develop. At the same time, unemployment in rural areas of South Africa is extremely high, which results in severe poverty conditions in these areas.

Nature's Wealth: The Economics of Ecosystem Services and Poverty, ed. P. J. H. van Beukering, E. Papyrakis, J. Bouma and R. Brouwer. Published by Cambridge University Press, © Cambridge University Press 2013.

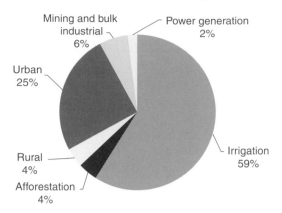

Figure 15.1 Water requirements by sector in South Africa for year 2000

The agricultural sector in South Africa is the largest consumer of water, using 59% of total water demand (Figure 15.1). Large-scale farmers primarily use 95% of irrigation water and small-scale farmers use the remainder (Schreiner and Van Koppen 2002). Afforestation requires 4% of the total water requirement, while rural and urban populations require 4% and 25%, respectively. Mining and bulk industrial, and power generation use 6% and 2%, respectively.

Depending on the poverty measure used, between 45% and 55% of all South Africans lived in poverty in 2001 (DME 2002). Even though average income statistics imply that South Africa is a middle-income country, most of the population experience serious absolute poverty or is vulnerable to poverty (Klasen 2000, May 2000, Woolard 2002). Poverty in South Africa is concentrated among the African and Coloured race groups (Table 15.1). Aliber (2002), quoting Schlemmer's work based on the All Media and Products Surveys (AMPS), shows that overall poverty has been increasing since 1993, for all ethnic groups. A poverty line of ZAR 400 per capita per month in 1989 was used.

Almost 50% of the population is income poor, spending less than ZAR 350 per adult equivalent per month, and about 70% of the poor live in rural areas (Schreiner and Van Koppen 2002). At least half of the rural population depend on remittances or pensions and grants. Water as a resource is essential to transform society towards social and environmental justice and poverty eradication (Schreiner and Van Koppen 2002). Rural people require water for drinking, hygiene and cooking and for productive purposes such as farming, livestock, forestry, fisheries and small-scale

Table 15.1 *Proportion of households below poverty line, by year and population group*

Year	Africans	Coloureds	Indian	White
1989	51%	24%	6%	3%
1993	50%	26%	8%	3%
1996	57%	22%	9%	3%
1997	55%	21%	6%	4%
2001	62%	29%	11%	4%

Source: Aliber (2002) quoting Schlemmer.

industries in order to deal with income poverty. A mere 24% of rural people have access to piped water on site and only 15% have access to sanitation.

The South African government is exploring ways to address water scarcity problems by introducing water resource management charges on the quantity of water used in sectors such as irrigated agriculture, mining and forestry. This implies that the water authorities would like to control water demand rather than trying to expand water supply. The additional charge on water used by economic sectors might improve allocation efficiency, conserve water resources and possibly impact positively on poverty alleviation. It translates into more water available for drinking, hygiene and productive activities, which can contribute to poverty reduction if revenues from this charge are transferred to the poor.

15.2 Data and methodology

15.2.1 Double dividend

The double dividend theory has been a popular research theme for at least 15 years, after papers from Pearce (1991) and Bovenberg and De Mooij (1994). The main reason for this popularity is its seeming promise of a free ride: environmental taxes could diminish pollution at no costs or even lead to additional benefits, if the revenues of the tax are used to reduce other distortionary taxes on, for instance, labour or capital. These additional benefits – the second dividend – are usually an increase in welfare or GDP or a decrease in unemployment. In this study, our focus is on GDP and poverty. In South Africa the latter is not linked to employment in general, but specifically to employment of the poor and to prices of commodities bought by poor households.

The literature suggests that for double dividends to possibly be attained there have to be initial inefficiencies in the tax system. Earlier studies that used simple one-factor models and efficient taxation found no double dividends (see Bovenberg and de Mooij 1994, Fullerton and Metcalf 1997, Goulder *et al.* 1997).

Later studies included more factors of production and therewith the possibility of the initial tax inefficiencies. These occur when one factor of production is relatively over-taxed compared to others. The later studies had mixed results and established a few important prerequisites for a double dividend. Most importantly, the burden of the environmental tax has to fall mainly on the under-taxed factor of production, and in line with this, the revenues from the tax have to be used to cut the tax rate of the over-taxed factor (Goulder 1994). General tax rules suggest that: the higher the tax rate, narrower the tax base and more elastic a production factor supply is, the higher is the likelihood that the factor will be over-taxed. The possibility of attaining a double dividend therefore depends on the initial structure of the tax system and on the characteristics of the envisioned green fiscal reform.

We propose that a tax or surcharge is levied on water use to achieve an environmental dividend, namely a reduction in water demand. Our first 'target variable' is therefore the quantity of water demanded in the economy, and if this could be reduced, we would obtain a first dividend. The typical second target variable studied in the double dividend literature is real GDP. The decrease in GDP that results from a 1 Rand increase in total tax revenue is referred to as the marginal excess burden (MEB) of a tax, that is:

MEB = decrease in real GDP/increase in real government income[1]

As both the numerator and the denominator have the same unit of measurement (Rand), the MEB measure is without dimension. By comparing MEBs of different scenarios one can identify the scenarios that produce a second dividend, i.e. an increase in GDP while maintaining total government revenue constant.

Could appropriate ways of recycling revenue from water-related environmental taxes result in double or triple dividends? We implement a computable general equilibrium (CGE) model to calculate and compare

---

[1] For readers familiar with the term MCPF (marginal cost public funds), the MCPF is equal to 1 + MEB.

the MEBs of water-related environmental taxes with the MEBs of recycling the revenue, in order to find possible double or even triple dividends.

15.2.2 Tax instruments for reduced water consumption

In South Africa, according to the National Water Act (DWAF 1998), the government is regarded as the public trustee of the nation's water resources. Under previous water legislation, pricing of water did not generally take into account the real cost of managing water, the cost of water supply and the scarcity value of water (King 2004). The principle behind the current water pricing policy in South Africa is that payment for water should not only recover full costs of supply, but should also be at a level reflecting its scarcity, except for water required to meet basic human needs, which is currently set at 25 litres of water per day per person. The South African government is introducing a water resource management charge to recover some of the costs for water management and to reflect water scarcity in the country. This means that the government is moving towards full economic costs of water by taking into account the supply cost and opportunity cost of water.

The following scenarios, which represent potential adjustments in water pricing, and which are based on suggestions proposed by water authorities and experts in South Africa, are evaluated:

1. A surcharge of 10c per m^3 water used by forestry (which translates to a 400% increase; see Table 15.2);
2. A surcharge of 10c per m^3 water used by irrigated agriculture (which translates to a 200% increase; see Table 15.2); and
3. A surcharge of 10c per m^3 water used by all mining industries (which translates to a 5% increase; see Table 15.2).

15.2.3 Recycling schemes

Three recycling or hand-back schemes are analysed: (1) a decrease in direct tax, (2) a general decrease in indirect taxes and (3) a decrease in taxes on food price. These recycling schemes are as politically sensitive as the environmental tax instruments discussed above, since business usually prefers a reduction of progressive direct taxes, while the labour unions prefer a reduction in regressive indirect taxes (Du Toit and Koekemoer 2003, p. 49). We implement the first recycling scenario via a uniform ordinary change in *ad valorem* rates of the direct tax on capital and labour. The second two scenarios are implemented via reductions in commodity taxes levied on purchases by households (VAT). In scenario two the tax reduction lowers prices of all

Table 15.2 *Average water tariffs (2002) and the semi-elasticity for water demand*

	Water tariff (ZAR/m³)	% change due to an increase of ZAR1	Elasticities	Semi-elasticities	Taxable water (million m³)
Irrigated field	0.05	2000.0	−0.25	−500.0	7152.0
Dry field	0.05	2000.0	−0.15	−300.0	0
Irrigated horticulture	0.05	2000.0	−0.25	−500.0	3400.0
Dry horticulture	0.05	2000.0	−0.15	−300.0	0
Livestock	0.05	2000.0	−0.15	−300.0	191.1
Forestry	0.025	4000.0	−0.40	−1600.0	1673.0
Other agric.	0.05	2000.0	−0.15	−300.0	24.8
Coal	2.12	47.2	−0.32	−15.3	40.3
Gold	2.12	47.2	−0.32	−15.3	284.8
Crude, petroleum and gas	2.12	47.2	−0.48	−22.6	0.7
Other mining	2.12	47.2	−0.32	−15.3	368.3
Food	4.00	25.0	−0.39	−9.8	376.4
Textiles	4.00	25.0	−0.33	−8.3	104.4
Footwear	4.00	25.0	−0.33	−8.3	0
Chemicals and rubber	2.12	47.2	−0.15	−7.2	59.4
Petroleum refineries	2.12	47.2	−0.48	−22.6	92.0
Other non-metal minerals	2.79	35.8	−0.32	−11.6	44.0
Iron and steel	2.79	35.8	−0.27	−9.8	56.2
Non-ferrous metal	2.79	35.8	−0.27	−9.8	14.0
Other metal products	2.79	35.8	−0.27	−9.8	60.0
Other machinery	4.00	25.0	−0.25	−9.5	37.3
Electricity machinery	4.00	25.0	−0.38	−9.5	6.2
Radio	4.00	25.0	−0.38	−9.5	0
Transport equipment	4.00	25.0	−0.38	−9.5	20.4
Wood, paper and pulp	2.12	47.2	−0.59	−27.8	157.5
Other manufacturing	4.00	25.0	−0.38	−9.5	13.0
Electricity	2.12	47.2	−0.80	−37.7	207.9
Water	2.12	47.2	−0.60	−28.3	5906.5
Construction	4.00	25.0	−0.38	−9.5	167.1
Trade	4.00	25.0	−0.19	−4.8	491.4
Hotels	6.11	16.4	−0.19	−3.1	319.8
Transport services	6.11	16.4	−0.19	−3.1	497.1
Community services	6.11	16.4	−0.19	−3.1	175.8
Financial institutions	6.11	16.4	−0.19	−3.1	281.3
Real estate	6.11	16.4	−0.19	−3.1	662.0
Business activities	6.11	16.4	−0.19	−3.1	26.2

Table 15.2 (*cont.*)

	Water tariff (ZAR/m^3)	% change due to an increase of ZAR1	Elasticities	Semi-elasticities	Taxable water (million m^3)
General government	6.11	16.4	−0.19	−3.1	524.8
Health services	6.11	16.4	−0.19	−3.1	331.3
Other service activities	6.11	16.4	−0.19	−3.1	198.7

Water tariff data: own analysis based on various unpublished Department of Water Affairs and Forestry, water board and municipal data.
Elasticities: CSIR 2001, DBSA 2000, Renzetti 1992, Le Maitre *et al.* 2000, Veck and Bill 2000.

consumer goods by an equal percentage. In scenario three only food becomes cheaper. All three scenarios would be simple to administer.

15.2.4 Target variables

Four target variables are calculated by the model to compare the different scenarios in terms of the three dividends: (1) environment via water consumption, (2) economy via GDP and employment, and (3) equity via total consumption by the poor. Changes in each target variable are expressed per change of government revenue, so that different policy scenarios can easily be compared to each other on the basis of equal extra tax revenues. These target variables have been chosen to see which water charge and tax-recycling scenarios would best yield environmental, economic and equity dividends.

15.2.5 The model

We use a computable general equilibrium (CGE) model for all our simulations. It is called 'UPGEM' (University of Pretoria general equilibrium model) and is based on the structure of the ORANI-G[2] model (Horridge 2000) written and solved using the GEMPACK[3] suite of

[2] Orani was the name of the original model builder's wife.
[3] GEMPACK is a suite of general-purpose economic modelling software especially suitable for formulating and solving general equilibrium models.

software (Harrison and Pearson 1996). The model has a theoretical structure that is typical of a static CGE model, and consists of equations describing producers' demands for produced inputs and primary factors; producers' supplies of commodities; demands for inputs to capital formation; household demands; export demands; government demands; the relationship of basic values to production costs and to purchasers' prices; market-clearing conditions for commodities and primary factors; and numerous macroeconomic variables and price indices. Conventional neoclassical assumptions define all private agents' behaviour in the model. Producers minimize cost while consumers maximize utility, resulting in the demand and supply equations of the model. The agents are assumed to be price takers, with producers operating in competitive markets, which prevent the earning of pure profits.

In general, the static model with its overall Leontief production structure allows for limited substitution on the production side, but more substitution in consumption. It has CES substructures for (1) the choice between labour, capital and land, (2) the choice between the different labour types in the model, and (3) the choice between imported and domestic inputs into the production process. In the short-run simulations reported here, we do not allow for substitution in production between water and other inputs. Household demand is modelled as a linear expenditure system that differentiates between necessities and luxury goods, while households' choices between imported and domestic goods are modelled using the CES structure.

In the model commodity composites, a primary-factor composite and 'other costs' are combined using a Leontief production function. Consequently, they are all demanded in direct proportion to total production. Each commodity composite is a CES function of a domestic good and the imported equivalent. The primary-factor composite is a CES aggregate of land, capital and composite labour. Composite labour is a CES aggregate of occupational labour types. Although all industries share this common production structure, input proportions and behavioural parameters may vary between industries (Horridge 2000).

The primary model database is the official 1998 SAM of South Africa, published by Statistics South Africa (SSA 2001). This SAM divides households into 48 groups (12 income by 4 ethnic), and distinguishes 27 sectors. For the purpose of this study, we split the energy intensive as well as the agriculture sectors further to arrive at 39 sectors.

The model's closure rules reflect a short-run time horizon. The capital stock in each sector is assumed fixed, while the rate of return on capital is allowed to change. The South African labour market is characterized by large unemployment of unskilled labour, and a shortage of skilled labour. The model differentiates between 11 different labour groups that are classified as either skilled or unskilled. Skilled labour is treated as human capital in inelastic short-term supply. The supply of unskilled labour is assumed to be perfectly elastic at fixed post-tax real wages (i.e. nominal post-tax wages deflated by the economy-wide CPI (the consumer price index)). The distinction between skilled and unskilled labour supply reflects the South African labour market realistically and allows for investigating the effect of certain policies on employment of unskilled labour. The supply of land is also assumed to be inelastic (Van Heerden *et al.* 2006).

It is assumed that aggregate investments, government consumption and inventories are exogenous and unaffected by the change in environmental taxes under consideration. Consumption spending by each of the 48 representative households follows labour income earned by each household, and the trade balance is endogenous. This specification allows us insight into the effect of the suggested policies on South Africa's consumption and competitiveness. All technological change variables and all tax rates are exogenous to the model. Finally, the nominal exchange rate is the numeraire in each simulation.

15.3 Results and discussion

15.3.1 Environmental effects

The first of a double or triple dividend is the environmental dividend reaped. Table 15.3 shows that all the simulations yield the first dividend, irrespective of the recycling method used. The first dividend here is a net decrease in the amount of water consumed per unit of real government revenue. An additional charge on water consumption always leads to a decrease in water demand. All that is needed for the environmental dividend to occur is that the increase in water consumption that results from a direct or indirect tax break is less than the decrease due to the water charge. The model results as shown in Table 15.3 indicate that this is expected to be the case.

The decrease in water consumption as a result of the water charges (second column) greater in absolute terms than the increases in water consumption because of the three possible tax breaks (third row), thereby yielding the environmental dividend in all the combinations of taxes and recycling schemes. For example, a tax on forestry results in a decrease in

Table 15.3 *Marginal change in water consumption due to tax change, and an indication of scenarios that result in a water dividend*[a]

| | | Recycling scheme | | |
| | | Direct tax break | Indirect tax break | Food tax break |
Water tax		0.000014	0.000013	0.000015
Tax on forestry water	−0.1749	Δ	Δ	Δ
Tax on mining water	−0.0007	Δ	Δ	Δ
Gold mining	−0.0007	Δ	Δ	Δ
Coal mining	−0.0005	Δ	Δ	Δ
Other mining	−0.0007	Δ	Δ	Δ
Tax on irrigated agriculture	−0.0328	Δ	Δ	Δ
Field crops	−0.0340	Δ	Δ	Δ
Horticulture	−0.0305	Δ	Δ	Δ

[a] The numbers represent the percent change in water use per million ZAR tax revenue. In the first column, a water tax is levied; the numbers are the reduction in water use. In the first row, the tax is recycled; the numbers are the percentage change increase in water consumption per million ZAR of government expenditure. If the sum of the two effects is negative, water use falls, and a 'Δ' is given to portray a positive water dividend (we are trying to save water).

water consumption by the economy of 0.1749% for each ZAR 1 million tax revenue that is collected. All the numbers in the table are in terms of percentage changes per amount of revenue collected. If a tax break is given, such as the food tax break, economic activity increases, and more water is consumed. For the food tax break, total water consumption increases by 0.000015% for each ZAR 1 million tax revenue that is collected. A combination of a tax on forestry and a food tax break, involving the same amount of revenue, will result in a net decrease in water demand.

It should be noted that the environmental effects are sensitive to the assumed elasticities of water demand in each industry (see Table 15.2). If the semi-elasticity for water demand from forestry is divided by four, the environmental effects of a water tax on forestry will be the same as the effects on agriculture.

15.3.2 Economic effects

The second dividend is the effect on the total economy, and the concept of MEB defined above is used to determine that. The MEBs for all eight

Table 15.4 *Marginal excess burdens of different tax instruments, and an indication of scenarios that result in a GDP dividend*[a]

Water tax		Recycling scheme		
		Direct tax break 0.201	Indirect tax break 0.137	Food tax break 0.167
Tax on forestry water	−0.208	–	–	–
Tax on mining water	−0.165	Δ	–	Δ
Gold mining	−0.298	–	–	–
Coal mining	−0.154	Δ	–	Δ
Other mining	−0.062	Δ	Δ	Δ
Tax on irrigated agriculture	−0.152	Δ	–	Δ
Field crops	−0.144	Δ	–	Δ
Horticulture	−0.170	Δ	–	–

[a] The numbers represent the percent change in gross domestic product per million Rand tax revenue. In the first column, a water tax is levied; the numbers are the reduction in GDP. In the first row, the tax is recycled; the numbers are the increase in GDP. If the sum of the two effects is positive, GDP increases, and a ' Δ ' is given.

water charge policy measures as well as the three recycling measures are compared in Table 15.4. A double dividend is indicated by a Δ in the table, when the increase in real GDP per unit of real government revenue lost as a result of a tax break (recycling policy) is larger than the decrease in real GDP per unit of real government revenue collected from a new water charge.

A charge on water consumed by the mining industry made real GDP decrease by 16.5 cents per Rand of real government revenue collected from the tax, while recycling via a direct tax break, indirect tax break and food tax break causes GDP to increase by 20.1, 13.7 and 16.7 cents per Rand of the real government revenue forsaken by government, respectively. The first and third combination render net gains to the economy. However, if only gold mining were paying the water charge, it would not render a double dividend, and neither would coal mining with the indirect tax break as the method of recycling, hence the negative signs in Table 15.4.

An additional charge on water consumption by irrigated agriculture renders double dividends with two of the recycling schemes, whether the tax is levied on field crops only, on horticultural crops, or on both. The damage done in terms of the MEB is smaller with field crops than horticulture. However, additional water charges on forestry do not yield a double dividend.

The percentage change in total employment per unit of real government revenue collected was also calculated, and the Δ's and minuses follow exactly the same pattern as in Table 15.4. That is, employment and GDP per unit of real government revenue are closely related to each other. The explanation is simply that the total production function in the model has Leontief and CES characteristics in terms of intermediate and primary inputs, so that GDP and employment will always move in the same direction as a result of an exogenous shock.

15.3.3 Poverty effects

The criterion used to measure an improvement in poverty levels is the percentage change in total real consumption of the three poorest household groups in the economy, by race. The model has 12 household groups and four race groups and it calculates consumption for each group by commodity, as well as total consumption. The results of total consumption for the poorest household group are given in Table 15.5, but the results for the poorest three groups are similar, so that the table would be representative of all three.

Some policy combinations render a net improvement for one race group while they have a detrimental effect on another. We have a 'Δ' in the table if all African poor persons benefit from a strategy, since they comprise 90% of all poor persons. In the mining sector the only water charge that could be recycled in a way that would benefit all four race groups within the poorest groups of households is a tax on water consumption by mining industries other than gold and coal. The table shows that the initial three scenarios proposed by policymakers (scenarios 1, 2 and 3 in Table 15.5, as discussed in Section 15.2.3) could be expanded into subscenarios, by breaking the mining sector up into gold, coal and other mining, and the agricultural sector into field and horticultural crops.

The water charges proposed affect the subsectors quite differently. For example, if all mining industries' water consumption were to be taxed, poor Africans would become much worse off than when only other mining industries pay the tax. In this case, any recycling scheme would render the poverty dividend. However, recycling tax revenue through a decrease in taxes on food renders a poverty dividend in combination with all the water charge scenarios except with gold mining. With irrigated

Table 15.5 *Marginal change in poverty (%), and an indication of scenarios that result in a poverty dividend*[a]

	Water tax	Recycling scheme		
		Direct tax break 0.133	Indirect tax break 0.091	Food tax break 0.365
Tax on forestry water (1)	−0.159	–	–	Δ
Tax on mining water (2)	−0.207	–	–	Δ
Gold mining (2a)	−0.390	–	–	–
Coal mining (2b)	−0.179	–	–	Δ
Other mining (2c)	−0.065	Δ	Δ	Δ
Tax on irrigated agriculture (3)	−0.143	–	–	Δ
Field crops (3a)	−0.112	Δ	–	Δ
Horticulture (3b)	−0.169	–	–	Δ

[a] The numbers represent the percent change in real consumption of the poorest household group per billion Rand tax revenue. In the first column, a water tax is levied; the numbers are the reduction in consumption. In the first row, the tax is recycled; the numbers are the increase in consumption. If the sum of the two effects is positive, poverty decreases, and a 'Δ' is given.

agriculture it helps to differentiate between water charges on field crops and on horticultural crops separately. It shows that a tax on irrigated horticultural crops has a more severe influence on the consumption of the poorest groups in that at least one group is made worse off with this tax, while with irrigated field crops at most one group is made worse off.

15.4 Conclusions and policy recommendations

15.4.1 Main lessons

The eight key commodities that Africans spend most of their income on are – in order of importance – food, petroleum, real estate, textiles, electricity, transport services, other manufacturing and agricultural goods. Taxes that negatively affect consumer prices of these commodities will harm the poor more. The direct impact of extra water charges on forestry is first an increase in the cost of the forestry industry and hence its prices, and second on wood, paper and pulp, which is part of other manufacturing.

The agricultural sector is the largest intermediate supplier to the food industry, and food is the most important commodity to all households – rich and poor. Other manufacturing is also high on poor consumers' priority list, and these two channels turn out to be significant in having the detrimental effect on the poor. Extra water charges on mining do not generally have a direct effect on households. They do buy coal for heating purposes, but the taxes on gold or other mining goods only influence consumers through the downstream effects on industries who buy the outputs of the mining industries.

The results in Table 15.5 also take recycling into consideration and the effects described above should be compared to the increases in consumption of various commodities due to recycling. Comparing the general tax break on all commodities with the break on food only, it is clear that the poor benefit more from the food tax break than they lose with the water taxes. As they have food as 60% of their consumption basket this result is not surprising. A general break is less distortionary to the economy because relative prices do not change much, and hence does not render net benefits to the poor, per unit of government revenue collected.

Extra water charges on irrigated agriculture directly increase the cost of field and horticultural production. Field and horticultural products comprise a large proportion of agricultural commodities, and an increase in their prices directly affects the prices of industries buying them as intermediate inputs. The four largest demanders of agricultural goods are food, other manufacturing, petroleum and textiles, all important to poor households. Hence, recycling via either a direct or indirect tax break does not outweigh the negative effects of the tax on irrigated agriculture. Only recycling through a subsidy on food benefits all consumers, including the poor.

15.4.2 Policy recommendations

The simulation results from our modelling exercises are summarized in Table 15.6. The first Δ in each cell shows the first dividend, namely the environmental effect, which is positive in all cases. The second Δ or – shows whether a double dividend has been achieved with the policy combination or not, while the third Δ or – shows the achievement of a double dividend or not. There are quite a number of policy combinations that render double dividends, but in this paper we are particularly interested in poverty reduction and environmental benefits.

Table 15.6 *Summary of results*

Water charge	Direct tax break	Indirect tax break	Food tax break
Charge on forestry	Δ – –	Δ – –	Δ – Δ
Charge on mining	Δ Δ –	Δ – –	Δ Δ Δ
Gold mining	Δ – –	Δ – –	Δ – –
Coal mining	Δ Δ –	Δ – –	Δ Δ Δ
Other mining	Δ Δ Δ	Δ Δ Δ	Δ Δ Δ
Charge on irrigated agriculture	Δ Δ –	Δ – –	Δ Δ Δ
Field crops	Δ Δ Δ	Δ – –	Δ Δ Δ
Horticultural crops	Δ Δ –	Δ – –	Δ – Δ

As mentioned above, the environmental dividend is obtained with all the policy combinations, suggesting that recycling methods do not stimulate the economy enough to offset the decreases in water demand that are achieved through the water taxes. The recycling method that works the best for poverty reduction is clearly the tax break on food, simply because food makes up most of the poor's consumption basket. It seems fairly easy to obtain the environment–poverty dividends from this recycling option, and we should therefore take the next steps and also try to achieve the economic dividend at the same time.

If we wish to obtain a 'triple' dividend, that is, alleviate poverty and improve upon real GDP, there are eight possible combinations of taxes and recycling schemes, as shown in Table 15.6. The direct tax break offers two combinations, the indirect tax break one, and the food tax break five.

The simulation results presented in this chapter are satisfactory. The largest water user is irrigated agriculture and it is an important result that a tax on water used by irrigated agriculture would render the desired triple dividends for all four race groups, either through a direct tax break or a food tax break. The environmental dividends per unit of government revenue are the strongest with a tax on forestry, but we found that both the economic and poverty dividends are difficult to achieve, and hence we would not recommend this policy option. Furthermore, the environmental results found depend strongly on the assumptions that we have made about water demand elasticities.

Even though triple dividends could be obtained by introducing a charge on water consumption in the mining sector, we also do not recommend this as a policy option for South Africa, since this part of the mining industry does not consume a significant amount of water.

In conclusion, we would like to recommend an additional water charge on irrigated field crops in conjunction with a general decrease in direct tax rates, or a decrease in taxes on food products consumed by the poor. A more detailed analysis with more specific charges needs to be carried out to further substantiate this conclusion.

References

Aliber, M. A. (2002). Overview of the incidence of poverty in South Africa for the 10-year review. Unpublished paper commissioned by the President's office, Human Sciences Research Council, Pretoria.

Bovenberg, A. L. and De Mooij, R. A. (1994). Environmental levies and distortionary taxation. *American Economic Review*, **84**(4): 1085–1089.

CSIR (2001). Water Resource Accounts for South Africa. Report number ENV-P-C 2001–050, Statistics South Africa and The Department of Environment Affairs and Tourism.

Development Bank of Southern Africa (DBSA)(2000). *Environmental Impacts of the Forestry Sector in South Africa with Specific Reference to Water Resources.* Midrand, South Africa: Development Bank of Southern Africa.

Department of Minerals and Energy (DME)(2002). *Digest of South African Energy Statistics.* Compiled by C. J. Cooper. Pretoria, South Africa: Department of Minerals and Energy.

Department of Water Affairs and Forestry (DWAF) (1998). National Water Act. Act No 36 of 1998. Pretoria, South Africa: DWAF.

Department of Water Affairs and Forestry (DWAF) (2002). *National Water Resource Strategy*, 1st edn. Pretoria, South Africa: DWAF.

Du Toit, C. and Koekemoer, R. (2003). A labour model for South Africa. *South African Journal of Economics*, **71**(1): 49–76.

Fullerton, D. and Metcalf, G. E. (1997). Environmental taxes and the double dividend hypothesis: did you really expect something for nothing? Working Paper No. w6199, National Bureau of Economic Research (NBER). Available at www.nber.org.

Goulder, L. H. (1994). Environmental taxation and the 'double dividend': a reader's guide. Working Paper No. w4896, National Bureau of Economic Research (NBER). Available at www.nber.org.

Goulder, L. H., Parry, I. W. H. and Burtraw, D. (1997). Revenue-raising versus other approaches to environmental protection: the critical significance of pre-existing tax distortions. *The RAND Journal of Economics*, **28**(4): 708–731.

Harrison, W. J. and Pearson, K. R. (1996). Computing solutions for large general equilibrium models using Gempack. *Computational Economics*, **9**: 83–127.

Horridge, M. (2000). ORANI-G: a general equilibrium model of the Australian economy. CoPS/IMPACT Working Paper Number OP-93. Centre of Policy Studies, Monash University, Melbourne, Australia. Available at http://www.monash.edu.au/policy/elecpapr/op-93.htm.

King, N. (2004). The economic value of water in South Africa. In J. Blignaut and M. De Wit (eds.), *Sustainable Options* Cape Town: UCT Press.

Klasen, J. (2000). Measuring poverty and deprivation in South Africa. *Review of Income and Wealth*, **46**(1): 33–58.

Le Maitre, D. C., Versveld, D. B. and Chapman, R. A. (2000). The impact of invading alien plants on surface water resources in South Africa: a preliminary assessment. *WaterSA*, **26**(3): 397–408.

May, J. (ed.) (2000). *Poverty and Inequality in South Africa, Meeting the Challenge*. Cape Town and London: David Phillip and Zeb Press.

Pearce, D. W. (1991). The role of carbon taxes in adjusting to global warming. *Economic Journal*, **101**(407): 938–948.

Renzetti, S. (1992). Estimating the structure of industrial water demands: the case of Canadian manufacturing. *Land Economics*, **68**(4): 396–404.

Schreiner, B. and Van Koppen, B. (2002). Catchment management agencies for poverty eradication in South Africa. *Physics and Chemistry of the Earth*, **27**: 969–76.

SSA (Statistics South Africa)(2001). *Social Accounting Matrix* 1998. Pretoria, South Africa: Statistics South Africa.

Van Heerden, J. H., Gerlagh, R., Blignaut, J. *et al.* (2006). Searching for triple dividends in South Africa: fighting CO_2 pollution and poverty while promoting growth. *Energy Journal*, **27**(2): 113–142.

Veck, G. A. and Bill, M. R. (2000). Estimation of the residential price elasticity of demand for water by means of a contingent valuation approach. Water Research Commission Report No: 790/1/00, Marlow, UK.

Woolard, I. (2002). An overview of poverty and inequality in South Africa. Working paper, Department for International Development, South Africa. Available at http://datafirst.cssr.uct.ac.za/resource/papers/woolard_2002.pdf.

Part V
Land-related ecosystem services

ELISSAIOS PAPYRAKIS AND
S. MANSOOB MURSHED

Humans have always depended on land-related ecosystem services to secure survival and improve well-being. Human-induced changes in land use have typically allowed populations all over the world to expand agricultural land, food production and human settlements. Extensive deforestation and mining secured access to timber and mineral resources, which consequently facilitated increases in production and consumption. This modification of land cover, and the natural environment more broadly, for the sake of human benefit has nevertheless not always coincided with improvements in living standards and poverty alleviation. Increases in food availability at least until the Industrial Revolution were for the most part accompanied by subsequent increases in population size rather than any significant welfare improvements. More recently, structural changes favouring manufacturing and technological advancements allowed for significant productivity gains (and improvements in nutrition, life expectancy and living conditions) despite high population growth rates, at least for the more developed economies.

There are multiple ecosystem services derived from land, which extend beyond the direct use of land resources (e.g. forest, arable land, pastureland; see MEA 2005). Such services often relate to very complex interactions taking place within and across ecosystems, both at a local and global scale. Deforestation, for instance, is known to interfere with the hydrological cycle resulting in reduced water availability as well as flood control

Nature's Wealth: The Economics of Ecosystem Services and Poverty, ed. P. J. H. van Beukering, E. Papyrakis, J. Bouma and R. Brouwer. Published by Cambridge University Press, © Cambridge University Press 2013.

(Nelson and Chomitz 2007). Large-scale habitat destruction reduces biodiversity and inhibits crop pollination (Fearnside 2005). Unsustainable agricultural practices (e.g. by use of artificial fires and inappropriate fertilizers) commonly result in reduced agricultural yields, gradual soil erosion and pollution. Services also fall in the cultural domain and can, hence, come in non-material form (Mandondo 1997). These can include recreational activities supported by local ecosystems, as well as spiritual/religious affinity with the natural environment. Unsustainable land use can also have significant implications both for climate change mitigation, as well as adaptation. Excessive deforestation reduces the potential for forest carbon sequestration, while no-tillage agriculture is nowadays promoted as more resistant to adverse weather shocks (in addition to its supplementary function as a carbon sink).

The multiplicity in services derived from ecosystems suggests that there is an equally rich list of trade-offs amongst resource users and management options (UNEP-WCMC 2007). Some of these trade-offs affect communities of close geographic proximity, while others take place at a more macro scale. For instance, mining activities are often linked to both localized air pollution and deforestation, as well as more remote impacts as a result of contamination of surface and ground water. Trade-offs take place not only across space but also across time. Unsustainable land management often has repercussions for the provision of ecosystem services in the future (and hence for the welfare of future generations) to the extent that ecosystems do not recover swiftly or fully from anthropogenic disturbances. Much of this damage (for instance, in the case of biodiversity loss) is often irreversible.

The trade-offs associated with different land-use management scenarios are often far from clear and require extensive interdisciplinary research to identify and quantify them. Different types of ecosystem services are often interlinked (land, water, air, biodiversity) and resource users are often unaware of the complexities involved. Upstream users of resources, for instance, may have little knowledge of the environmental externalities their activities generate upon downstream communities. Resource users often fail to fully comprehend the links between individual behaviour and collective damage, typically resulting in overharvesting and environmental degradation (i.e. the 'tragedy of the commons'). Even when trade-offs are clearly mapped, individuals often fail to adjust behaviour in order to limit damage inflicted on relatively distant ecosystems and communities (transboundary pollution is a typical example).

The link between the provision of ecosystem services and poverty is also rather complex and dynamic in nature (Dasgupta *et al.* 2005). Most of the world's poor reside in rural communities and are hence directly dependent on land-related activities (e.g. timber extraction, farming, herding). For many of these rural communities the provision of food is very closely linked to the status of surrounding ecosystems, as they are often very loosely connected to markets outside their vicinity. The health of local dwellers is also often dependent on the health of ecosystems, with the latter regulating the supply of natural medicines. The causality between poverty and land use runs in both directions. While the poor depend on land-related ecosystems for their livelihoods, any excessive use of resources will deplete the natural resource base and aggravate poverty (in the absence of alternative economic activities). For this reason, sustainable land management ensures that local systems maintain their function as an important source of rural income.

There are multiple reasons that explain why the poor often resort to unsustainable land management (Broad 1994). They often have a high time preference (i.e. they heavily discount the future), which weighs down any negative side-effects of unsustainable harvesting in their decision-making. They often have limited access to credit, savings and other productive assets that would allow for a smoothing of consumption over time. For this reason any sustainable land management that even temporarily restricts resource use and reduces rural income is likely to face fierce opposition from local communities. At the same time, their economic dependence on land enhances their vulnerability to ecosystem degradation. This is particularly the case for those of poor health and little purchasing power, who often face constraints in migrating to alternative resource-rich areas or importing resources from elsewhere. Many other factors interact with poverty to result in unsustainable land use. Population growth is commonly mentioned as a key driver behind environmental pressure and is closely associated with land and food scarcities (Harte 2007). The absence of strong local institutions that clearly define tenure and user rights (and guarantee their enforceability) has often led to over-exploitation of land resources. Privatization of resources, though, is not always a better solution than communal management, since poor individuals often fail to gain access to restricted resource use. A final factor, that has been receiving much attention due to the increased interest in climate change, is the variability in weather conditions. Changing rainfall patterns and consequent water scarcity will exacerbate land degradation, particularly in those areas where currently the majority of the poor are located. For this reason any direct impact of poverty on land degradation is likely to be further reinforced by the predicted changes in climatic conditions.

Organization of Part V

Part V of the book brings together four contributions that collectively verify the strong nexus between poverty alleviation and land management. The section highlights the strong dependencies of rural communities on land use and identifies key factors that support the sustainable management of land-related ecosystems. The section is methodologically pluralist and a number of different research techniques have been employed to uncover the linkages between poverty and land degradation. Part V is organized as follows.

Chapter 16 examines the extent of dependence of rural communities in the Jhabua district in India on common-pool natural resources. Data are collected for a sample of 536 households in 60 villages in the area and were matched with remote-sensing information. The analysis is methodologically novel by estimating an asset-based measure of income (land, livestock, capital) that is likely to remain more stable over time (and hence a better proxy of household wealth). The authors estimate imputed values for disaggregated economic activities and determine the extent of economic dependence on common property resources (e.g. fuelwood, fodder, seeds/ leaves). The analysis infers that dependence on common-property resources is equally prevalent amongst rich as well as poorer households. This is an important finding, suggesting that poverty alleviation alone (in the form of higher income levels) is not sufficient to ease pressure on common resources. The study also concludes that richer and poorer households depend on different types of common-pool resources, with the former relying more on flowers/seed collection to supplement income and the latter more on fodder and construction wood (while fuelwood appears to be an important income source for all income cohorts). Seed and flower collection appears to be a low-return time-intensive activity and hence appeals mainly to the poorer households.

Chapter 17 focuses on the role of land tenure security in improving agricultural sustainability and increasing food supply. The chapter hence examines the mediating role of property rights (institutions) in enhancing agricultural productivity and alleviating poverty in rural communities. The study is based on data collected from a survey of 457 households from 18 villages in rural Kenya (and was supplemented by a community questionnaire circulated to key informants in each village). Data on investment in soil and water conservation (e.g. tree planting, terracing with grass strips, mulching) were also gathered for 684 plots located in the villages. A factor analysis was conducted in order to determine the key variables to be included in the empirical analysis. The study concludes that while tenure security does not appear to affect expenditure per person (i.e. the adopted measure of welfare) directly, it indirectly raises income by encouraging investment in soil and water conservation (with the latter being strongly and positively correlated with per capita expenditure). The empirical model hence supports a strong poverty–environment nexus, with sustainable land management being strongly correlated with increased income. Population density is also negatively associated with welfare, signifying a close link between population pressure and environmental degradation. Results also confirm that investment in soil and water conservation of a more permanent (rather than seasonal) character benefits the most from increased tenure security. Supplementary productive assets (livestock, farm equipment) as well as plot proximity to household are also found to be positively correlated with investment in conservation. The results hence point to the need for a comprehensive land use policy that incorporates secure land property rights as a means to simultaneously achieve poverty alleviation and sustainable land use.

Chapter 18 is concerned with another type of degradation of common-pool land resources, which commonly affects rural livelihoods in many parts of the developing world; namely the extent of pastureland degradation as a result of overgrazing. The study focuses on two separate regions in Mongolia (Ugtaal, Gurvansaikhan) where livestock herding constitutes a key source of rural income for local communities. It analyses herder behaviour through a sophisticated game-theory framework, utilizing data collected from 60 households in the two areas. The study collects information on the herding strategies of respondents (as well as on their views of other herders' behaviour) with respect to recent changes in their livestock. These are subsequently combined with data on herders' income, allowing the estimation of payoff functions for herding-related activities. Although different types of equilibriums arise in the two sites, in both cases payoffs are at their minimum when all herders simultaneously opt for increases in their livestock (while at their maximum when a herder decides to increase his or her livestock, while the rest of the group keeps livestock size constant). This suggests that government interventions in terms of restricting access to pastureland could be welfare maximizing for the livestock-dependent communities. One of the key inferences of the study is that food short-ages (as a result of declines in livestock numbers attributed to adverse weather conditions) could be prevented by adopting semi-commercial herding, where meat and milk are exchanged for grain in urban markets (a caloric terms of trade table provides such conversion rates by combin-ing nutritional information on food products with their corresponding price levels over time).

Chapter 19 adopts a more macro perspective by utilizing extensive household surveys and census data covering multiple regions in rural Uganda. The study combines welfare estimates with biophysical informa-tion in order to explore contemporaneous links between resource degra-dation and poverty levels between 1992 and 1999. It makes an important methodological contribution by generating simultaneous estimates for changes in poverty and land use through poverty-mapping techniques. The study concludes that large declines in poverty have been particularly observed in areas experiencing degradation of local wetlands (as a result of land conversion for the purpose of subsistence farming). A similar pattern between forest degradation and declines in poverty also emerges when overlaying poverty estimates and biophysical data. Larger declines in pov-erty are observed in areas where changes in land use (i.e. forest and wetlands degradation) supported commercial rather than subsistence farming. While resource degradation has raised average household income in Uganda, this constitutes a rather short-term windfall rather than any sustainable long-term solution to poverty. The authors of the study conclude that initial conditions play a major role in shaping subsequent patterns of change. High initial levels of (widespread) poverty, jointly with population pressure, encourage land degradation as a means to supplement rural income. In the absence of community-level management that regulates the use of wetlands and forest resources, ecosystem degradation will only temporarily provide a relief to impoverished rural communities.

The picture that emerges from Part V points to a strong interrelation-ship between land-related ecosystem services and poverty levels. This is by no means a straightforward linear relationship, but one that is rather complex and mediated by a long list of factors (e.g. property rights, governance, market access, initial conditions). The need for establishing secure user rights that limit overharvesting of local resources is a common recommendation of all four studies. Despite their variety in methodological approaches and geographical focus, all four studies hence conclude that sustainable land use is the only viable solution to poverty alleviation.

References

Broad, R. (1994). The poor and the environment: friends or foes? *World Development*, **22**: 811–822.

Dasgupta, S., Deichmann, U., Meisner, C. and Wheeler, D. (2005). Where is the poverty-environment nexus: evidence from Cambodia, Lao PDR, and Vietnam. *World Development*, **33**: 617–638.

Fearnside, P. M. (2005). Deforestation in Brazilian Amazonia: history, rates and consequences. *Conservation Biology*, **19**: 680–688.

Harte, J. (2007). Human population as a dynamic factor in environmental degradation. *Population and Environment*, **28**: 223–236.

Millennium Ecosystem Assessment (MEA) (2005). *Ecosystems and Human Well-being: Biodiversity Synthesis*. Washington DC: World Resources Institute.

Mandondo, A. (1997). Trees and spaces as emotion and norm laden components of local ecosystems in Nyamaropa communal land, Nyanga District, Zimbabwe. *Agriculture and Human Values*, **14**: 353–372.

Nelson, A. and Chomitz, K. M. (2007). The forest–hydrology–poverty nexus in Central America: a heuristic analysis. *Environment, Development and Sustainability*, **9**: 369–385.

UNEP-WCMC (2007). *Biodiversity and Poverty Reduction: The Importance of Biodiversity for Ecosystem Services*. Cambridge: United Nations Environmental Programme, World Conservation Monitoring Centre (UNEPWCMC).

16 · Income poverty and dependence on common resources in rural India

URVASHI NARAIN, SHREEKANT GUPTA
AND KLAAS VAN 'T VELD

16.1 Introduction

Despite the recent rapid expansion of the IT software and services industry in India (which often draws widespread media attention in the West), the country still remains a predominantly agrarian economy. According to India's 2001 census, more than 70% of the population lives in rural areas, and mostly in conditions of extreme poverty. The economy of these poor, rural households is closely connected to the village natural resource base – its forests, grazing lands and water resources. Whether households can make a living from agriculture depends, in large part, on the amount of water available for irrigation. The availability of fodder on village grazing lands affects the income that households derive from livestock rearing. The amount of fuelwood in village forests determines how much time households must devote to collecting fuelwood, and thereby also how much time remains for other income-generating activities.

Given the simultaneous existence of high levels of poverty and dependence on local common-pool resources, two questions arise: (1) can improved natural resource management form the basis of poverty alleviation policies in rural India; and (2) can common-pool resource stocks serve as a public asset for poor households, substituting for the private assets that they lack? Attempts to answer these questions have given rise to a growing literature on poverty–environment interactions (for reviews, see Barbier 2005, Duraiappah 1998, Horowitz 1998, Reardon and Vosti 1995).

Some of this literature (recently reviewed by Beck and Nesmith 2001, Kuik 2005, Vedeld *et al.* 2004) explores how the degree of households'

Nature's Wealth: The Economics of Ecosystem Services and Poverty, ed. P. J. H. van Beukering, E. Papyrakis, J. Bouma and R. Brouwer. Published by Cambridge University Press, © Cambridge University Press 2013.

dependence on common-pool natural resources varies with household income, where *dependence* is usually defined as the share of overall income derived from natural resource use. A common finding in this literature is that dependence on resources declines with income (Adhikari 2003, Cavendish 2000, Jodha 1986, Reddy and Chakravarty 1999). However, the literature only provides conjectures, unsupported by hard data, as to why the reported regularity is obtained. Reddy and Chakravarty (1999) merely note that the poor have less land and theorize that this explains their higher dependence on forest resources. Cavendish (2000) suggests that the decline in dependence with income may in part be due to cash constraints: poorer households are less able to purchase food and must therefore collect it from the commons instead. Jodha (1986), in contrast, provides a fairly detailed discussion, suggesting three specific reasons: (1) common-pool resources act as a substitute for the private assets that poor households lack – instead of acquiring fuel and fodder from private lands, for example, land-poor households collect these resources from common lands; (2) poor households have surplus labour that is well suited to resource extraction, an activity where labour is usually the only input; and (3) returns to extraction from the commons are often low and are therefore unattractive to the rich. However Jodha (1986) does not support these statements with any data.

Our study, using purpose-collected data from 536 households in 60 Indian villages, examines the relationship between rural household incomes and natural resources in greater detail and brings information on household characteristics to bear to understand why certain trends emerge between resource use, dependence and income. We thereby focus on the question of how private holdings of productive assets (land, livestock, farm capital and human capital) affect households' use of the commons, and in particular, whether, as Jodha (1986) suggests, the natural assets of common-pool resources serve as a substitute for the private assets that poor households lack.

A key finding, one that contradicts conventional wisdom, is that dependence on natural resources does not decline with rising income. Instead, we find evidence of a U-shaped relationship – that is, dependence on natural resources first decreases and then increases with income. Similarly, among households with livestock, time spent grazing increases strongly and monotonically with income. This is true, despite the higher land holdings of richer households, indicating that contrary to Jodha's (1986) conjecture about the importance of private assets to resource dependence, provision of these resources from private land is not an important substitute for provision from the commons for most households. In fact, the private asset of livestock, which richer households own more of as well, appears to act as a complement instead.

The rest of the chapter is organized as follows. Section 16.2 describes the study site and the methodologies used to estimate household incomes. Section 16.3 presents descriptive statistics on these income measures and our results on resource dependence. Section 16.4 concludes with some policy recommendations.

16.2 Data and methodology

16.2.1 Study site and field work

The study site for our project is the Jhabua district in the Indian state of Madhya Pradesh. Jhabua is a hilly region of 68 000 ha in western Madhya Pradesh. According to the Human Development Report published by the state government in 1998, of the total land area, 54% is classified as agricultural land, 19% as forestland and the rest as land not available for cultivation. Jhabua is one of the poorest districts in the state, with a Human Development Index of only 0.356, the lowest out of 45 districts in the state. Only 26.3% of the men and 11.5% of the women in the district are literate, life expectancy of an average person is only 51.5 years, and 30.2% of the district's rural population and 41.6% of its urban population are classified as living below the poverty line. Agriculture is the main occupation of households, with 90.6% of the workforce employed in this sector (Government of Madhya Pradesh 1998). Furthermore, agriculture in the district is predominantly rainfed. Households in this region usually supplement their incomes through livestock rearing and with various products from the local (often degraded) forests − most notably fuelwood, construction wood, tendu (*Diospyros melonoxylon* Roxb.) leaves and mahua (*Madhuca indica*) flowers and seeds.

The local socio-economic characteristics make Jhabua a suitable study site (i.e. its level of poverty and dependence on agriculture and natural resources more broadly, e.g. fodder, construction wood and other forest products). Moreover, since high dependence on rainfed agriculture, livestock income and supplementary resource income characterize the economies of large parts of rural, semi-arid India, the results of this study plausibly generalize to areas beyond Jhabua.

Data for the study were collected from 536 households in 60 villages in the district of Jhabua, covering the period June 2000 to May 2001.

A random sample of households for the survey was generated through a two-stage sampling design: first, a stratified random sample of villages was selected, and then a stratified random sample of households in those selected villages.

The village sample frame comprised 89 villages in the district of Jhabua where the Madhya Pradesh Groundwater Department has monitored the groundwater level thrice yearly (pre-monsoon, post-monsoon and winter) since 1973. From the sample frame of 89 villages, a stratified sample of 64 villages was selected to maximize variability in the forest stock. For the latter, we used data from the Madhya Pradesh Forest Department's 1998 inventory of all forest 'compartments' (the smallest forest management unit of area) in Jhabua. For each compartment, the inventory gives area and total volume of trees in cubic meters. Summing the volume over all compartments within a 5-km radius from the centre of a village provides a measure of the total forest biomass available to the village as a whole, which was then divided by the number of village households. The resulting measure of per-household biomass was used as the basis for stratification. Unfortunately, political unrest in Jhabua at the time of the survey made it impossible to complete the survey in four villages, leaving 60 villages in all.

Household sample frames were constructed for each of the sample villages from village landownership records and from the Madhya Pradesh state government's village-level list of households living below the poverty line (BPL). A random sample of households was selected from three strata: BPL, land-poor (owning less than 3 ha of land) and land-rich (owning 3 ha or more of land), with oversampling of BPL and land-rich households. Finally, remote-sensing images were used to obtain village-level measures of forest and fodder biomass.

16.2.2 Different components of household income

To determine the extent to which households in rural Jhabua use common-pool natural resources for their livelihood, we calculated the income that each household obtained from seven major sources: (1) agriculture, (2) livestock rearing, (3) common-pool resource collection, (4) household enterprise, (5) wage employment, (6) transfers and (7) financial transactions. Income from each source was calculated as the difference between total revenue obtained and total input costs incurred, where these totals include both market transactions and imputed values for non-market activities. For example, the revenue obtained from common-pool resource collection includes imputed values for resources collected but not sold by the household. Similarly, the input costs incurred for livestock rearing includes imputed values for fodder collected from the commons and then fed to own livestock. For income sources 1–4, no cost is imputed for a household's own labour inputs, however; in this sense, the income from these sources is 'gross' income.

Common-property income, in turn, comprises income from the main resources collected from village commons: wood for fuel, wood for construction, fodder, mahua flowers used to make local liquor, mahua seeds used to make cooking oil, tendu leaves used to make cigarettes and animal dung used as agricultural fertilizer and cooking fuel. Here, as noted above and based on information from the field, we assumed that markets exist for all the products that were gathered from village commons. Table 16.1 gives some indication of how prevalent trade is in the most common of these common resources. For example, 61% of all sample households were found to trade (either sell some fraction of what they gathered or buy supplemental amounts) in fodder and another 17% in fuelwood. The last two rows of the table show the number of villages in which at least one sample household respectively gathered or traded these products. Trade in fodder occurred in all 60 villages, and trade in the other three products was widespread as well.

Once income from different sources was calculated, we made these numbers comparable across households by dividing the income obtained by the number of adult-equivalent units in the household.

16.2.3 Current and permanent income

The top panel of Table 16.2 shows the composition of current per capita income for the current-income quartiles. We notice a large disparity

Table 16.1 *Sample prevalence of trade in the most important common-property outputs*

Common-property output	Fodder	Fuelwood	Construction wood	Mahua flowers
No. of households that:				
– gathered	359 (66%)	333 (61%)	129 (24%)	102 (19%)
– traded	335 (61%)	94 (17%)	98 (18%)	96 (18%)
No. of villages that:				
– gathered	60 (100%)	58 (97%)	53 (89%)	43 (72%)
– traded	60 (100%)	37 (62%)	42 (70%)	41 (68%)

between the mean current per capita income of households in the bottom three quartiles, and that of households in the top quartile. The mean household in the lowest quartile lost Rs 2024 over the course of the survey year, while the mean household in the top quartile earned Rs10 383. The large losses from agriculture are explained by the fact that the survey year was the fifth consecutive drought year in Jhabua.

Also shown in Table 16.2 is resource dependence, defined (as is common in the literature) as the percentage of overall current income derived from resources. For the second to the fourth income quartiles, dependence clearly declines monotonically, consistent with other findings in the literature. For households in the bottom income quartile, however, the dependence measure is ill defined because most (106 of 129) of these households have either negative or zero overall incomes. Moreover, if the

Table 16.2 *Current per capita household income in Rs by major sources and income quartiles for whole sample*

Current income quartiles	Lowest 25%	25%–50%	50%–75%	Highest 25%
Income from agriculture	−1329	79	530	2113
Income from livestock rearing	−374	-96	7	24
Income from resource collection	320	369	483	996
Fuelwood	72	144	152	477
Construction wood	2	16	3	146
Fodder	159	143	257	258
Other resources	86	66	71	115
Income from household enterprise	45	109	171	1320
Income from wage employment	590	1197	1994	5017
In-village casual labour	106	158	258	433
Off-village casual labour	451	967	1395	1435
Private and public jobs	33	72	340	3150
Income from transfers	156	120	202	1476
Relatives	11	36	55	245
Friends	1	1	2	315
NGOs	0	2	0	27
State	144	81	146	888
Income from financial transactions	−1432	−612	−499	−562
Total current income	−2024	1166	2889	10383
Resource dependence	n/a	40%	16%	14%

measure is calculated for only those households in the bottom quartile that have positive overall incomes, dependence is extremely high – 354% on average. The reason is that for nine of these households, resource income greatly exceeds overall income.

Viewing households in the bottom quartile as highly dependent on resources seems inappropriate, however, because these households are by no means asset-poor. Table 16.3 shows that they cultivate on average as much land per capita as households in the top income quartile, and more than households in the second and third income quartiles. The per capita value of land owned by these households is also considerably above that of households in the second and third income quartiles, though below that of households in the top quartile. Similarly, households in the bottom quartile have on average more farm capital and livestock than households in the top three quartiles.

All else being equal, households rich in private assets should be considered less dependent on common resources because their assets serve as a buffer to unexpected income losses. A household that owns gold jewellery can sell it to make up for losses it may have incurred in agriculture. A household without assets, on the other hand, may have no other option but to rely on the local forest and sell fuelwood, for example. There is evidence that households in the bottom quartile engage in such buffering. These households typically took on new debt and sold jewellery over the course of the survey year to make up for their income losses (see Table 16.3).

To account for the difference in buffering capability that stems from differences in private asset holdings, we define what we call the household's 'permanent income' – income that households can expect to derive from their asset holdings over the long run. For incomes derived from

Table 16.3 *Asset holdings per capita by current income quartiles for whole sample*

Current income quartiles	Lowest 25%	25%–50%	50%–75%	Highest 25%
Amount of land cultivated (ha)	0.36	0.20	0.24	0.35
Value of land owned (Rs)	29833	14046	18116	36841
Value of farm capital (Rs)	7260	1904	2738	4324
Value of livestock (Rs)	3136	2500	2598	2643
Financial assets (Rs)	−3865	−1462	−1293	−1494
Total asset value (Rs)	36364	16988	22159	42314
Net investment (Rs)	−383	160	77	1472

Table 16.4 *Relationship between current and permanent income quartiles for whole sample*

	Permanent-income quartiles			
Current income quartiles	Lowest 25%	25%–50%	50%–75%	Highest 25%
Lowest 25%	40%	25%	18%	17%
25%–50%	44%	28%	20%	7%
50%–75%	15%	37%	34%	15%
Highest 25%	1%	9%	28%	61%

private assets (land, livestock, farm capital and financial assets), we combine current-year returns on these assets with their annualized end-of-year value; for incomes derived from natural resources, wages, household enterprise and transfers, we simply extrapolate current-year income.

Table 16.4 shows that the correlation between current- and permanent-income quartiles is not very high. For example, although 61% of households that fall into the top permanent-income quartile also fall into the top current-income quartile, 15%, 7% and 17% of these households fall into the third, second and bottom current-income quartiles, respectively. This indicates that income in one particular year may not give an accurate picture of the household's expected long-run income, reinforcing our argument that resource dependence should be calculated in terms of the latter.

16.3 Results and discussion

16.3.1 Permanent-income statistics

As shown in Table 16.5, households in the lowest permanent-income quartile earn Rs 2420 per capita on average, while households in the top quartile earn Rs 16 275. According to the Madhya Pradesh Directorate of Economics and Statistics, the average per capita income in the state for both rural and urban households combined was Rs 11 244 in 1999–2000. Although this figure is not directly comparable to our measures of permanent per capita income, it does suggest that our sample captures a significant amount of income variability.

Permanent income from most sources – agriculture, resource collection, household enterprise, wage employment and transfers – increases monotonically from the first to the fourth income quartile. Income

Table 16.5 *Permanent per capita household income in Rs by major sources and income quartiles for whole sample*

Current income quartiles	Lowest 25%	25%–50%	50%–75%	Highest 25%
Income from agriculture	1557	2470	3725	7983
Income from livestock rearing	126	186	214	180
Income from resource collection	189	326	549	1103
Fuelwood	103	149	223	370
Construction wood	2	2	41	122
Fodder	23	71	196	526
Other resources	60	104	89	85
Income from household enterprise	51	145	229	1217
Income from wage employment	598	1482	2096	4618
In-village casual labour	127	271	217	340
Off-village casual labour	419	1044	1479	1308
Private and public jobs	52	167	400	2970
Income from transfers	133	153	189	1475
Relatives	31	33	36	248
Friends	0	3	0	313
NGOs	2	0	0	27
State	100	117	153	887
Income from financial transactions	−235	−206	−304	−302
Total current income	−2024	4555	6698	16275

from livestock rearing increases from the first to the third quartile and then decreases, while income from financial transactions shows no clear trend.

After income from agriculture, income from wage employment was the largest source of income for the households in all four quartiles. However, for the first three quartiles, the wage income mostly came from off-village casual employment. Households in these quartiles earned about 70% of their total wage income from such seasonal migratory labour. In contrast, households in the top quartile earned only 29% of their wage this way, with 64% coming from regular jobs in the private or public sector. The main source of transfer income for households in all four quartiles was the state; examples include subsidies to deepen wells and for school meals. Interestingly, even though such government transfers are aimed at helping the poorest of the poor, households in the top quartile received substantially higher transfer incomes than households in the bottom three quartiles.

16.3.2 Income from the commons

As for income derived from common-pool resource collection, which is the main focus of this study, Table 16.5 and Table 16.6 show that, consistent with similar findings by Cavendish (2000) and Adhikari *et al.* (2004), use of all resources combined increases monotonically with income, although the same is not true for all resources considered individually. The average household in the bottom quartile earned Rs 189 per capita from natural resources (mostly from collecting fuelwood and resources other than wood and fodder), while the average household in the top quartile earned Rs 1103 per capita (mostly from collecting fuelwood and fodder).

In contrast to the findings of Jodha (1986), Reddy and Chakravarty (1999) and Cavendish (2000), however, Table 16.6 shows that dependence on common-pool resources does not decrease with income. Instead, dependence follows a U-shaped relationship with income, declining at first but then increasing. Among collecting households (400 households in all, dispersed across all 60 villages in the sample), the poorest derive about 12% of their total income from resources. Dependence decreases to 9% for households in the second income quartile, and then increases again to 11% for the third income quartile and to 13% for households in the fourth quartile. In short, wealthier households depended on the commons as much as the poorest ones.

For the subsample of collecting households, the U-shaped relationship is explained by a combination of trends in dependence on individual

Table 16.6 *Dependence on resources (%) by income quartile for collecting households*

| Collecting households | Permanent-income quartiles | | | |
	Lowest 25%	25%–50%	50%–75%	Highest 25%
Fuelwood	6.0	4.1	4.3	4.4
Construction wood	0.1	0.0	8.0	1.7
Fodder	1.2	1.9	3.9	5.8
Other resources	4.4	2.8	1.9	1.1
All resources	11.6	8.9	10.9	13.0

resources. As shown in Table 16.6, whereas increasing use of construction wood and fodder accounts for the increase in overall dependence at higher incomes, decreasing use of other resources (mahua flowers and seeds, tendu leaves, gum and dung) accounts for the decrease in overall dependence at lower incomes. As with overall resources, dependence on fuelwood follows a U-shaped relationship with income. In other words, at the income extremes, high resource dependence of the poorest collecting households is mostly due to their high dependence on fuelwood (6.0%) and other resources (4.4%), whereas the high resource dependence of the richest households is mostly due to their high dependence on fodder (5.8%) and again fuelwood (4.4%).

As the top panel of Table 16.7 indicates, households in all income quartiles meet the bulk of their consumption of fuelwood from the commons, supplemented to a small extent by collection from private trees and by market purchases. The initial decline in dependence on fuelwood with income therefore appears to be largely driven by consumption patterns: a tabulation (not shown) of per-capita consumption of fuelwood as a share of income in fact shows that this share falls monotonically with income, suggesting that fuelwood is an inferior consumption good. The slight increase in dependence on fuelwood for the top income quartile appears to be driven by the fact that only households in this quartile sell some of the fuelwood they collect. Apparently, poorer households do not consider collecting fuelwood for sale a productive use of their time. Also, higher-income households' heavy reliance on the commons for fuelwood provision, despite their higher land holdings, indicates that the private asset of land in this case does not substitute to any significant extent for the public asset of the commons.

Similarly, the high dependence of the rich on construction wood is driven by their higher consumption demand. Although much of this higher demand is met through higher market purchases, a significant fraction is met through higher collection from the commons as well, and only a small fraction through higher private provision (see second panel of Table 16.7). Also, the high dependence of the rich on fodder collection is driven by their larger animal holdings and therefore greater demand for fodder. Here, too, higher demands for fodder are not met from private sources.

Resources other than wood and fodder are collected by 302 households in our sample, and dependence on these other resources decreases almost monotonically with income. This suggests that collecting these resources is a relatively low-return activity, and one that the rich move away from as more productive uses for their labour become available.

Table 16.7 *Collection and consumption of resources by permanent-income quartile for households that collect each resource*

Permanent income quartiles	Lowest 25%	25%–50%	50%–75%	Highest 25%
Fuelwood:				
Collection from commons	213	291	501	962
Sale from commons	0	0	0	226
Collection from private sources	26	60	51	74
Sale from private sources	21	36	28	7
Market purchase	31	17	25	10
Total consumption	250	333	550	813
% of households that collect fuelwood	48%	51%	45%	38%
Construction wood:				
Collection from commons	67	61	406	1420
Sale from commons	0	0	0	0
Collection from private sources	136	1	428	655
Sale from private sources	0	0	0	404
Market purchase	0	14	24	3364
Total consumption	203	76	858	5035
% of households that collect construction wood	3%	3%	10%	9%
Fodder:				
Collection from commons	297	685	1136	2884
Sale from commons	0	7	177	0
Collection from private sources	386	848	827	1021
Sale from private sources	0	202	72	0
Market purchase	132	239	614	619
Total consumption	815	1564	2327	4524
% of households that collect fodder	8%	10%	17%	18%
Grazing:				
Days spent grazing	41	48	50	50
% of households that graze animals	74%	85%	90%	76%

Contrary to the findings in the literature, then, dependence on resources does not decline with income. Instead, dependence follows a U-shaped relationship with income. That said, the rich and the poor are not dependent on the same type of resources. The poorest households are particularly dependent on fuelwood and resources other than wood and fodder, whereas the richest households are particularly dependent on construction wood and fodder. Factors that appear to underlie these findings are differences in consumption (the rich consume much more construction wood), asset holdings (the rich have more animals) and time constraints (the rich do not bother to collect resources other than wood and fodder).

16.3.3 Grazing and water

So far, we have described only income from resources that are collected by households directly – that is, by hand. Fodder, however, is also gathered indirectly, when households let their animals graze in common grazing lands. Lacking a reliable way of converting time spent grazing to a monetary value, we instead consider time spent grazing one's animals as a proxy for grazing income. As with fodder collection by hand, time spent grazing is found to increase with income, again for the simple reason that it is the rich who have larger animal holdings (see the last panel of Table 16.7).

Unfortunately, largely because of the difficulties of pricing water, we have been unable to consider how dependence on water varies with household income. Given that one of the main uses of water is irrigation, however, we would expect land to act as a complement to common water resources, which would tend to further increase the overall resource dependence of the rich.

16.3.4 Non-collectors and non-grazers

Not all households in our sample either collect resources by hand or graze their animals on the commons. The bulk of non-collecting households are in the top income quartile. For households at this top end, Jodha's three conjectures do appear to apply: non-collecting rich households have (1) higher private landholdings, whose fodder and wood substitute for resources from the commons; (2) fewer children and therefore less surplus labour to devote to resource collection; and (3) more human capital, which likely makes extraction from the commons a low-return activity. That said, we find that most non-collecting rich households do graze their animals in the village commons and thus do engage in indirect collection. Non-grazers, a still-smaller sample of households than non-collectors, are at the very top end of the income distribution, and for these households, too, we find evidence consistent with Jodha's conjectures. Non-grazers tend to be more educated and to have smaller families. They also have much less land and livestock and instead much higher income from non-agricultural sources, such as home enterprises and jobs in the private or public sector.

16.4 Conclusions and policy recommendations

A main conclusion of our study is that, contrary to conventional wisdom, rural households do not appear to turn to the environment solely in times of desperation. The relatively rich households in our sample, which tend to have a broader set of options for earning a livelihood, regard the forests and other resources as a profitable source of income. As a result, they derive a comparable share of their income from natural resources as do poor households, though the rich and the poor do depend on different resources.

This finding suggests that improving efficiency in natural resource management may have a lasting impact on rural economies, beyond reducing poverty. If dependence on resources decreases with income – as the conventional wisdom has it – then efforts to improve the natural resource base of rural villages would boost the incomes of the poorest of the poor. However, as these households made their way out of poverty, they would turn to sources of income other than those based on natural resources and would no longer benefit from efforts to improve their environment (at least in monetary terms). If, on the other hand, both the poor and the rich are dependent on these resources – as our study suggests is the case – then households will continue to draw on natural resources to earn a living even as household incomes improve.

Our study suggests also that the poorest of the poor will benefit the most, and most immediately, by efforts that improve the returns to collection of 'minor' forest products. In Jhabua these products include tendu leaves and mahua seeds, on which the poorest households are the most dependent. To benefit a larger population, over the longer run, efforts need to be made, along with improving the natural resource base, to improve livestock holdings, which we find to be an important complementary asset to natural resources.

References

Adhikari, B. (2003). Property rights and natural resources: socio-economic heterogeneity and distributional implications of common property resource management. Working Paper 1–03, South Asian Network for Development and Environmental Economics (SANDEE), Kathmandu, Nepal.

Adhikari, B., Di Falco, S. and Lovett, J. C. (2004). Household characteristics and forest dependency: evidence from common property forest management in Nepal. *Ecological Economics*, **48**(2): 245–257.

Barbier, E. B. (2005). Rural poverty and resource degradation. In *Natural Resources and Economic Development*. Cambridge: Cambridge University Press, Chapter 8, pp. 286–320.

Beck, T. and Nesmith, C. (2001). Building on poor people's capacities: the case of common property resources in India and West Africa. *World Development*, **29**(1): 119–133.

Cavendish, W. (2000). Empirical regularities in the poverty–environment relationship of rural households: evidence from Zimbabwe. *World Development*, **28**(11): 1979–2003.

Duraiappah, A. K. (1998). Poverty and environmental degradation: a review and analysis of the nexus. *World Development*, **26**(12): 2169–2179.

Government of Madhya Pradesh (1998). The Madhya Pradesh human development report, Technical report, Government of Madhya Pradesh, India.

Horowitz, J. (1998). Review of 'The environment and emerging development issues'. Volumes I and II. Edited by P. Dasgupta and K.-G. Mäler. UN/WIDER, 1997. *Journal of Economic Literature*, **36**(3): 1529–1530.

Jodha, N. S. (1986). Common property resources and rural poor in dry regions of India. *Economic and Political Weekly*, **21**(27): 1169–1181.

Kuik, O. (2005). The contribution of environmental resources to the income of the poor: a brief survey of literature. Poverty Reduction and Environmental Management (PREM) programme, mimeo.

Reardon, T. and Vosti, S. A. (1995). Links between rural poverty and the environment in developing countries: asset categories and investment poverty. *World Development*, **23**(9): 1495–1506.

Reddy, S. R. C. and Chakravarty, S. P. (1999). Forest dependence and income distribution in a subsistence economy: evidence from India. *World Development*, **27**(7): 1141–1149.

Vedeld, P., Angelsen, A., Sjaastad, E. and Kobugabe Berg, G. (2004). Counting on the environment: forest environmental incomes and the rural poor. Paper 98, World Bank Environment Department, Washington DC.

17 · Tenure security and ecosystem service provisioning in Kenya

JANE KABUBO-MARIARA,
VINCENT LINDERHOF,
GIDEON KRUSEMAN AND
ROSEMARY ATIENO

17.1 Introduction

Poverty in Kenya is most severe in rural areas where an estimated 65% of the population resides, deriving their livelihood from the natural resource base. About 60% of this rural population was estimated to fall below the poverty line in the year 2000 (Republic of Kenya 2003). One of the most important features of Kenyan agriculture is the large subsistence sector, which makes agriculture even more important for food security. However, over the last three decades, soil erosion and land degradation have become major environmental concerns and present a formidable threat to food security and sustainability of agricultural production.

The proximate causes of land degradation are numerous and militate against the ongoing efforts at poverty alleviation. Efforts at poverty alleviation have failed to bring progress and development despite decades of development assistance. A growing population in combination with poor initial resource endowments, and macro-economic policies biased against agriculture have not only failed to alleviate poverty, but also led to a deterioration of the natural resource base on which rural livelihoods depend. It is generally accepted that development hinges on the dimensions of ecological sustainability, economic feasibility and social acceptance. However, a number of critical dimensions in the development context are unfavourable, yet they constitute the main issues related to sustainable development. These so-called development domain dimensions include agro-ecological potential, population density, market access and institutional setting. Less-favoured areas are typically characterized by

Nature's Wealth: The Economics of Ecosystem Services and Poverty, ed. P. J. H. van Beukering, E. Papyrakis, J. Bouma and R. Brouwer. Published by Cambridge University Press, © Cambridge University Press 2013.

a combination of low agricultural potential and/or poor market access; often existing in an institutional setting that is not conducive to alternative viable development pathways (Pender *et al.* 1999).

Soil and water conservation (SWC) investments cannot be seen in isolation from development domain dimensions that frame the livelihood strategies of households in a specific area. Although land tenure can be hypothesized to play an important role, it is not the only factor, and is often subordinate to market access and relative resource endowments in general (Bruce *et al.* 1994).

There is a no easy solution for the complex situation of less-favoured areas. To improve the lot of the poor, there is need for a combination of appropriate technology, an institutional setting that helps households to cope with existing market and government failures, and a set of policy measures that induce behaviour leading to both higher levels of household welfare and improved management of the natural resource base (World Bank 2003).

Poverty, agricultural stagnation and resource degradation are interlinked with overgrazing and certain modes of unsustainable crop cultivation (WCED 1987). Pender *et al.* (1999) also acknowledge the existence of a downward spiral of resource degradation and poverty. In their view, natural resource degradation contributes to declining agricultural productivity and reduced livelihood options, while poverty and food insecurity in turn contribute to worsening resource degradation by households. Livelihoods in many resource-poor farming and pastoral systems have been sustained by land-use practices that have tended to perpetuate poverty, soil erosion and other land degradation phenomena.

In Kenya, conservation and sustainable utilization of the environment and natural resources form an integral part of national planning and poverty reduction efforts. The Poverty Reduction Strategy Paper (PRSP) recognizes that weak environmental management, unsustainable land-use practices and depletion of the natural resource base have resulted in severe land degradation. This seriously impedes increases in agricultural productivity and must be addressed in order to check its adverse impact on poverty. This study examines some of the PRSP concerns and attempts to fill in research and policy gaps identified by the PRSP. It aims to gain insights into tenure security issues, providing guidelines for poverty reducing land conservation practices.

This study contributes to the literature by analysing the impact of tenure security and development domain dimensions on household welfare and SWC investments in Kenya. The study addresses the following questions: (1) What variables underpin investments in soil and water conservation? (2) What factors determine household poverty? (3) What is the link between tenure security, SWC investments and household poverty? (4) To what extent do poverty and SWC investments depend on development domain dimensions? (5) Is there any link between environment and poverty?

The rest of the chapter is organized as follows: Section 17.2 presents the study site and the data, as well as the theoretical context of our analysis. It then proceeds with a discussion of the conceptual framework and relevant methodology. Section 17.3 presents and discusses the empirical results and Section 17.4 concludes.

17.2 Data and methodology

17.2.1 Study site and data

This study is based on data collected from a survey of 457 households from three districts and 18 villages in Kenya. The household data was collected in November and December 2004[1] using a structured household questionnaire. The sample comprised 188 households from Maragua district, 151 from Murang'a district and 118 from Narok district. In addition to the household survey, a community questionnaire was administered to selected key informants (elders) in each village. The village survey gathered information on product and input prices, markets, village infrastructure and was meant to supplement information collected from households.

Since the household survey collected very extensive information on most variables of interest, factor analysis was applied to some of the data to determine the key variables for the empirical analysis. These variables relate to plot-level data on tenure security and soil quality measures and community-level data for village institutions and market access. For SWC investments, information is based on 684 plots, on which a combination of seasonal and permanent investments had been made. The three most frequently observed SWC investments in the survey were tree planting

---

[1] For detailed information on the household survey, including choice of study districts and their characteristics, sampling procedure and summary statistics, see Kabubo-Mariara *et al.* (2006).

(28% of SWC investments), terracing with grass strips (26%) and grass strips (23%). Other investments included mulching, stone terraces, soil terraces, fallowing and ridging.

17.2.2 Theoretical context

Institutional economics and the sustainable development literature, as well as historical evidence, inform us about the existence of co-evolutionary processes of three key state variables, namely the environment, institutions (the constellation of property rights arrangements) and technology. These state variables are subject to pressures related to the development domain dimensions. Population growth is a fairly autonomous process that increases pressure on natural resources. The ongoing globalization with macro-economic effects in terms of liberalization and structural adjustment – and with its local impacts in terms of market access, changes in service provision and development of rural infrastructure – not only affects the relative profitability of SWC investments but also household welfare. Every day new technological innovations are introduced, while existing ones adapted to local circumstances are adopted. The development and adaptation of technology, however, cannot be seen in isolation from other pressures. Property rights arrangements related to agricultural lands are also subject to evolution, partly as a result of changing formal rules regarding land tenure and sustainable land management, and partly as a result of endogenous processes of co-evolution.

Available evidence further shows that development domain dimensions have important implications for the poverty–environment nexus as well as for the adoption and levels of soil and water conservation. This study analyses the impact of four development domain dimensions, as well as household characteristics, on household welfare and SWC investments. In this context, land tenure is the most important dimension of the institutional setting. The empirical literature on tenure security, however, shows arbitrariness in the choice of tenure security indicators due to differences in definition and measurement of property rights and tenure security. Our survey collected data on all aspects of land rights – ranging from access to duration of ownership – on both actual and expected land rights, and on modes of acquisition. From the survey data, we used factor analysis to derive a vector of tenure security variables in order to overcome this arbitrariness (see Kabubo-Mariara et al. 2006).

17.2.3 Conceptual framework

Over the last two decades, economists have reached a consensus on how to analyse agricultural household behaviour (see for instance Sadoulet and de Janvry 1995 and Singh *et al.* 1986). The basic analytical framework in use is the agricultural household model, where a household is assumed to engage in activities using scarce resources in order to attain specific goals and aspirations, taking into account that they are constrained by external environmental and socio-economic circumstances. In such a model, the main equation relates to the utility function, which is a function of a vector of consumption goods, leisure and household characteristics. The utility function takes into account the cumulative distribution function of states of nature that captures the inherent risk and uncertainty of rural livelihood systems in terms of prices, weather and in some cases tenure. The inclusion of SWC technology implies a longer time horizon, which requires the inclusion of a subjective time preference as a discount rate, comparable to the general formulation of optimal control models. Utility is maximized subject to a cash income constraint, which is a function of vectors of market purchased and household produced consumption commodities, market and farm-gate commodity prices, factor prices and exogenous income. In addition, the household faces a set of resource constraints, which imply that the household cannot allocate more resources to activities than is available in terms of the total stock. The household also faces a production constraint reflected by a technology function that depicts the relationship between inputs and outputs conditional on farm characteristics, soil quality and technology level.

Standard farm household theory postulates that consumption and production decisions for farm households in developing countries are often non-separable. Consequently, the complexity of the interacting components of the farm household and the problem of unobservable key variables imply that there are serious consequences for econometric estimation of structural models (see Kruseman 2001).

Solving the agricultural household model can be done in a number of ways: the first is to estimate the full structure of the model, by estimating each equation separately and then using the quantified model to simulate responses, as is common in bio-economic modelling. The alternative is to estimate reduced form equations of the household model, traditionally considered the most appropriate way of dealing with these types of complexities. The coefficients in the reduced form equations capture the sum of both direct and indirect effects. Because of this characteristic, it is imperative to include all relevant explanatory variables in the analysis, even if their coefficients are insignificant for the analysis being undertaken. This approach, however, deserves some attention. If we derive the first-order conditions for the agricultural household model and combine and collapse the resulting equations, the end result is a system of equations where the dependent variables consist of the choice variables of the household (production structure, investment, consumption, resource allocation) and on the right-hand side all the exogenous factors (household characteristics, farm characteristics, institutional characteristics). However, we have to be very careful about causality and attribution in the intertemporal context.

17.2.4 Empirical implementation

The principal goal of this study is to analyse the impact of tenure security on household welfare and SWC investments. According to economic theory, households faced with uncertain outcomes with respect to income streams will diversify their portfolio of activities in order to ameliorate the threat to their welfare by the failure of individual activities. The overriding objective of the household will therefore be to maximize its welfare. Closely linked to welfare is vulnerability, which is related to the activities and investment decisions that a household will undertake. Insecurity of land tenure, for example, adds to vulnerability and will determine how much investment a household can undertake on the farm (Ellis 2000).

In addition to the farm household model (Sadoulet and de Janvry 1995, Singh *et al.* 1986), household welfare analysis is founded on the standard economic theory of consumer behaviour (Deaton and Muellbauer 1980, Glewwe 1991). Because household welfare is unobservable, consumption expenditure can be used as a proxy for welfare. The expenditure variable can, however, be scaled down to take into consideration differences such as household size, so that the dependent variable collapses into per capita rather than aggregate expenditure. This allows for comparison of welfare levels across households with different composition and across regions with different prices. Following standard econometric practice, we can then estimate a reduced form model of per-capita expenditure or income combining all the various structural relationships that affect welfare.

For policy analysis, it is important to include variables influenced by government interventions. Therefore, we include a vector of standard explanatory variables (see for instance, Glewwe 1991) but also introduce farm characteristics, institutional variables and other development domain dimensions. Using the survey data, we examine the correlation between welfare, soil and water conservation and land tenure security. To find out if there is a relationship between the endogenous variable related to welfare and SWC beyond the relationship between the exogenous variables that determine both issues, we analyse the regression model residuals. The residuals enter into the welfare equation as error correction terms (ECM) in order to eliminate any possible bias.

Our deviation from standard poverty analysis is to introduce institutional factors and SWC investment variables into the welfare model. The institutional factors, such as the presence of special interest groups and extension agencies, and the choice of SWC investments are (partly) endogenous, and they have to be explained themselves. In particular, the presence of interest groups and the willingness to engage with ('listen to') extension agents affect the willingness to invest.[2] In order to disentangle partly endogenous effects we use a step-wise estimation approach (see Kabubo–Mariara *et al.* 2006 for more details).

Institutional factors are indicators of how well households/villages organize themselves to enhance household welfare as depicted in Figure 17.1. To capture the impact of the membership of the household in special interest groups/village institutions, we specify probit models for three different interest groups: membership in income generating groups, loans groups and benevolent groups. In addition, we are interested in analysing the impact of extension services on household welfare. We specify probit models for the willingness to listen to extension services in general, and the willingness to listen to extension in natural resource management.

We then specify the willingness to invest in natural resource management at household level. This willingness to invest is defined as investments made up to 5 years ago and depends on the willingness to listen to natural resource management (NRM) extension, and the willingness to listen to extension in general as well as the awareness of the presence and membership of NRM and other special interest groups. However, since these variables are endogenous, we apply a nested methodology for capturing these variables. We then use the residual terms of each of the probit equations as explanatory variables in the willingness to invest equation. The reason for using these residual terms for estimating the

---

[2] Village institutions in this context refer to village groups/committees, which are often organized by villagers for different purposes. Most of them are based on different church denominations (see Kabubo–Mariara *et al.* 2006).

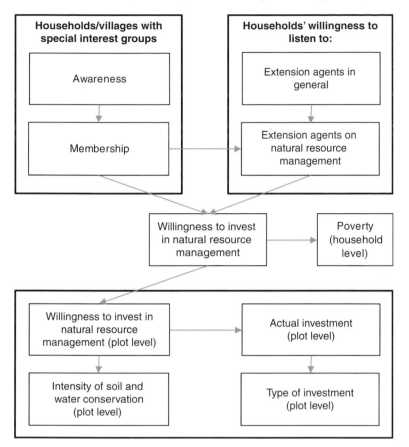

Figure 17.1 Link between village institutions, soil and water conservation investments and poverty

endogenous variables is that the terms are orthogonal to other independent variables in the equation at hand.[3] The truly independent variables still capture both the direct and indirect effects, while the residual terms capture the idiosyncratic effect of the endogenous variable (Kabubo-Mariara *et al.*, 2006).

Due to our approach, the residuals are a set of identical equations, explained by the same set of factors. In principle, membership in special

---

[3] Note that the predicted residuals are not necessarily independent from the error term of the equation in which the endogenous variable appears as explanatory variable. If they are dependent this estimation approach will yield biased estimators, although the bias will be small.

interest groups and willingness to listen to extension services (including extension in NRM) are related and therefore we correct for correlation of variances. We do this by applying factor analysis on the residuals of each of the probit results for each equation to find a common variance factor(s). These factors are then included in the final estimation model of the willingness to invest in SWC at the household level (Kabubo-Mariara *et al.* 2006). To capture the effect of membership in interest groups, willingness to listen to extension and the willingness to invest in SWC on household welfare, we enter the residual terms into the welfare function (see Figure 17.1). This way we are able to capture the direct and indirect effects of both the exogenous and potentially endogenous variables.

The adoption of SWC investments is based on the same framework as household welfare analysis but the analysis is at the plot level (Figure 17.1). We limit the analysis to plots that have had investments in the last two seasons. We distinguished between seven main types of investments: grass strips, mulching, tree planting, general terraces, soil terraces, grass stripped terraces, and all other investments (fallowing, ridging and crop rotation). We categorized these investments into permanent and seasonal investments, depending on how they were adopted, and used the probit model to explain the adoption of these investments. In addition, we explain the number (intensity) of investments at the plot level using the Tobit model. Results for individual SWC conservation investments are presented in Kabubo-Mariara *et al.* (2006).

17.3 Results and discussion

17.3.1 Household welfare analysis

This subsection presents the regression results linking household welfare, tenure security and SWC investments. For tenure security, we used five variables constructed using factor analysis. The first variable captures owned plots (which are often inherited), which have been owned by the family for a long period of time and which can be sold without permission. The second variable captures family land that can be sold or bequeathed with or without permission. The other variables include land registered in the family name, rented out and the right to rent out land without permission. Our results (Table 17.1) suggest that tenure security may not be an important determinant of household welfare, but we note that this may be due to inclusion of both direct and indirect effects of all endogenous variables.

Table 17.1 *Reduced form estimates of household welfare (per-capita expenditure)*

Variable	Coefficient	Standard error	t-value
Investment in SWC			
Number of SWC structures on farm	0.195	0.047	4.16
Soil quality and topography			
Coarse soils	−0.108	0.037	−2.91
Workability	0.071	0.024	2.91
Red vs. black soils	0.058	0.032	1.79
Undulating terrain	0.071	0.024	2.91
Tenure security and related factors			
Land registered in household head or spouse	−0.070	0.042	−1.68
Right to sell family land with permission	−0.070	0.042	−1.68
Plot area (farm size)	−0.059	0.026	−2.29
Distance to plot	−0.004	0.002	−2.82
Village characteristics			
Population density	−0.002	0.001	−3.54
Market access	−1.054	0.521	−2.02
Murang'a district dummy	−0.297	0.160	−1.85
Narok district dummy	−4.472	1.819	−2.46
Household characteristics and assets			
Age of household head	−0.036	0.019	−1.93
Age of household head squared	0.0002	0.0002	1.48
Child under 5-years-old in a household	−0.113	0.045	−2.52
Children 6- to 16-years-old in a household	−0.077	0.035	−2.17
Number of adult women in a household	−0.165	0.049	−3.4
Masaai tribe dummy	2.120	1.014	2.09
Lagged value of livestock (log)	0.146	0.055	2.68
Lagged value of farm equipment (log)	0.153	0.047	3.26
Error correction terms (residuals)			
Listened to extension services	0.132	0.071	1.85
Membership in benevolent groups	0.132	0.071	1.85
Willingness to invest in SWC	0.199	0.091	2.2
Constant	5.795	0.849	6.83
Observations	454		
R^2	0.33		
$F_{(44, 409)}$	4.62		

Notes: The actual regression is based on 44 explanatory variables (see Kabubo-Mariara *et al.*, 2006). For convenience, we only present the results for a selected number of relevant variables.

The impact of SWC investment on welfare is captured by two variables: the number of SWC investment structures present per plot[4] and a residual for the willingness to invest in SWC. The total number of SWC investments on land used by a household has a large positive and significant impact on per capita expenditure, implying a poverty–environment link. The willingness to invest in SWC is also positively associated with household welfare. The SWC variables are jointly significant determinants of welfare, further suggesting a poverty–environment link.

Population density is inversely related to expenditure, implying that poverty is concentrated in regions of high population density. Market access has an unexpected negative and significant sign, which can be explained by the fact that market access is the factor score of the factor related to distance to markets. The expected correlation between population density and market access and also the testing of both direct and indirect effects probably accounts for these results. The district dummies for Murang'a and Narok exhibit significantly negative coefficients, which implies that households located in these districts are likely to be poorer than those located in Maragua district. This result seems to correspond to the observed distribution of per capita expenditure across the three districts.

We further investigate the impact of two different categories of farm characteristics: soil quality and topography, both indicators of development domain dimensions. In addition, we include acreage and distance to plot. The results indicate that soils with easy workability are positively correlated with per capita expenditure, but coarse soils exhibit an inverse relationship with per capita expenditure. Relative to black soils, red soils have a significant positive impact on per capita expenditure. Undulating land is found to be inversely and significantly correlated with per capita expenditure, because it is more prone to soil erosion than flat land. Other important determinants of household welfare are some household characteristics including household composition and assets (initial livestock, equipment and education).

17.3.2 Soil and water conservation investments

In addition to the welfare analysis, we determine the impacts of household, farm and community characteristics as well as development domain

---

[4] The number of SWC investment structures includes all SWC investments that have been present for 1 year and which have not been abandoned. Thus, actual SWC investments made in the current year are excluded.

dimensions on SWC investment decisions at the plot level.[5] Table 17.2 presents the regression results for both seasonal and permanent investments as well as for the intensity of adoption. The results suggest that most household characteristics do not influence the adoption decision. Although this result is not uncommon in the literature (see for instance, Gebremedhin and Swinton 2003), it could be due to the fact that the coefficients capture the sum of direct and indirect effects.

The Murang'a and Narok district dummies suggest that location is an important determinant of the decision to adopt SWC. Specifically, there is a higher probability of adoption if a household is located in Murang'a relative to Maragua district. The likelihood of adoption is lower in Narok district but the coefficients are insignificant (note that Murang'a is more hilly and undulating than Maragua district, while Narok is relatively flat in terms of terrain). Given this diversity of agro-ecology in the three districts, the regional dummies reflect the unobserved relative importance of different development domains.

The coefficients of most tenure security variables exhibit the expected signs and are significant, which confirms the importance of tenure security in adoption of SWC investments. The marginal impacts of the tenure security factors show that bundles of rights that relate to owned plots, which have been owned by the family for a long period of time and which can be sold without permission, have the largest positive impact on investment in soil and water conservation; the impact is highest for permanent rather than for seasonal investments. Bundles of rights that relate to land rented out have a large negative impact on adoption of SWC investment. This is because a household has no control over investments on plots that it has rented out and it will not make any new investments so long as the land is rented out. The magnitudes of coefficients for the intensity of SWC investments show the same impact on the two separate bundles of rights. The differences in coefficients for permanent and seasonal investments support the finding that tenure security favours long-term SWC investments more than short-term investments (see Gebremedhin and Swinton 2003).

Soil depth is positively correlated with adoption of SWC investments. Moderate slopes and undulating terrain favour adoption of permanent

[5] To derive the parameter estimates of the final estimation models, we use several steps, including step-wise regression and factor analysis of residuals in order to take care of any possible covariance and joint determination of the variables at hand (see Kabubo-Mariara *et al.* 2006 for more details).

Table 17.2 *Regression results for adoption of soil and water conservation investments (plot level)*

Variable	Permanent investments — Marginal effects (probit model)	Seasonal investments — Marginal effects (probit model)	Intensity of SWC — Coefficients (Tobit)
Tenure security and related factors			
Land registered in household head or spouse	0.0657 (4.88)***	0.0426 (2.81)***	0.4392 (5.83)***
Family land registered in extended family	0.0090 (0.77)	−0.0031 (−0.21)	0.0151 (0.22)
Right to sell family land with permission	0.0035 (0.32)	0.0284 (2.10)**	0.1392 (2.18)**
Rented out land	−0.0469 (−2.53)**	−0.0508 (−2.55)***	−0.3656 (−3.73)***
Lent out land	−0.0321 (−2.43)**	−0.0323 (−1.97)*	−0.2053 (−2.67)***
Plot area (farm size)	0.0017 (1.35)	0.0014 (0.85)	0.0150 (2.03)**
Distance to plot	−0.0028 (−2.97)***	−0.0051 (−2.62)***	−0.0099 (−2.01)**
Soil quality and topography			
Moderate vs. fine texture	−0.0510 (−3.89)***	−0.0330 (−2.10)**	−0.3570 (−4.60)**
Soil depth	−0.0117 (−0.78)	−0.0393 (−2.29)**	0.0413 (3.53)***
Red vs. black soils	−0.0266 (−2.07)**	−0.0030 (−0.19)	−0.2244 (−2.74)***
Fertile to average fertile	0.0608 (5.27)***	0.0317 (2.11)**	−0.1481 (−1.97)**
Steep slope	−0.0056 (−0.48)	0.0348 (2.60)***	0.0264 (0.39)
Moderate slope	−0.0510 (−3.89)***	−0.0330 (−2.10)**	0.3982 (5.62)***
Flatness of slope	0.0071 (3.70)***	0.0015 (0.60)	−0.1391 (−1.91)**
Undulating terrain	−0.0117 (−0.78)	−0.0393 (−2.29)**	0.0973 (1.47)

Household characteristics and assets			
Age of household head	-0.0063 (-1.54)	0.0129 (1.70)*	-0.0030 (-0.09)
Age of household head squared	0.00004 (1.18)	-0.0001 (-1.69)*	-0.0001 (-0.23)
Children 6 to 16 years old in a household	0.0294 (3.51)***	-0.0028 (-0.26)	0.1208 (2.37)***
Household head years of schooling	0.0070 (2.07)**	-0.0014 (-0.33)	0.0129 (0.66)
Lagged value of livestock (log)	0.0438 (3.45)***	0.0078 (0.49)	0.2229 (2.95)***
Lagged value of farm equipment (log)	0.0173 (1.79)*	0.0093 (0.77)	0.0619 (1.10)
Previous soil conservation structures	-0.1293 (-7.71)***	-0.145 (-7.01)***	-1.089 (-10.39)***
Error correction terms (residuals)			
Willingness to invest in SWC	0.1669 (6.94)***	0.0516 (1.89)*	0.9692 (7.00)***
Constant			-1.2329 (-1.12)
Observations	684	684	684
LR chi2(42)	169.07	137.11	236.58
Pseudo R2	0.2715	0.2001	0.1802
Log likelihood	-226.847	-273.999	-538.12

Note: The actual regression is based on 42 explanatory variables (see Kabubo-Mariara *et al.* 2006). For convenience, we only present the results for a selected number of relevant variables. Robust *t*-statistics in brackets; *, **, *** denote significance at 10%, 5% and 1%, respectively.

investments. Undulating terrain is also positively correlated with the probability of adopting seasonal investments. Conversely, steep slopes lower the probability of adoption of all SWC investments. There is also less likelihood of adoption of most SWC investments on flat land. Plot size increases the likelihood of adoption, but the impact is insignificant. Distance to plot is inversely and significantly correlated with adoption of all SWC technologies, implying that all technologies are more likely to be adopted on home plots rather than distant plots.

The error correction variables for extension services and membership in village institutions have no important impact on adoption of SWC. However, willingness to invest is positively correlated with all investment decisions, both permanent and seasonal, implying the need to provide incentives for SWC. Comparing the intensity of adoption with the adoption of seasonal and permanent investments, the results largely suggest that the intensity of adoption is determined by the same set of factors that influence the decision to invest in SWC.

17.4 Conclusions and policy recommendations

This chapter investigates the impact of tenure security on household poverty and on SWC investment for rural areas in three districts in Kenya. The key hypothesis tested is that tenure security affects both SWC investments and household welfare, but that SWC investment also affects household welfare. In analysing these relationships, we also test the impact of household, farm and village characteristics including development domain dimensions that condition the link between poverty and the environment. An innovation of this study is the use of factor analysis to construct key variables for tenure security, soil quality and topography, institutional presence and market access for the empirical analysis. In addition, we are not aware of other studies that directly link household welfare, investment in SWC and tenure security.

The results suggest that the total number of conservation structures in place is also an important determinant of household welfare, implying a poverty–environment link. Farm characteristics, particularly soil quality and topography, are important determinants of household welfare. Since these variables are directly linked to the environmental status and agro-ecological potential of land, their impact on welfare also supports existence of the poverty–environment link. The district dummies suggest existence of district specific direct and indirect effects on household welfare and therefore suggest unobserved heterogeneity in determinants of welfare. These results also suggest that among other factors, the poverty–environment link is also conditioned by the agro-ecological potential. We conclude that development domain dimensions together with other farm, household and village characteristics are important correlates of household welfare.

Consistent with the literature, our analysis of SWC investments supports the importance of tenure security in determining adoption and the intensity of SWC investments. In addition, the results support the importance of household assets, farm characteristics (soils and topography), presence of village institutions and development domain dimensions (market access and population density) in adoption of SWC investments. Soil quality and topography, as well as location (agro-ecological diversity) are particularly important determinants of SWC investment. The impact of household assets (livestock and farm equipment) on SWC investment implies a poverty–environment link, because households poor in assets are less likely to invest in soil and water conservation. The intensity of adoption is also lower for households poorer in assets than their richer counterparts. The results further suggest that the factors affecting the intensity of investment are the same as the factors determining the decision whether or not to invest in SWC.

The results suggest a number of policy interventions for alleviating poverty and providing incentives for soil and water conservation. The importance of development domain dimensions suggests the need for geographical targeting with incentives oriented towards specific development domains that takes into account diversity on market access, population density and agro-ecology. In addition policies that provide incentives for boosting household assets and village institutions would positively impact on both investment in SWC and household welfare.

The results point to the need for a comprehensive land-use policy that will facilitate land-use management and tenure security for environmental protection as a way of increasing agricultural productivity and hence enhancing rural livelihoods. In addition, incentives that encourage SWC investments are also important in improving land productivity.

References

Bruce, J. W., Migot-Adholla, S. E. and Atherton, J. (1994). The findings and their policy implications: institutional adaptation or replacement. In J. W. Bruce and S. E. Migot-Adholla. Dubuque (eds.), *Searching for Land Tenure Security in Africa*. Dubuque, IA: Kendall/Hunt Publishing Company, pp. 251–265.

Deaton, A. and Muellbauer, J. (1980). *Economics and Consumer Behavior*. Cambridge: Cambridge University Press.

Ellis F. (2000). *Rural Livelihoods and Diversity in Developing Countries*. New York: Oxford University Press.

Gebremedhin, B. and Swinton, S. M. (2003). Investment in soil conservation in Northern Ethiopia: the role of land tenure security and public programs. *Agricultural Economics*, **29**: 69–84.

Glewwe, P. (1991). Investigating the determinants of household welfare in Cote d'Ivoire. *Journal of Development Economics*, **35**: 307–337.

Kabubo-Mariara J., Linderhof, V., Kruseman, G., Atieno, R. and Mwabu, G. (2006). Household welfare, investment in soil and water conservation and tenure security: evidence from Kenya. PREM report 06/06. Institute for Environmental Studies, VU University, Amsterdam, the Netherlands.

Kruseman, G. (2001). Household technology choice and sustainable land use. In N. B. M. Heerink, H. van Keulen and M. Kuiper (eds.), *Economic Policy Reforms and Sustainable Land Use in LDCs: Recent Advances in Quantitative Analysis*. Heidelberg, Germany: Physica-Verlag, pp. 135–150.

Pender, J., Place, F. and Ehui, S. (1999). Strategies for sustainable agricultural development in the East African Highlands. Environmental and Production Technology Division Working Paper 41. IFPRI, Washington DC.

Republic of Kenya (2003) *Geographic Dimensions of Well-Being in Kenya: Where are the Poor*, Vol. **1**. Nairobi, Kenya: Central Bureau of Statistics, Ministry of Planning and National Development.

Sadoulet, E. and de Janvry, A. (1995). *Quantitative Development Policy Analysis*. Baltimore, MD: Johns Hopkins University Press.

Singh, I., Squire, L. and Strauss, J. (1986). *Agricultural Household Models: Extensions, Applications, and Policy*. Baltimore, MD: Johns Hopkins University Press.

World Bank (2003). *World Development Report 2003*. Oxford: Oxford University Press.

World Commission on Environment and Development (WCED). (1987). *Our Common Future*. Oxford: Oxford University Press.

18 · *Pastureland degradation and poverty among herders in Mongolia*

SEBASTIAAN M. HESS, AUYRZANA
ENKH-AMGALAN, ANTONIUS J. DIETZ,
TUMUR ERDENECHULUUN, WIETZE LISE
AND BYAMBA PUREV

18.1 Introduction

After the collapse of communist rule in Mongolia in 1991, the demise of the livestock collectives resulted in private (household-based) livestock ownership and unclear range management institutions. Between 1991 and 1998 the livestock sector rapidly expanded, partly assisted by relatively good weather conditions, and partly by many new entrants in the livestock economy as a result of de-industrialization of the urban economy. In 1990 Mongolia had 25.9 million domesticated animals. In 1998 this number had grown to 32.9 million, an increase of 27%.

Between 1990 and 1998 the weather conditions were indeed rather favourable. Compared to the 1980s rainfall was higher, and winters were less severe (Batjargal *et al.* 2000). The carrying capacity of the Mongolian grazing lands improved, and the growing livestock population could, on average, be accommodated by these improved grazing conditions. However, changes in livestock mobility and range management styles, as well as unclear grazing institutions under privatized livestock regimes, already created carrying-capacity tensions in some areas. Where water wells were no longer maintained, some grazing areas were abandoned, resulting in condensed grazing in other areas.[1]

---

[1] According to CPR (2003), out of 41 600 wells operational in 1990, only 30 900 were still operational in 2000.

Nature's Wealth: The Economics of Ecosystem Services and Poverty, ed. P. J. H. van Beukering, E. Papyrakis, J. Bouma and R. Brouwer. Published by Cambridge University Press, © Cambridge University Press 2013.

These early signs of overuse in some places turned into a national disaster when weather conditions worsened after 1998. Between 1999 and 2002, winter conditions were very severe and Mongolian herders and livestock were faced with consecutive *dzuds*; the Mongolian term for winters when ice and snow prevent animals from foraging on pastures. Furthermore, the spring and summer rainfall was also disappointingly low during these years. This meant grass was less abundant in the autumn when the animals usually renew their fat reserves for the coming winter. Together this resulted in estimated losses of 12 million animals nationwide; out of an estimated 190 000 herding households in 1998, 11 000 families lost all of their animals (Danker 2004, p. 26). In December 2002 the total number of animals had gone down to only 24 million, back to the level of the late 1980s.

Part of the reason these losses were so extreme is because the capability to provide emergency feed for the animals (which used to be the responsibility of the collectives) had been eroded. Most district authorities did not have enough feed stored, nor was there enough transport capacity to distribute the stocks that were available to the distressed herders and animals. The livestock sector found itself in a transitory state between the collectives of the old system and a mature free-market system with its own emergency relief mechanisms.[2] Because 70 years of communist rule had also destroyed the old communal institutions this transitory period basically meant the range lands were in a state of open access, illustrated by the large number of new, inexperienced entrants in the livestock sector. The disasters of the late 1990s and the early years of the new century are therefore a case example of Hardin's 'Tragedy of the Commons' (1968), or more fittingly, a 'Tragedy of Open Access' (Bromley and Cernea 1989).

This tragedy is often associated with the prisoners' dilemma game, in which actors make choices about their natural resource use in isolation and end up in a suboptimal situation. If the situation can indeed be characterized as a prisoners' dilemma, and had the actors taken their decisions cooperatively, both they and the environment would have been better off. Establishing institutions that facilitate such cooperative decision-making can alleviate the Tragedy.

---

[2] According to the National Statistical Office of Mongolia (unpublished data, 2003) the land for natural hay production for winter storage decreased from 1.2 million to 0.8 million hectares (out of *c.* 129 million hectares of natural pasture), between 1989 and 2002. Green fodder and silage production more or less disappeared, and manufactured feed production was more than halved.

In this chapter, one of the main objectives is to use game theory to characterize the type of game that is being 'played' in the Mongolian livestock sector. Depending on the type of game, different actions can be taken to address the problem of overgrazing and herder poverty.

A second objective of this chapter is to provide a better insight into the dynamics of the pastures' carrying capacity. The traditional approach of having a fixed carrying capacity is not sufficient to explain the large numbers of animal losses that occurred. It is shown that when we allow the carrying capacity to be dynamic by letting it depend on weather conditions, the cycle of animal losses and increases can be better explained.

A final objective is to establish whether a further developed market system might have prevented part of the disaster or could prevent others in the future. To accomplish this we look at the so-called *Caloric Terms of Trade* (CToT) that the herders faced during both good and bad years. This indicator describes at which caloric rate the herders can trade their animal products against other food sources. If these rates are favourable, especially during bad periods, more trading could relieve the pressures faced by the herders.

The remainder of this chapter is organized as follows. Section 18.2 starts with a description of the study sites and Section 18.3 presents our research methods. Results and their discussion are provided in Section 18.4 followed by conclusions and policy recommendations in Section 18.5.

18.2 Study site

Two areas were studied. The first area, Ugtaal *sum* (district) in the Tov *aimag* (province) is in the north (in an area often referred to as the 'forest steppe'). This area has more rainfall and more severe winter conditions. The second area, Gurvansaikhan *sum* in Dundgovi *aimag*, is in the south close to the Gobi desert. These two districts were selected for the following reasons:

- They have different levels of pasture degradation.
- They are in different ecological regions, with varying ecological conditions and land-use patterns.
- They vary in terms of how they were impacted by the 1999–2002 *dzuds*.

18.2.1 Ugtaal

Ugtaal is located in Mongolia's steppe region and was created in 1924. The *sum* centre is closer to the capital (at 155 km distance) than to its own *aimag* centre (at 177 km distance). It covers a land area of 154 800 ha, out

Table 18.1 *Livestock number in Ugtaal* sum

	1999	2000	2001	2002	2003
Camel	2	9	2	4	6
Horse	8 402	7 858	6 949	6 296	5 490
Cattle	7 088	6 758	3 836	3 238	2 630
Sheep	38 011	41 358	28 810	24 901	19 737
Goat	12 712	16 319	14 105	14 987	13 357
Total	150 804	151 644	113 174	101 909	85 998

Note: Total is expressed in sheep units (SU): 1 sheep = 1 SU, 1 horse = 7 SU, 1 cattle/ yak = 6 SU, 1 camel = 5 SU, 1 goats = 0.9 SU.
Source: *Sum*'s livestock census data, 1999–2003. Livestock data are collected in the autumn of each year.

of which 110 700 ha is pasture. The remainder consists mostly of arable land and forests, but there is also some land reserved for haymaking. According to the 2003 *sum* statistics, 23 000 hectares of pastures had been degraded. The *sum*'s population stood at 2816 as of January 2004, with a total of 715 households.

The main economic activity is livestock herding and crop farming. Livestock was severely affected by the *dzuds* in 2000–02. In 2000 the number of livestock reached its maximum level of 152 000 sheep units. At the end of 2003, there were only 86 000 sheep units left in the *sum* (a decrease of 43%). The available data on livestock are presented in Table 18.1.

There were also corresponding socio-economic tensions. In 2002 there were 105 households who lost all their animals. On the other hand, the share of households with more than 200 animals increased from 6% in 1997 to almost 10% in 2003; a clear case of asset polarization.

18.2.2 Gurvansaikhan

The second study site is located in the Gobi region. The district centre lies 331 km south of Ulaanbaatar and the distance to the *aimag* centre is 71 km. Gurvansaikhan is much less densely populated than Ugtaal. Its population is just a bit smaller, standing at 2690 (673 households) in January 2004, but the *sum* covers 542 000 ha, consisting almost entirely of pastureland (99%).

As with most other *sums* in Mongolia, the primary economic activity is livestock herding. Gurvansaikhan was hit especially hard by the 1999– 2000 *dzud* and drought. Fifty per cent of all horses died, and losses among

Table 18.2 *Livestock numbers in the Gurvansaikhan* sum

	1999	2000	2001	2002	2003
Camel	1 794	1 465	1 369	1 324	1 303
Horse	13 475	6 701	7 099	7 628	8 321
Cattle	10 157	1 867	1 673	2 174	2 862
Sheep	73 275	48 161	54 100	58 277	65 731
Goat	57 674	31 222	41 963	49 724	61 549
Total	289 419	141 695	158 443	176 089	203 059

Note: Total is expressed in sheep units (SU): 1 sheep = 1 SU, 1 horse = 7 SU, 1 cattle/
yak = 6 SU, 1 camel = 5 SU, 1 goat = 0.9 SU.
Source: *Sum*'s livestock census 2000–03. The animal numbers are collected in the
autumn of each year.

cattle amounted to a staggering 82%. In total sheep units, numbers went
down from 289 000 to 142 000. However, from 2000 onwards, livestock
numbers increased again to a level of around 203 000 sheep units in 2003
(see Table 18.2).

Household herds are generally larger in Gurvansaikhan, but we see the
same trend as in Ugtaal with the share of households with larger numbers
of animals increasing from 2000 onwards. The two districts also differ in
one other important factor: migration. In Ugtaal the number of households decreased between 2000 and 2003, whereas Gurvansaikhan showed
an increase, although there was some out-migration in 2000–01.

18.3 Data and methodology

As mentioned in the introduction, there were three main objectives in the
project. In this section the methods used to reach these objectives will be
described.

18.3.1 Carrying-capacity dynamics and market-based population
supporting capacity

The concept of 'carrying capacity' generally deals with the relationship
between land (pasture) and livestock. To get an idea about the carrying
capacity of the land for livestock, range management scientists use 'rules of
thumb' to assess whether there are too many animals on the land, and
hence a high probability of land degradation due to overutilization.

Overutilization can result in further degradation (erosion, diminishing biomass production, desertification), and hence a downward spiral of deteriorating conditions for livestock production. Often these rules of thumb are rather crude and static. Current thinking in range management circles recognizes that livestock management on collective levels needs to be based on more complex models and therefore take more notice of the variability in range conditions. These include:

- The availability and accessibility of range lands, whilst also looking at the distribution of water points, and the relative differences in security, social and legal barriers and labour availability for movements to remote areas;
- The relative usefulness of different types of biomass for livestock utilization, which partly depends on range management institutions;
- The weather conditions, with lower-than-average (spring and summer) rainfall translating into more than proportional decreases of feed availability (including hay production for winter storage). Moreover, the differences in severity and length of winters, particularly in Mongolia, directly affects range lands' stress levels.

In our quantitative analysis we will focus on the third aspect to start making the concept of carrying capacity more dynamic, and try to establish the cause of environmental degradation in Mongolia.

We also include the concept of 'population supporting capacity' in our analysis. In its most simple form it translates the calculated carrying capacity, or the observed numbers of animals, into the number of people that can be fed from the land on a subsistence basis. Apart from the size of the livestock, information on meat and milk production of these animals is incorporated in the analysis and the associated caloric values are compared to people's caloric requirements.

To move away from this simple form of population supporting capacity, which holds the restrictive assumption of full subsistence herding, we also look at CToT. This indicator gives the caloric rate at which animal products can be traded against other food products such as grain and potatoes. In most parts of the world these terms of trade are favourable for the herder, but they also vary with pasture conditions, as these influence the supply of animal products. Using primary and secondary data on prices and caloric values of both animal and non-animal food products, we can show the changes in the CToT over the cycle of the disasters that struck Mongolian herders.

18.3.2 The inter-herder game in Mongolia

Through game theory analysis we aim to highlight the unsustainable nature of herder behaviour. To picture the kind of game we will try to estimate, let us consider the following situation: it is November and winter is about to start. Imagine a delineated winter grazing area in a Mongolian region. The summer grass-growth season is over and the winter pasture is restored and ready for grazing. This is the time for the herders to decide on the off-take in their herds by slaughtering a number of their animals.

This situation can be formalized by identifying strategies and players in a game as follows. The herders are playing a game about which herd size to maintain. Herders' decisions, however, have certain consequences. When they choose to maximize their herd sizes the risk of negative impacts of a possible *dzud* or drought, or diseases increases. On the other hand, when they choose to slaughter a larger proportion of their herds, they may not fully benefit from the (winter) grazing opportunity. Since they compete for the same pastures, they also have to take the strategies of other herders into account.

We can demonstrate the mechanism of this game by a hypothetical two-person, two-strategy representation. We assume two *identical* herders, Herder 1 and Herder 2, where Herder 2 represents all other herders contesting for the same winter pasture. In the analysis, Herder 1 is 'played' in turn by every herder in our sample.[3]

Both one's own strategy and that of other herders will determine the survival rate of the herd and hence, one's payoff. According to the strategies adopted, payoffs can be classified in different groups: when both herders keep their livestock constant they both obtain x; when herders increase their livestock they both obtain y; when one herder increases his herd and the other keeps his herd constant, the herd increasing herder obtains a and the herd maintaining herder obtains b. Table 18.3 shows the resulting payoff matrix.

The type of game herders face when competing for grazing pastures is determined by looking at actual herder behaviour. In a field survey carried out in the winter of 2003–04 we collected data on the strategy of herder households, their view of other herders' strategy and the payoff of each household. In both Ugtaal and Gurvansaikhan 60 herding households

---

[3] Hence, we assume a 1 versus n–1 persons game (see also Lise 2001, Lise *et al.* 2001) in which Herder 1 interprets the actions of other herders as a simultaneous move. One of the earliest accounts of games among herders is the herdsman game by Muhsam (1973), which is also analysed in a 1 versus n–1 person setting.

Table 18.3 *Payoff matrix for the game between two symmetric herders*

		Herder 2:	
		Keep herd constant	Increase herd size
Herder 1:	Keep herd constant	*x,x*	*b,a*
	Increase herd size	*a,b*	*y,y*

were interviewed in an even split between 'poor' (<200 animals) and 'rich' households (>300 animals).

We base herders' strategies on information about offtake and offspring, indicating the growth rate of the herd. The perceived strategy of other herders was proxied by the perceived environmental conditions of the herders. We assume that a negative perception is linked to a herder's view that the other herders are maximizing their herds, since this will have a negative effect on the pastures, and vice versa. Finally, for the payoff we used net income from the herd, including costs of fodder purchases, veterinary services, sales of animals and animal products. Unfortunately, own consumption of animal products by the household could not be included in the net income because the response to these questions in the survey was incomplete. For details on the survey and the constructed variables the reader is referred to Lise *et al.* (2006).

Knowing both the strategy of Herder 1 and his perception of the strategy of the other herders, we can place his payoff in one of the four cells in Table 18.3. By calculating the average payoff in each group and determining the order between the groups we will be able to establish which type of game is played on the windy plains of Mongolia.[4]

Since we hypothesized a prisoners' dilemma game we expect to find the following relationship between the payoff groups:

$$a > x > y > b, \qquad (18.1)$$

which can be conceptualized as such: if Herder 2 (all other herders) maintains his/her herd, there will be more grass left for Herder 1 who can feed his herd without purchasing additional and expensive fodder. If Herder 2 maximizes his/her herd, it is still better for Herder 1 to do the same, as he/she may not be sure whether the pasture will be in good condition, once he/she accesses it in the winter. Both herders, however,

---

[4] For more details on the technical aspects of the estimation process the reader is referred to Lise (2001).

are best off with a mutual constant herd size, so that they are in a better position to face a possible *dzud*, since the impact of a *dzud* will be much more severe in the case where they both start with a large herd. The same reasoning is true for droughts, as it is much more difficult to maintain a large herd during a drought than a smaller more mobile herd.

18.4 Results and discussion

18.4.1 Carrying capacity and market-based population

In order to make the concept of carrying capacity more dynamic, we looked at weather data for the 1990s and early years of the twenty-first century. We calculated an aridity index for all years, based on the precipitation data for the vegetative period,[5] divided by a proxy for evapotranspiration for the same period.[6]

In Ugtaal rainfall was mostly in the semi-arid range prior to 1995 (>0.25) and in the arid range after 1995. Rainfall in Gurvansaikhan was in the arid range for the whole period. If we compare the severity of drought and *dzud* conditions for 1998–2002 with the period as a whole we find that Ugtaal in 1998 was a drought and severe *dzud* year. In 2000 there was another severe *dzud*, followed by a somewhat warmer summer in 2001 and another severe drought in 2002. In Gurvansaikhan 1998 was not a problem year, but 1999 and especially 2000 were very problematic drought years, made worse by an additional *dzud* in 2000. The year 2002 was also a severe drought year.

These drought and *dzud* conditions are summarized in Table 18.4. We also add a tentative assessment of the variations in carrying capacity in sheep units based on both the aridity index for the two areas and on a hypothetical carrying capacity model; this combines the aridity index with

[5] We define vegetative periods for grasslands as all 10-day periods between the first and the last measurement of an average temperature of 5°C for that 10-day period, based on data provided by the Meteorological Service of Mongolia for Ugtaal and Gurvansaikhan. For both areas the vegetative period is between 140 and 180 days, normally between somewhere in April and somewhere in September.

[6] This was based on the assumption that the average temperature for the vegetative period (see note 5) × 100 gives an adequate evapotranspiration assessment. With the existing evapotranspiration levels in summer, and spread of rainfall during the year, we estimate that Mongolia has arid conditions (P/ETP<0.25) when annual rainfall is below 250 mm. and semiarid conditions (0.25<P/ETP<0.40) when rainfall is between 250 and 400 mm. This means that we assume that for Mongolia as a whole the annual evapotranspiration is in the range of 1000 mm.

Table 18.4 *Drought and dzud conditions in Ugtaal and Gurvansaikhan in 1998–2002 compared to average situation in 1990–2002, and assessment of theoretical carrying capacity compared with actual livestock numbers.*[7]

Variable	Ugtaal					Gurvansaikhan				
	1998/99	1999/00	2000/01	2001/02	2002/03	1998/99	1999/00	2000/01	2001/02	2002/03
Temperature vegetative period	−	+	+	+	++	+	+	++	+	++
Precipitation vegetative period						--	--	--	--	--
Aridity assessment	0.16	0.20	0.25	0.17	0.09	0.10	0.04	0.04	0.07	0.04
Aridity	−					--	--	--	--	--
DROUGHT	Yes	No	No	No	Yes!	No	Yes!	Yes!	No	Yes!
Temp. Oct–Mar	−		+		NA				NA	NA
Snow depth Oct–Mar	++		+		NA	+			NA	NA
DZUD	Yes!	No	Yes!	No	No	(Yes)	No	Yes	No	No
SU/ha	1.1	0.8	1.3	1.2	0.5	0.5	0.3	0.2	0.4	0.5
Theoretical carrying capacity[8]	132	96	156	144	60	265	159	106	212	60
Carrying capacity based on sample areas	114	96	128	136	54	90	60	80	130	106
Actual SU	156	145	147	109	98	276	281	139	154	173

Note: The actual livestock numbers and carrying capacity is expressed in 1000 sheep units (SU).

[7] In this theoretical assessment of the carrying capacity the accessibility of rangelands is not taken into account. As was stated earlier on, a shortage of water points, social and legal boundaries, a shortage of labour availability, etc. can reduce the amount of actually usable pastures. As we know, livestock mobility reduced and rangeland management (e.g. the maintenance of wells) deteriorated after 1990. This resulted in condensed grazing in some areas, particularly around *sum* and *aimag* centres, where some social services are provided, and around the remaining water points. The actual carrying capacity of the two *sums* is therefore probably lower.

[8] We use 120 000 ha of realistic pastureland for Ugtaal and 530 000 ha of realistic pasture land for Gurvansaikhan.

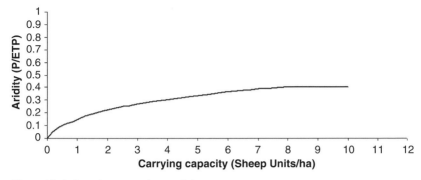

Figure 18.1 Carrying-capacity model

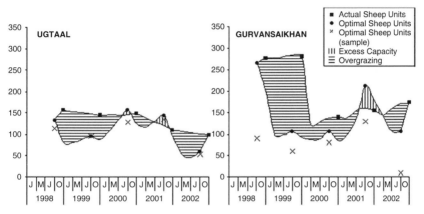

Figure 18.2 Estimated optimal livestock numbers (sheep units × 1000) in Ugtaal and Gurvansaikhan

environmentally sustainable sheep unit numbers.[9] Figure 18.1 shows the carrying-capacity model and the results for the two *sums* are depicted in Figure 18.2.

---

[9] This is based on Dietz (1987: 83), which in turn was based on an analysis of carrying capacity assessments for African rangelands, derived from aridity indexes. The most sophisticated source was KSS (1982: 46–47). The model is based on an empirically derived assumption that if aridity (P/ETP) is equal to 0.1, 0.25 and 0.4, the carrying capacity would be 0.5, 2.5 and 10 sheep units per hectare, respectively. This is based on the overall assumption that one sheep unit would have a live-weight of 30 kg, and a total annual feed consumption of 300 kg, with less than 15% of all biomass production consumable in the hyper-arid area, between 15 and 25% in the arid area, and between 25 and 40% in the semi-arid area.

During the worst rainfall years in Gurvansaikhan (i.e. 2000 or 2002) the potential sheep units per hectare can be estimated at around 0.2. During high rainfall (e.g. 1996) the carrying capacity increases to 0.7 sheep units per hectare. The average 'static' measure for 1990–2002 as a whole is 0.4 sheep units/ha. Actual sheep units decreased from 289 000 in 1999 to 142 000 in 2000. However, after 2000 livestock numbers increased again to a level of 176 000 sheep units in 2002. In Gurvansaikhan stock numbers were (far) in excess of optimal (theoretical) carrying capacity in 1999 and 2000, due to the adverse conditions. Indeed the slump in livestock numbers brought an adjustment to much lower levels. In 2001 weather conditions slightly improved and so did livestock numbers, which for a few months were below the optimum carrying capacity. However, soon the actual numbers exceeded the optimum numbers again.

For Ugtaal the worst rainfall years (i.e. 2002) have a carrying capacity of close to 0.5 sheep units/ha, and the best rainfall years close to 3 sheep units/ha, with an average 'static' figure of 1.4 sheep units/ha. In actual sheep units, the situation deteriorated from about 150 000 in 1999–2000 to 100 000 in 2002. Compared to the theoretically derived assessment of sustainable numbers of sheep units the situation between late 2000 and early 2002 was close to the optimum level, with some excess grazing capacity around October in both years. In the period before late 2000 and after early 2002 there were more sheep units than the theoretical carrying capacity. In 2003 the carrying capacity was locally judged to be exceeded by 20% and patches of degraded pasture were visible.

In Table 18.4 and Figure 18.2 we also add an assessment of optimal numbers of sheep units, based on actual grass yields in a few sample areas in the two areas. These were collected by the Range Management Department of the Government of Mongolia. Extrapolation of these data for the area as a whole has been done rather conservatively, which is why we regard the overall figures as being too low. However, the trend based on grass yield samples does resemble the theoretical trend based on our aridity assessment very closely.

18.4.2 Market-based population supporting capacity

For both areas we may hypothesize that the deterioration of livestock numbers and of local food production conditions caused a food crisis, which could only be solved by importing food from elsewhere. One possibility would be to sell livestock and buy grain, if the caloric exchange rates were favourable. Let us look at the evidence.

Table 18.5 *Caloric terms of trade in Ugtaal and Gurvansaikhan, 1998 and 2002*

| | Ugtaal | | Gurvansaikhan | |
	1998	2002	1998	2002
Beef t/kg	550	(900)	380	700
Mutton t/kg	600	(850)	400	700
Horse milk t/l	NA	NA	450	550
Wheat flour t/kg	320	380	350	400
Rice t/kg	450	420	420	400
Beef/wheat	1.7	2.4	1.2	1.8
CToT beef/wh	3.2	4.6	2.3	3.4
Beef/rice	1.2	2.1	0.9	1.8
CToT beef/rice	2.3	4.0	1.7	3.4
Mutton/wheat	1.9	2.2	1.1	1.8
CToT mut/wh	3.4	4.0	2.0	3.2
Horse milk/wheat	NA	NA	1.3	1.4
CToT milk/wh	–	–	9.6	10.4

For both areas we can estimate the trends in CToT, based on data on price levels for various products, adjusted to local circumstances: horse milk only has 487 Cal/l, beef 1872 Cal/kg and mutton 2029 Cal/kg, but wheat flour and rice are both estimated to have 3600 Cal/kg and hence the horse milk/wheat-rice conversion factor is 7.4; the mutton/wheat-rice conversion factor is 1.8, and the beef/wheat-rice factor is 1.9. Table 18.5 compares 1998 with 2002 for both *sums*.

Looking at the findings for the CToT for these two case study regions we can conclude that in all cases the CToTs improved during the livestock crisis, as expected. However, in Ugtaal levels were always higher than in Gurvansaikhan, probably reflecting the difference in distance to Ulaanbaatar, which with 1.2 million inhabitants (out of the current 2.5 million Mongolians) is the primary centre of demand. However, CToT levels in and around Ulaanbaatar are much better, both in 1998 and in 2002, than in either of the two case study regions.[10] The meat–wheat exchange ratio (CToT) in 1998 was almost four times better around the capital city than in Ugtaal and almost six times better than in Gurvansaikhan. In 2002 the relative situation of Gurvansaikhan had slightly improved. When looking at the horse milk–wheat exchange

[10] For the calculations of the CTOT in Ulaanbaatar see Dietz *et al.* (2005).

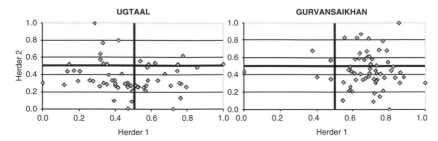

Figure 18.3 Scatter plot of strategies in Ugtaal and Gurvansaikhan

rate (only data for Gurvansaikhan) the difference with the situation around Ulaanbaatar is less extreme.

18.4.3 Game estimation

In order to obtain insights into the assignment of payoffs to the four payoff groups, the strategies of Herder 1 and Herder 2 are plotted in Figure 18.3 (the strategies are normalized between 0 and 1).[11] The strategy of Herder 1 is represented by the growth in herd size; low growth is associated with high values, and vice versa, since low growth suggests that the herder scores high on cooperative behaviour. The strategy of Herder 2 is represented by the perception of the environment; high scores stand for positive perceptions, and low ones for negative perceptions, since a positive perception implies that the herder assumes that his/her competitors adopt cooperative behaviour. Figure 18.3 shows the results, where observations in the lower left corner are assigned to payoff group y, those in the upper right to group x, and upper left and lower right are assigned to groups a and b, respectively.

The interpretation of Figure 18.3 already leads to an interesting finding, namely that there is more herd size maximizing behaviour in Ugtaal than in Gurvansaikhan. This is shown in the figure by the concentration of data in the lower left-hand side in Ugtaal (increasing herd size, quality of the environment perceived to be low) and the right-hand side in Gurvansaikhan (more constant herd size, irrespective of perceived environmental quality).

---

[11] Preliminary results from our game analysis have previously appeared in Lise *et al.* (2006). Due to improvements in the estimations and the data, the results presented here deviate significantly from those in this prior publication. We believe the current results offer a better interpretation of the actual games played in Ugtaal and Gurvansaikhan.

Table 18.6 *The estimated herder game using primary data, when the strategy is to choose herd growth rate.*

	a	b	x	y	Payoff order	Name of the game
Ugtaal	655.0	455.3	268.9	248.1	a>b>x>y	Non-coordination game
	(8)	(19)	(6)	(27)		
Gurvansaikhan	1330.1	942.8	1126.9	862.0	a>x>b>y	Chicken's game
	(2)	(36)	(19)	(3)		

Note: the number in the brackets denotes the number of observations within the payoff group. Payoffs are expressed in thousand Tugrik (the local Mongolian currency), which was equivalent to €0.74 in January 2004.

The game estimation results are presented in Table 18.6. These results suggest a so-called non-coordination game for Ugtaal and a chicken's game for Gurvansaikhan. This would mean that there are two equilibriums, both characterized by one herder maintaining his/her herd size and the other herder maximizing his/hers. These outcomes suggest that the highest payoffs correspond to a mix of herders, with some striving for low growth and others maximizing their animal numbers. This is difficult to interpret, especially since we simplified our analysis by assuming a two-person game whereby the whole group of herders compete for the same pastures. In any case, maintaining herd numbers leads to a higher payoff than maximizing the size of the herd, so there are no incentives for everyone to simultaneously maximize herd size. The difference in payoff between the two strategies is 20 000 Tugriks in Ugtaal and 265 000 Tugriks in Gurvansaikhan. However, neither of these strategies constitutes an equilibrium.

Finding a non-coordination game in Ugtaal suggests that there is no voluntary institutional solution to the problem. In view of the lack of institutional solutions, one could think of more coercive government intervention that forces herders to maintain a certain herd size. An alternative would be to create ownership or user rights of the land, which, when adequately enforced, gives clear incentives to take better care of the pastures.

Finding the chicken's game with this particular payoff structure in Gurvansaikhan suggests that in the long run, when the chicken's game is repeated, there could be an equilibrium in which herders keep their herd size constant.[12] This is only the case, however, if they attach a relatively high value to future income, i.e. if they have a low discount rate. The discount rate will be lower when current income can (easily) satisfy basic needs; otherwise this naturally takes precedence over any longer term considerations. If current income is not high enough, raising it or providing income security could lower the discount rate and increase the chance of reaching an equilibrium in which herds are smaller.

These games were estimated using current payoffs. Of course, expectations about future availability and quality of grazing pastures will also affect herders' decisions about herd size. Offering them better information about the long-term detrimental consequences of continued overgrazing will affect these expectations and could lead to alternative equilibriums.

18.5 Conclusions and policy recommendations

After communist rule collapsed the weather became one of the main regulatory factors of animal numbers in Mongolia in the late 1990s and early years of the twenty-first century. Favourable conditions in the years up to 1998 allowed large increases and many new entrants in the sector, leading to overgrazing and land degradation. In a way, nature corrected the overuse when severe winters and droughts hit Mongolia in the latter part of the period and brought animal numbers back to their old level. This correction mechanism worked, but had disastrous consequences for the people who were dependent on their herds for food and income. Letting nature handle things also has the drawback that land degradation caused by overuse can cause permanent damage to the land.

Fortunately, there are alternatives. Our game estimation results tentatively suggest that the current conditions in Ugtaal provide incentives to some herders to maximize their herds and to others to keep growth low. If this result holds when we relax some of the simplifying assumptions, it would suggest that top down coercive measures would be required to force all herders in lower growth, more sustainable, herding strategies. This could for instance be achieved through fines or taxes. A different approach would be to assign land or use rights to individual herders or herder groups. The game indicates that such strategies would mean lower overall payoffs.

This could partly be compensated by a move towards more market-based livestock rearing. Our analysis on the terms of trade herders faced

---

[12] See Lise (2005) on the estimation of repeated games.

during the period shows that smaller herds do not necessarily have to mean lower incomes as well. Better use could be made of both the urban and export markets. These markets can be provided with livestock products in exchange for grains, which then become more important in pastoralists' diets.

Increasing herders' payoff from the herding business by taking a more market-based approach would also be very important in Gurvansaikhan. By making it easier to provide for current needs, herders can give more consideration to future effects, and this would lower their discount rate. Our game estimation results for Gurvansaikhan suggest this will increase the chance of ending up in an equilibrium where everyone aims for low growth in animal numbers.

Government agencies can support a move towards more market-based livestock rearing and the accompanying change in diets. This can be done, for instance, by stimulating food trade (e.g. giving credit and training to grain providers) and by stimulating dietary changes (e.g. by modifying school dinners or by including recipes in the popular media). In Mongolia some changes are already visible. Although the official figures about the composition of the Mongolian diet (National Statistical Office of Mongolia, unpublished data, 2003) are rather doubtful,[13] it is quite clear that cereals have indeed become important during the last few years. One can expect further developments along this road of ever more market-oriented pastoralism. Our comparison of the CToT between different regions also suggests that market access is very important. Improving infrastructure could provide market access to more distant regions and increase the profit herders could make from marketing their animal products.

A more market-based approach could be combined with a move towards intensification of the sector, at least for those regions that are close to a large market with the ability to supply additional forage, such as concentrated feed. This could result in further reductions of the pressure on pastures, which is especially large near these large centres.

Even a system with smaller herds, where a larger share of animal products is marketed, needs well-functioning institutions for successful management of water sources and emergency relief during disasters. These institutions do not have to be set up by the government alone.

---

[13] Calculated diets for the 1999–2002 periods consist of an average of 507 000 Cal/capita, which is much below the necessary food intake of the 'required 950 000 Cal/cap'. Probably a lot of consumption is not measured or taken into account. So the figures about the composition of food intake should also be interpreted with care.

Considering Mongolia's old communal society, herder cooperation should be actively stimulated by creating more herder groups in which agreements about water source and land management are made. If herders cooperate and have faith in such institutions, the need to maximize the herd as a form of insurance will also diminish.

References

Batjargal, Z., Dagvadorj, D. and Batima, P. (eds.) (2000) *Mongolia National Action Programme on Climate Change*. Ulaanbaatar, Mongolia: National Agency for Meteorology, Hydrology and Environment Monitoring and JEMR Publishing.

Bromley, D. W. and Cernea, M. M. (1989). The management of common property natural resources: some conceptual and operational fallacies. World Bank Discussion Paper 57. World Bank, Washington DC.

Centre for Policy Research (CPR) (2003). PREM Mid-term Report. CPR, Ulaanbaatar.

Danker, M. (2004). *Hörs-Gazar*. Cry of a woken bull: changes in pastoral life and pastureland management in Mongolia. MSc Thesis, PREM programme, VU University, Amsterdam.

Dietz, A. J. (1987). Pastoralists in dire straits. Survival strategies and external interventions in a semi-arid region at the Kenya/Uganda border: Western Pokot 1900–1986. PhD thesis, University of Amsterdam (published as Netherlands Geographical Studies).

Dietz, A. J., Enkh-Amgalan, A., Erdenechuluun, T. and Hess, S. M. (2005). Carrying capacity dynamics, livestock commercialisation and land degradation in Mongolia's free market era. PREM working paper 05/10. Institute for Environmental Studies, Department of Economics and Technology.

Hardin, G. (1968). The tragedy of the commons. *Science*, **162**: 1243–1248.

Kenya Soil Survey (KSS) (1982). *Exploratory Soil Map and Agro-Climatic Zone Map of Kenya 1980*. Nairobi: KSS.

Lise, W. (2001). Estimating a game theoretic model. *Computational Economics*, **18**: 141–157.

Lise, W. (2005). A game model of people's participation in forest management in Northern India, *Environment and Development Economics*, **10**: 217–240.

Lise, W., Garrido, A. and Iglesias, E. (2001). A game model of farmer's demand for irrigation water from reservoirs in Southern Spain. *Risk Decision and Policy*, **6**: 167–185.

Lise, W., Hess, S. M. and Purev, B. (2006). Pastureland degradation and poverty within herder communities in Mongolia: data analysis and game estimation. *Ecological Economics*, **58**: 350–364.

Muhsam, H. V. (1973). A world population policy for the world population year. *Journal of Peace Research*, **10**: 91–99.

NSOM (2001). National Statistical Office of Mongolia, and World Bank, Mongolia, *Participatory Living Standard Assessment 2000*, Summary report prepared for the Donor Consultative Group Meeting, Paris, 15–16 May 2001.

19 · *Changes in welfare and the environment in rural Uganda*

VINCENT LINDERHOF, PAUL O. OKWI,
JOHANNES HOOGEVEEN AND THOMAS
EMWANU

19.1 Introduction

The link between poverty and the environment is both context and location specific (Barbier 2000, Chomitz 1999, Ekbom and Bojo 1999). Poverty in remote, inaccessible areas with unfavourable natural conditions often coexists alongside relative affluence in more favourable locations close to major cities and markets. Information on the spatial distribution of poverty and environmental degradation is of interest to policymakers and researchers for a number of reasons. It can be used to quantify disparities in welfare across regions or districts and identify areas that are falling behind. It facilitates the targeting of programmes whose purpose is, at least in part, to alleviate poverty; i.e. programmes related to environmental conservation, education, credit, health and food aid. It may also shed light on the geographic factors associated with poverty, such as degraded forest areas and reclaimed wetlands.

Environmental quality is an important determinant of welfare as it affects health, security, energy supplies and housing quality (Dasgupta and Maler 1995). Typically the relation is quite complex (Roe 1998), but it is clear that especially for poor people environmental quality is a key determinant of welfare as their livelihoods so often depend on natural resource use (e.g. Cavendish 1999, 2000).

There is an extensive literature regarding the correlation between environmental quality and poverty, as well as the importance of the environment as a source of income for the poor. However, little is known about changes in poverty in relation to changes in the natural

Nature's Wealth: The Economics of Ecosystem Services and Poverty, ed. P. J. H. van Beukering,
E. Papyrakis, J. Bouma and R. Brouwer. Published by Cambridge University Press,
© Cambridge University Press 2013.

environment over time in developing countries like Uganda. This study attempts to contribute to the environment–poverty nexus literature by exploring data for over 900 sub-counties in rural Uganda. The study specifically looks at how changes in poverty between 1992 and 1999 are associated with changes in the natural environment, and how initial conditions affect changes in poverty and environmental degradation.

The systematic study of the relationship between changes in poverty and changes in environmental quality requires poverty and biophysical information for small geographical areas. For Uganda, biophysical information is available for 1992 and 1999 (Forest Department 2002). Poverty maps (that do not incorporate, though, biophysical information) are also available for 1992 and 1999 (see UBOS and ILRI 2004).

In our study, we emphasize the link between poverty and environment. Okwi *et al.* (2005) have shown that the inclusion of biophysical information in estimating poverty for small geographic areas through poverty mapping techniques significantly improves poverty estimates. Therefore, we recalculate poverty for 1999 following the methodology outlined in Emwanu *et al.* (2006) while incorporating the biophysical information as suggested in Okwi *et al.* (2005). In combination this results in a panel database of poverty indicators and land use for rural Uganda in 1992 and 1999 (for more details see Okwi *et al.* 2006).

The chapter is organized as follows. Section 19.2 briefly summarizes the data and methodology on the estimation of poverty indicators for rural Uganda in 1992 and 1999. Section 19.3 presents and discusses the biomass maps for rural Uganda and the results on poverty estimates. Finally, Section 19.4 concludes and discusses policy implications.

19.2 Data and methodology

19.2.1 Methodology

In Uganda, the availability of high-resolution datasets on land use provides a strong foundation for the construction of poverty–biomass maps. Although several approaches have been developed to design poverty maps, there has been less effort to develop detailed poverty–environment maps. In our approach, we use statistical estimation techniques (small area estimation) to overcome the typical limitations in the geographic coverage of household welfare that surveys provide, and the lack of welfare indicators in the census data. In addition, we include environmental information to address local impacts. Note that we use biophysical information in

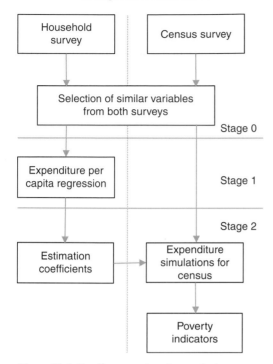

Figure 19.1 Small-area estimation technique

the statistical analysis of estimating poverty indicators at the level of households and their local areas.

We adopt the approach developed by Elbers *et al.* (2003) and modified by Emwanu *et al.* (2006). In addition, we include biophysical information as suggested in Okwi *et al.* (2005).

Figure 19.1 shows the concept of the small-area estimation method, which is typically divided into three stages. In stage 0, variables are identified that relate to income or consumption and that are present in both the household and census survey. Based on the household survey, stage 1 estimates a measure of welfare (per capita consumption), as a function of these household characteristics using regression analysis. At this stage, biophysical variables are included to minimize the variation of the location effect as part of the residuals (see Okwi *et al.* 2005). In contrast with Elbers *et al.* (2003) who regress location residual separately, we include those biophysical variables directly in the welfare equation. Moreover, we emphasize that it is our objective to estimate the conditional expectation of consumption and not a causal relationship between poverty and the environment. Practically, our main purpose is to obtain a high goodness-of-fit of the regression equation (reasonably high R^2) rather than an interpretation of individual coefficients of the variables. Finally, in stage 2 the distributions of regression coefficients are used to simulate per capita consumption with the corresponding household characteristics in the census data. In particular, the estimation results of the coefficients from the household expenditure equation estimated in stage 1 are applied to the census data. Since we are using household-level census data, this combination produced estimates of per capita expenditure for each household. We simulated the level of consumption for each household in the census (see Elbers *et al.* 2003, Mistiaen *et al.* 2002). With these per capita consumption simulations, we can ultimately construct poverty indicators at different administrative unit levels, i.e. at the district, county or even sub-county level. Moreover, poverty indicators such as poverty incidence, poverty gap or poverty gap squared (see Foster *et al.* 1984), are accompanied with standard errors as well.

For the 1991 estimates, the household survey was sufficiently large to distinguish between four different rural regions (Central, East, North and West Uganda) and allow exploration of different per capita consumption regression equations. This means that the four first-stage regression equations for 1991 might have different sets of explanatory variables including biophysical variables.

For 1999 we adjusted the methodology due to limited data availability as suggested by Emwanu *et al.* (2006, see also subsection 19.3.2). From the 1999/2000 Uganda National Household Survey (UNHS), we have per capita consumption from a selected panel of households, who were also included in the 1991 Integrated Household Survey (IHS). However, there is no information on household characteristics in 1999 (Government of Uganda 1999). Therefore, we use the past household characteristics from 1991. Thus, in contrast with the approach for 1991, the panel is relatively small and covers almost exclusively rural areas. We therefore estimate only one (rural) per capita consumption regression equation. After identifying a set of common variables, a model for per capita consumption in 1999 was regressed on household characteristics from the 1999 IHS survey and biophysical variables for 1999. Although the estimation of consumption on past household characteristics is unusual, we emphasize that it is our objective to estimate the conditional expectation of consumption for 1999 and not a causal relation (see Emwanu *et al.* 2006). The model is only meaningful if its coefficients are estimated accurately (to limit the variance attributable to model error) and if a reasonably high R^2 (to assure disaggregation for small target populations) is obtained. As in Emwanu *et al.* (2006), the method of estimating future consumption is based on a single regression equation for all four regions, while Okwi *et al.* (2005) estimate regression equations for all four rural areas separately.

19.2.2 Data

For the estimation of the poverty indicators for rural Uganda, we use three different household datasets: the Integrated Household Survey (IHS) for 1991, the Uganda National Household Survey (UNHS) for 1999/2000, and the census survey for 1991. The IHS used a stratified sample of 10 000 households in both rural and urban areas, from which we only use the 6305 rural households. The survey questionnaire collected information on household and demographic characteristics, education, assets, employment, income and consumption (UBOS and ILRI 2004). This survey was based on four regions divided into rural and urban strata. From the 1999/ 2000, we use 1058 households present in both the IHS and the UNHS for the four rural strata, so that we derive updated welfare estimates for 1999.

The 1991 Population and Housing Census was conducted by the same institution (UBOS) and was meant to cover the entire population in both rural and urban areas. The Census covered information on household members, education, housing characteristics and access to basic utilities, although it did not collect information on income and consumption. The Census and Survey data have several common household variables such as household size composition, education, housing characteristics, access to utilities and location of residences.

The biophysical information is derived from the National Biomass Study (NBS) organized by the Ministry of Water, Lands and the Environment. NBS collected geo-referenced information on a variety of spatially referenced variables describing topography, land cover and land use, and roads for 1992 and 2000. The project developed its own classification system based on a combination of land covers and land uses. This information captures changes in land cover such as broadleaved tree plantation or woodlots, coniferous plantations, tropical high forests (normal and depleted/encroached), woodland, grassland, wetlands, water resources and land use, such as subsistence and commercial farmland and changes in landscape. In the NBS project, the country was split into 9000 plots with three sample plots at each intersection. However, due to influences of population density and agro-ecological zones on land cover and tree growth, some adjustments were made to the overall total sample plots. Topographic maps, land-cover maps (1:50 000) and Global Positioning Systems (GPS) were used to locate the field plots on the ground. There were four categories of data capture and processing: (1) mapping (spatial and its attributes); (2) biomass survey (filed plot measurements); (3) monitoring of biomass; and (4) land-cover change. This information details the woody biomass stock for each plot and can be used to assess the relationship between tree cover and poverty. The data is extremely rich in biophysical factors and also includes the distribution of infrastructure like markets, roads and schools. The GIS format of the data also allows us to explore the possibilities of merging the datasets using GIS variables.

19.3 Results and discussion

19.3.1 Results

The poverty-mapping method generates estimates for changes in poverty and the environment. In this section, we show how changes in poverty between 1992 and 1999 are related to changes in land use over the same period. Land-use change refers to any increase or decrease in the proportion of land area under any type of land use (e.g. wetlands being reclaimed for subsistence farming or urban areas growing to reclaim some subsistence farm areas).

Figure 19.2 shows the changes in poverty in Uganda at the sub-county level between 1992 and 1999. The results from the analysis of poverty changes are encouraging, with large and widespread decreases in poverty observed countrywide. The highest drops in poverty in rural areas between 1992 and 1999 can be seen in Central and parts of the Western region in the districts of Kibaale, Luwero, Bushenyi, Rakai, Mpigi and Kisoro. Poverty is observed to have increased in Arua, Moyo and Apac in the Northern region and Kasese district in the West. At the sub-county level, the maps demonstrate how almost all rural areas in Uganda benefited from the growth that took place during the 1990s. Poverty worsened in only 8% of Uganda's rural sub-counties during this period, whereas a number of land-use changes can be observed. A number of key patterns emerge when we overlay changes in poverty with changes in land use between 1992 and 1999.

Figure 19.3 and Figure 19.4 show the changes in land use particularly for wetland and forest resources. Overall there is evidence that the proportion of land under wetlands decreased (Figure 19.3). The maps

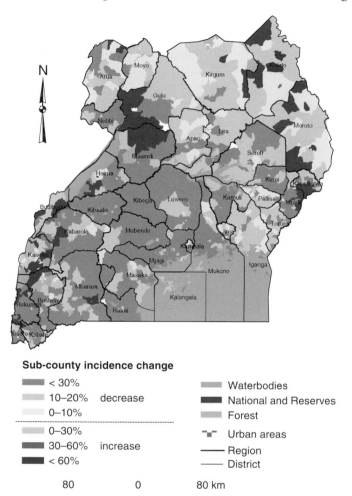

N

Sub-county incidence change

< 30%	
10–20%	decrease
0–10%	
0–30%	
30–60%	increase
< 60%	

Waterbodies
National and Reserves
Forest
Urban areas
—— Region
—— District

80   0   80 km

Figure 19.2 Changes in rural poverty in Uganda 1991–99 using the small-area estimation techniques including biophysical information (see colour plate section)

show that between 1992 and 1999, many of the wetlands were converted to wetland subsistence farms especially in the areas around the shores of Lake Kyoga and Lake Victoria. Wetland farming (particularly the growing of rice, sugar cane and yams) is becoming a major income source for many families strategically located in the neighbourhood of the wetland areas.

In Figure 19.5, changes in poverty are overlaid with changes in wetland quality. We find that negative changes in wetland cover are correlated with decreases in poverty. Households living within the vicinity of wetlands in most parts of the country generally experienced a decline in

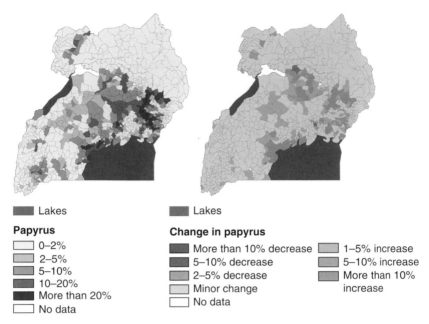

■ Lakes   ■ Lakes

Papyrus
- 0–2%
- 2–5%
- 5–10%
- 10–20%
- More than 20%
- No data

Change in papyrus
- More than 10% decrease
- 5–10% decrease
- 2–5% decrease
- Minor change
- No data
- 1–5% increase
- 5–10% increase
- More than 10% increase

Figure 19.3 Wetlands in 1992 and the changes between 1992 and 1999 (see colour plate section)

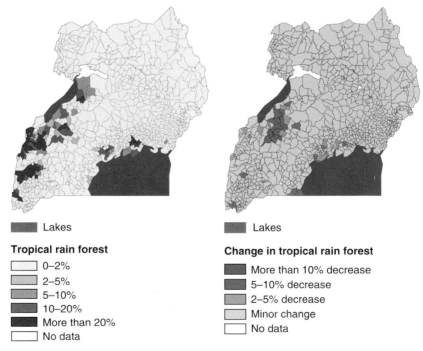

■ Lakes   ■ Lakes

Tropical rain forest
- 0–2%
- 2–5%
- 5–10%
- 10–20%
- More than 20%
- No data

Change in tropical rain forest
- More than 10% decrease
- 5–10% decrease
- 2–5% decrease
- Minor change
- No data

Figure 19.4 Tropical high forest 1992 and the changes between 1992 and 1999 (see colour plate section)

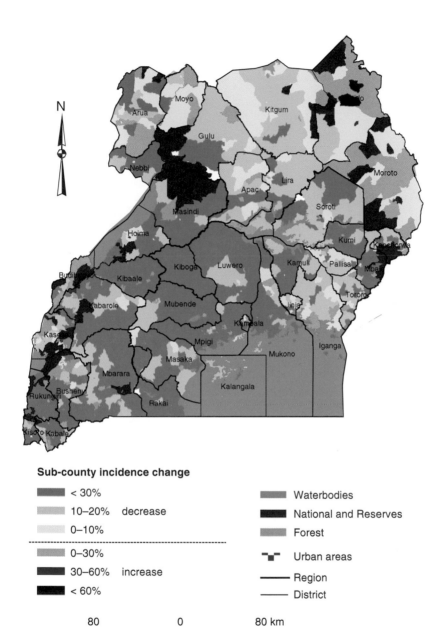

Sub-county incidence change

▬	< 30%
▬	10–20% decrease
▬	0–10%
- -
▬	0–30%
▬	30–60% increase
▬	< 60%

▬	Waterbodies
▬	National and Reserves
▬	Forest
⊤	Urban areas
—	Region
—	District

80   0   80 km

Figure 19.2 Changes in rural poverty in Uganda 1991–99 using the small-area estimation techniques including biophysical information

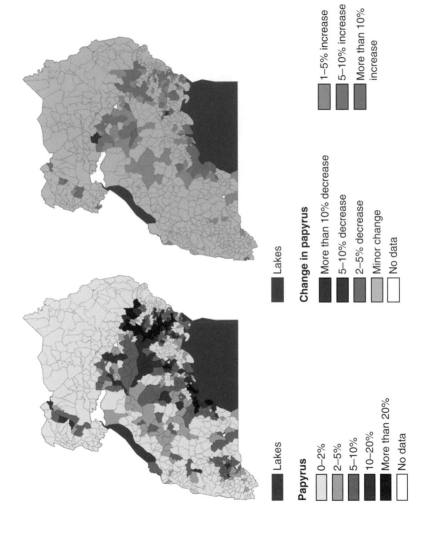

Lakes

Papyrus
- 0–2%
- 2–5%
- 5–10%
- 10–20%
- More than 20%
- No data

Lakes

Change in papyrus
- More than 10% decrease
- 5–10% decrease
- 2–5% decrease
- Minor change
- No data
- 1–5% increase
- 5–10% increase
- More than 10% increase

Figure 19.3 Wetlands in 1992 and the changes between 1992 and 1999

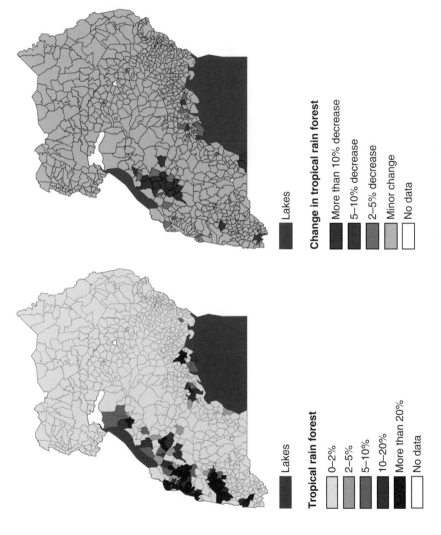

Lakes

Tropical rain forest

0–2%
2–5%
5–10%
10–20%
More than 20%
No data

Lakes

Change in tropical rain forest

More than 10% decrease
5–10% decrease
2–5% decrease
Minor change
No data

Figure 19.4 Tropical high forest 1992 and the changes between 1992 and 1999

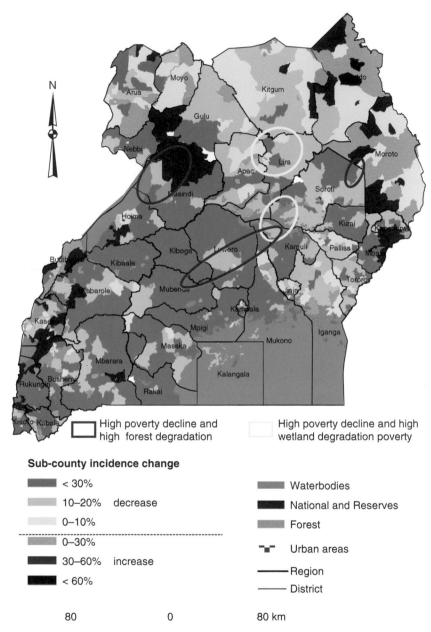

Figure 19.5 Overlay of changes in poverty and land use between 1992 and 1999

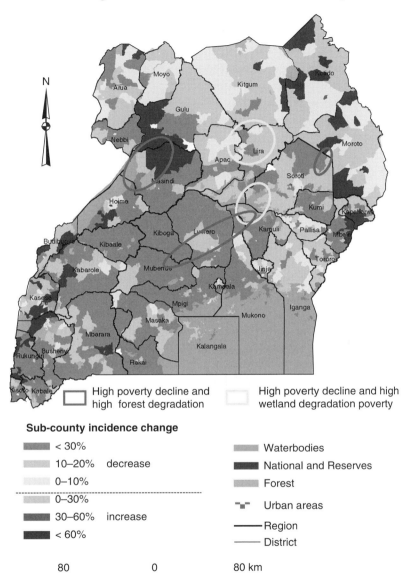

High poverty decline and high forest degradation

High poverty decline and high wetland degradation poverty

Sub-county incidence change

< 30%	
10–20%	decrease
0–10%	
0–30%	
30–60%	increase
< 60%	

Waterbodies

National and Reserves

Forest

Urban areas

Region

District

80   0   80 km

Figure 19.5 Overlay of changes in poverty and land use between 1992 and 1999 (see colour plate section)

poverty and increases in wetland degradation, indicating that reduction in poverty is possibly caused by increased extraction of wetland resources. For example, the least poor sub-county in 1999 in the Northern Region was Namasale in Kioga County, Lira district with a poverty rate of 37.5%.

The poorest sub-counties are Loyoro (98%), Lolelia (97%), Kaabong (97%) and Sidok (96%), all found in Dodoth County, Kotido district. Namasale sub-county is located on the shores of Lake Kyoga while the poorest sub-counties are located in the semi-arid parts of the region. Most of the households in Namasale sub-county participate in extraction of wetland resources such as hunting, fishing, papyrus harvesting, firewood collection, harvesting building materials, beekeeping, water collection, mining and livestock grazing. Interestingly, the Lake Kyoga area has between 6 and 11 of the extraction activities mentioned earlier taking place in the wetlands, compared to the sub-counties to the south of the Lake, where only one or two of these activities are reported (Wetland Division, Ministry of Lands Uganda). The sub-county of Namasale also recorded very high levels of use of wetlands (between 75–95%) and this probably explains the high degradation rate and low poverty levels. According to the Wetlands Division,[1] this area also witnessed among the highest impact of use of wetland resources. Other examples of this relationship are found in the Eastern and Central regions. In the Eastern region, the increase in rice farming in Bunyole, Busiiki and Pallisa could also explain part of the reduction in wetland areas and decline in poverty levels in this region. Note that the proportions of subsistence farmlands in wetland areas slightly increased in 1999. There is a seemingly high correlation between the increase in rice farming, reduction in wetland area and a decline in poverty in this region. According to the Wetland Division, more areas in the region have experienced increases in rice farming. This is likely to have serious implications on the wetlands in this region. Similarly, the sub-counties in the Central region located near Lake Victoria and other wetlands have experienced a far greater decline in poverty than others. This relationship points to a reclamation of wetlands. A key pattern that emerges, therefore, is that initial high poverty in this area has been reduced by the ability of the people to extract wetland resources but this is causing the wetlands to degrade at a high rate, which in turn raises questions about sustainability.

19.3.2 Discussion

Obviously, the initial level of either poverty or environmental quality largely determines the magnitude of changes in those indicators over time.

---

[1] Personal communication with Paul Mafabi, Wetland Division, Ministry of Lands and Natural Resources, Uganda

If a sub-county already has a low initial level of environmental quality, it is unlikely that it will experience a high level of environmental degradation. Similar reasoning holds for the initial level of poverty.

To further explore the implications of the relationship between changes in poverty and the environment at a sub-county level, we define four possible situations based upon initial conditions, see Table 19.1. The type of development observed in sub-counties can be categorized by four possible initial situations. High initial conditions refer to better environmental quality in terms of a high proportion of land use under wetlands or forests (i.e. the environment is less degraded). On the other hand, low initial conditions refer to poor environmental conditions or quality, such as low proportions of land under grassland or wooded area. In terms of poverty, high initial poverty would be used to describe

Table 19.1 *Relationship between initial conditions and changes in poverty in Uganda, 1992–99*

	Low initial poverty	High initial poverty
Low initial environmental quality	We do not find such an association except for urban environments. This itself is an indication that in rural environments low environmental quality is a huge determinant of poverty.	Example in text from Northern region shows poverty stagnant and so is environment. Possibly a trap. High poverty less degradation. (Qualification: Security issues could have a major role to play in this situation).
High initial environmental quality	Two scenarios: 1. In Central region shows high poverty decline and high degradation; (Qualification: possibly driven by population density) 2. In Mbarara, high poverty decline and less degradation (Qualification: possibly driven by lower population growth and better resource management.) Poverty reduction was observed in all sub-counties in western Uganda.	Δ poverty Δ environment Two scenarios: Kisoro, Bushenyi, Mbale and Tororo high poverty decline and high degradation. In Kumi high poverty and less degradation. (Qualification. Initial conditions in Kumi are different from the other areas).

situations where the poverty rate is higher than the national average while the reverse is true for low initial quality. Note that we can distinguish more than one scenario per initial situation.

19.3.2.1 Central Uganda

The Central region stood out as the least poor region both in 1992 and 1999 for rural areas, and has reasonably high land quality (Okwi *et al.* 2005), i.e. cell 3 (bottom left) of Table 19.1. The land-use maps show increasingly more degraded forest areas at the sub-county level. This region is mainly covered with subsistence farmlands and tropical forests. We observe that between 1992 and 1999 the proportion of forest cover has declined. The sub-counties found in Buikwe, Buvuma, Busiro and Mukono that have experienced increases in forest degradation have also experienced among the highest declines in poverty in the region. The decline in poverty within these sub-counties could be due to the increased use of forest resources such as timber, fruits, charcoal and expansion of farmland, especially near the Mabira forest reserve. As households seek to improve their welfare, they cut down trees to extract timber for sale and clear more land in order to increase the availability of farmland. This pressure is caused by high population growth within the area. Thus, we find that negative changes in forest area are correlated with improvements in welfare in this region.

19.3.2.2 Eastern Uganda

In the Eastern region, land use mainly changed in Butembe, Bunya and Bukooli in terms of forest cover (soft- and hardwood lots) and tropical high forests. Forest degradation increased and was highest in the wealthy counties near Jinja town and around Lake Victoria. Specifically, Bunya and Bukooli had the highest forest degradation rates in the region. These counties are also located along the shores of Lake Victoria and also have a high forest density. Apparently, communities living in these areas have used forest resources to significantly reduce their poverty levels between 1992 and 1999. Sub-counties located in Kongasis and Kween near Mt. Elgon also experienced increased degradation and decreased poverty, an indication that the population in these areas is harvesting forest resources from the Mt. Elgon reserve to improve welfare.

The land-use map shows that areas that have typically high subsistence farming generally attained lower reductions in poverty compared to degraded areas and commercial farms. Evidence consistently shows that poverty reduction was achieved to a lesser extent in areas that typically rely on subsistence farming. The data shows that although there was an increase in the area under subsistence farmland in almost all sub-counties, there was reportedly very low overall change in poverty levels. Moreover, agricultural production registered a high annual growth rate of about 5% on average between 1992 and 1999. But this overall picture masks great differences at lower levels. For example, variation was very high at the sub-county level in 1999 (13% to 71%) in the Eastern region. The poorest sub-county and the one that registered the least change in poverty at 66% was Gweri (a typical subsistence farming community) in Soroti county, Soroti district. By contrast, the richest sub-county with a major decline in poverty was Kakira in Butembe county, Jinja district with a poverty rate of 5%. Kakira sub-county is dominantly commercial sugarcane farmland implying that as expected, the gains from commercial agriculture are much higher than from subsistence farming. For the typical subsistence farming areas, most of the registered increases in output could have come from expanding the area farmed, which implies that other types of land uses were turned to subsistence farmland. But even then, there was a large increase in annual population growth of about 3.2% (Uganda Bureau of Statistics 2003) in this region. This increase, therefore, means that the economic output barely kept pace with the population growth, emphasizing the historical importance of subsistence agriculture in sustaining the basic welfare of rural households. The converse is true for the commercial farms, especially in Jinja district.

19.3.2.3 Northern Uganda

Poverty remained high in the grassland and wooded areas of the country. No clear patterns are observed in terms of increases or declines in other land-use types as they remained almost unchanged over the period.

In the Northern region, which is an example of low initial environmental quality and high initial poverty (top-right cell of Table 19.1), a large part of land cover consists of wood and grassland. There are a few pockets of major poverty decline observed. As mentioned above, these changes occurred in the wetland areas near Lake Kyoga. Even though there have been small increments in the plantation woodlots and wooded areas in the sub-counties, generally, the state of poverty remained high in these areas. For example, in the North, much of the land use remained unchanged after 1992, probably because most of the land-use activities have been hampered by conflict in that region. The high poverty incidence in the Northern region could also be due to the fact that this is one of the most semi-arid parts of Uganda, and the sandy soils make it difficult to practice intensive agriculture. In addition, this area is generally poorly served with roads and therefore access to markets is also difficult. In general terms, the typical grassland and wooded areas in the North continue to display the highest incidence of poverty.

19.3.2.4 Western Uganda

In parts of the Western region, which is better connected and less affected by security problems, grasslands have also remained intact, although a few areas that were formerly wooded and grasslands have been transformed into subsistence farmlands. The area generally has a mix of subsistence farming and cattle rearing. These few areas have also witnessed a substantial decline in poverty but with less environmental degradation.

In the Western region, there were pockets of high forest degradation between 1992 and 1999 in the mid-western sub-counties of Bujenje, Buliisa, Rujumbura Bujunje, Buruli, Kibanada and Bokunjo, where the proportion of hardwood plantation forests generally decreased. These are areas close to the mountainous parts of Rwenzori with difficult access to roads and markets. The overlay of changes in poverty and forest cover has revealed an important relationship that highlights the extreme vulnerability of forest areas. These maps have enabled us to identify forest degradation hotspots and how they are related to poverty changes. They show the roles of forest resources in reducing poverty among these communities, i.e. the role of poor communities in forest degradation.

19.4 Conclusions and policy recommendations

We have shown in this chapter that poverty declined in some land-use types, mainly wetlands, natural forests and commercial farming areas during 1992–99, while it remained fairly constant in the major woodland, grassland and other land use areas. These decreases in rural poverty could be attributed not only to the extraction of these resources, but also to other factors such as security, better soils and good climate. Increases in extraction of wetland and forest resources led to the reduction of poverty in some of these areas. We can also argue that changes in resource availability can go a long way in explaining changes in poverty. In particular, decreases in the proportion of land under wetlands and natural forests are correlated with a fall in poverty. However, the environmental degradation of wetlands and tropical high forest are, in some cases, possibly caused by high population pressures. Given these results, we need to ask ourselves about the challenges ahead for different land uses and areas in Uganda.

What are the challenges for wetlands? The availability of detailed spatial information about changes in the wetlands and changes in poverty allow us to formulate some hypotheses. The maps reveal how changes in poverty and changes in wetlands can go in various directions. The results are fascinating and also important for policymakers. These maps and the overlays have allowed us to test some hypotheses. For example, the overlay between poverty and wetlands reveals a striking relationship between declines in poverty levels and increased degradation of wetlands, indicating that higher degradation of wetlands may induce poverty reduction through increased extraction of wetland resources such as fish, beekeeping, water, grazing and wetland farming, among others. It could also be that high poverty levels are inducing increased degradation of wetlands.

What are the challenges for forest areas? In the case of forest degradation, we find an association between decreases in poverty and forest area, indicating that higher degradation leads to lower poverty levels or that higher poverty may lead to increased degradation of forests. A large number of rural poor people living within the vicinity of forest areas continue to depend on forest resources for survival. In contrast, households living in sub-counties with woodlands, grasslands and impediments such as steep slopes, rocks and poor soils appear to be caught in this vicious cycle and may be in a spatial poverty trap. In other types of land use, we find an inverse or no correlation. Thus, generally, we observe that in areas with wetland and forest cover, welfare improvement of households or economic progress is associated with degradation of these resources.

What are the challenges for environmental conservation and poverty reduction? As long as poverty levels remain high, people will continue to exploit natural resources unsustainably. This, in turn will increase pressure on the already strained natural resources. A large number of poor people who depend on these natural resources will end up with less natural resource assets and this will worsen their poverty levels leading to a vicious cycle of poverty.

In terms of policy, these overlays allow a better targeting of policy actions. Although these correlations are only a first attempt to relate changes in land use to changes in poverty, they are suggestive of the important role these play in determining poverty levels and dynamics. Government departments that are involved in natural resource management and poverty monitoring can utilize the land-use and poverty-change maps to identify 'poverty/environment hotspots' or intervention sites. For example, the Wetland Division in Uganda, which is targeting sustainable use of wetland resources, can identify for each wetland area the potentially high degradation areas, 'hotspots', and design interventions for wetland use. Farming in the wetland areas could be allowed but on a sustainable basis. Policymakers should devise means of monitoring and limiting the overexploitation of the wetlands by imposing area limits and restrictions on some activities such as tree extraction, which may seriously affect the quality of the wetlands. Activities such as fish farming could be encouraged to complement what is already available and community monitoring should be implemented.

Similarly, the National Forest Authority, which is the agency entrusted with the management of forest resources in Uganda, can identify key 'poverty–forest degradation hotspots' and design control systems. Efforts to protect forests may not produce meaningful results unless the underlying causes of deforestation are addressed. This is because other factors that cause deforestation, such as increased population pressure, may continue to raise demand for fuelwood even if governments enforce control measures on forest use. This is the exact situation faced in the Mt. Elgon and Mabira forest reserves in Eastern Uganda where there is tension between local communities and government enforcement agents due to high population growth rates and the need for more farmland. Therefore, reforestation or controlled use of forests alone is not enough. It has to accompany policies that aim at controlling the other causes of deforestation and improve the performance of management institutions.

This raises the need for increased community-level management of forests and wetlands. The Local Council system could be a good starting point for the formulation of environmental goods user groups. These groups should be trained to manage their environmental resources to ensure their sustainable use, for example, by restricting collection to only dry and fallen wood and limiting extraction of wetland products. Through enforcement of controlled use, forest and wetland resources would increase and collection would become easier hence better implications for welfare. Therefore, the challenge is to develop management institutions, which limit forest degradation and ensure a fair distribution of benefits.

Finally, these maps will serve to raise awareness among local governments, donors and researchers seeking to meet the Millennium Development Goals. In terms of future research, with more detailed information available, the causal relationship will be analysed in further detail. Another conclusion that we reached is that without further verification the updated results should not be used as indicators for the welfare in specific sub-counties, counties or districts. Despite the fact that the poverty indicators for 1999 are helpful in analysing the poverty–environment nexus in rural Uganda, their confidence interval is rather large, so that the 1999 poverty estimates have to be interpreted with some care. Hopefully, new household surveys and new census surveys will be co-ordinated and addressed regularly, so that more recent poverty maps for low administrative areas can be prepared in order to guide future land use and poverty alleviation policy.

References

Barbier, E. (2000). The economic linkages between rural poverty and land degradation: some evidence from Africa. *Agriculture, Ecosystems and Environment*, **82**: 355–370.

Cavendish, W. (1999). Poverty, inequality and environmental resources: quantitative analysis of rural households. Working Paper Series 1999–09. Centre for the Study of African Economies, University of Oxford.

Cavendish, W. (2000). Empirical regularities in the poverty-environment relationship of rural households: evidence from Zimbabwe. *World Development*, **28**: 1979–2003.

Chomitz, K. (1999). Environment poverty connections in tropical deforestation. Discussion notes prepared for the WDR workshop on Poverty and Development, 6–8 July, Washington DC.

Dasgupta, P. and Maler, K.-G. (1995). Poverty, institutions and the environmental resource-base. In J. Behrman and T. N. Srinivasan (eds.), *Handbook of Development Economics*, Vol. 3. Amsterdam: Elsevier Science Publishers.

Ekbom, A. and Bojo, J. (1999). Poverty and environment. Evidence of links and integration in the country assistance strategy process. Africa Region Discussion Paper no. 4, World Bank, Washington DC.

Elbers, C., Lanjouw, J. O. and Lanjouw, P. (2003). Micro-level estimation of poverty and inequality. *Econometrica*, **71**(1): 355–364.

Emwanu, T., Hoogeveen, J. G. and Okwi, P. O. (2006). Updating small area welfare indicators in the absence of a new census. *World Development*, **34**(12): 2076–2088.

Forest Department (2002). National Biomass Study Final Report. Kampala, Uganda.

Foster, J., Greer, J. and Thorbecke, E. (1984). A class of decomposable poverty measures. *Econometrica*, **52**: 761–766.

Government of Uganda, (1991). Uganda population and housing census. Uganda Bureau of Statistics, Government of Uganda.

Mistiaen, J. A., Ozler, B., Razafimanantena, T. and Razafindravonona, J. (2002). Putting welfare on the map in Madagascar. African Region Working Paper Series no. 34, The World Bank, Washington DC.

Okwi, P. O., Hoogeveen, J. G., Emwanu, T., Linderhof, V. and Begumana, J. (2005). Welfare and environment in rural Uganda: results from a small-area estimation approach. PREM-report 05/04. Amsterdam. IVM Institute for Environmental Studies, VU University.

Okwi. P. O., Linderhof, V., Hoogeveen, J. G., Emwanu, T. and Begumana, J. (2006). Welfare and environment in rural Uganda: panel analysis with small-area estimation techniques. PREM-report 05/04. Amsterdam. IVM Institute for Environmental Studies, VU University.

Roe, E. (1998). *Taking Complexity Seriously. Policy Analysis, Triangulation and Sustainable Development*. Boston, MA: Kluwer Academic Publishers.

Uganda Bureau of Statistics (2003). Statistical abstract. Uganda Bureau of Statistics, Government of Uganda.

Uganda Bureau of Statistics (UBOS) and International Livestock Research Institute (ILRI) (2004). *Where Are the Poor? Mapping Patterns of Well-being in Uganda: 1992 and 1999*. Norwood, MA: Regal Press.

Index

410 · Index